MODEL PLANTS and CROP IMPROVEMENT

MODEL PLANTS and CROP IMPROVEMENT

Edited by

Rajeev K. Varshney
Robert M.D. Koebner

Taylor & Francis
Taylor & Francis Group
Boca Raton London New York

CRC is an imprint of the Taylor & Francis Group,
an informa business

CRC Press
Taylor & Francis Group
6000 Broken Sound Parkway NW, Suite 300
Boca Raton, FL 33487-2742

Printed in the United States of America on acid-free paper
10 9 8 7 6 5 4 3 2 1

International Standard Book Number-10: 0-8493-3063-7 (Hardcover)
International Standard Book Number-13: 978-0-8493-3063-6 (Hardcover)

Library of Congress Cataloging-in-Publication Data

Model plants and crop improvement / editors, Rajeev Kumar Varshney and Robert Koebner.
p. cm.
Includes bibliographical references and index.
ISBN-13: 978-0-8493-3063-6 (alk. paper)
ISBN-10: 0-8493-3063-7 (alk. paper)
1. Crop improvement. 2. Plant molecular biology. I. Varshney, R. K. (Rajeev K.), 1973- II. Koebner, Robert.

SB106.I47M63 2006
631.5'233--dc22 2006017797

Visit the Taylor & Francis Web site at
http://www.taylorandfrancis.com

and the CRC Press Web site at
http://www.crcpress.com

Foreword

Agriculture is vital to our existence: plants provide the oxygen we breathe, the food we eat, many of the fibers for our clothes, and some of the materials used to build our homes, as well as fuel and fodder. Beyond these absolute necessities, a quarter of all medicinal drugs are derived from plant species. However, despite these many vital contributions of plants, far less is known about their biology and, particularly, their genetics than has been gleaned from the mouse, the fruit fly, or the major bacterial species that inhabit our intestines. We need to learn more about how plants grow and develop; how they produce useful chemicals; how they protect themselves from pests; and how they sense, respond to, and even alter our environment. In attempting to illuminate many of these important questions, the last 25 years or so have been very rewarding; development and application of DNA technologies have enabled a quantum leap in our ability to study such diverse topics as genome architecture, plant adaptation, and plant improvement.

To clarify the function of plant genes and to optimize crop improvement strategies, it was realized early that the generality of the central paradigm of molecular biology meant that sophisticated applications such as genome sequencing and functional genomics could be carried out in simple plant species acting as models for more complex ones. As a result of its small size, diploid genetics, small genome, and relatively short life cycle, thale cress (*Arabidopsis thaliana*) was accepted as the first model plant species, and its complete genome sequence was published in 2000. Since then, other plant species—in particular, rice, *Medicago*, *Lotus*, and poplar—have been promoted as complementary models. As DNA sequencing has become less expensive, full genome sequences of these second-generation models have been or are soon to be completed. Though technological and scientific advances reported over recent years continue to be important for basic research, consensus is limited as to whether an improved understanding of *Arabidopsis* or other models has contributed or can ever contribute materially to the breeding of commercial crops.

This book documents achievements (also failings) and prospects of model plant research in the context of its contribution to the advancement of crop science. The editors have commissioned a range of relevant and interesting reviews concerning model species research from leading authorities. I am sure the book will make a significant contribution toward enhancing knowledge on the model-crop paradigm and help practitioners of plant genetics and breeding.

William D. Dar
Director General
International Crops Research Institute for the Semi-Arid Tropics (ICRISAT)
Patancheru, India

Preface

The past two decades have seen significant research activity in model plant biology. In particular, the workhorse *Arabidopsis thaliana* has become a reference species for plant biologists and has taken its place in the genomic universe alongside yeast and the animal models worm, fly, zebra-fish, and mouse as well. The vision at the outset of plant molecular biology was that much of the biological, genetic, and later genomic insights gained from the dissection of this small dicot plant would prove transferable to higher plant species in general and to crop species in particular. However, because many of the world's staple crops are monocots, the age of the monocot/dicot divergence meant that this early optimism had to be tempered with the realization that a separate monocot model was probably essential. The choice fell on rice, which also enjoys a small genome size (though not quite as small as that of *Arabidopsis*) and is notably a crop plant in its own right (unlike the weed *Arabidopsis*), but is less well suited to model status in the context of its slow generation turnover and the large physical size of the adult plant.

Since then, the list of model species has continued to grow, capturing the uniqueness of the important legume–*Rhizobium* symbiosis and tackling the phenomena of perennial and juvenile characteristics of tree species. Although the given major reason for using all these models has been to simplify research, an important additional justification has always been the promise of the flow of discoveries and technologies to crop improvement. As a result, far more is now known about the biology and genetics of the models than of any single crop species (rice, of course, excluded).

The relevance of these models for crop improvement remains a horizon application that has yet to be tested adequately. In one scenario, greater use of models closely associated with respective crop breeding programs will be a winning combination because it will enable many more hypotheses to be tested than is possible using a crop species in isolation, thus streamlining discovery of solutions to crop problems. In the opposing scenario, the model research effort can be better described as expenditure rather than as investment.

The need for improvement in all crops is so urgent and the volume of information flowing from the models so large that closer associations between models and expanded crop biology programs are a priority. Therefore, as the postgenomics era dawns, it has become timely to consider achievements and failings of the model paradigm with respect to crop science and to ask how continued research in models can contribute to the goal of delivering the outputs of molecular biology to crops. We planned the present volume as a means to gather the opinions of "modelers" and "croppers," along with those working at the model–crop transition. The book includes chapters covering the application of discoveries and research in major models (i.e., *Arabidopsis* and rice) for crop improvement programs and provides overviews of other model species such as *Medicago*, *Brachypodium*, and *Chlamydomonas* and a critical assessment of their potential for understanding the molecular genetics of crops.

The editors are grateful to the contributing authors (see "Contributors" section), who not only reviewed the published research work in their area of expertise but also shared their unpublished results to bring the chapters up to date. We also appreciate their patience and cooperation in meeting deadlines and revising their manuscripts, when required. We also acknowledge the strong support of the many collaborators (see "Reviewers" section), who willingly reviewed the manuscripts and gave useful suggestions for improvement. As editors, we take responsibility for errors (we hope few in number!) that may have crept in as a result of our editorial work.

The cooperation and help received from David Fausel and John Sulzycki of CRC Press during various stages of the development and completion of this project are appreciated. Producing this book on the back of full-time research jobs has been demanding of our time, and we thank family and friends for their forbearance in putting up with these demands. RKV particularly acknowledges the help and support of his wife, Monika, who contributed directly to formatting the text, tables, and figures in several chapters of the book.

The editors hope that the book will prove useful for our target audience and that readers will bring any errors or omissions to our notice, as well as offer suggestions, so that any future update in such a quickly changing field will be facilitated.

Rajeev K. Varshney
International Crops Research Institute for the Semi-Arid Tropics (ICRISAT)
Patancheru, India

Robert M.D. Koebner
John Innes Centre
Norwich, U.K.

Editors

Rajeev K. Varshney, Ph.D., works at the International Crops Research Institute for the Semi-Arid Tropics (ICRISAT), Patancheru, India. Dr. Varshney has basic training in the areas of plant genetics, breeding, and biotechnology. After obtaining his Ph.D. from CCS University, Meerut, India, in January 2001, he took a position as a research scientist at the Institute of Plant Genetics and Crop Plant Research (IPK) in Gatersleben, Germany. Dr. Varshney recently joined the ICRISAT as a senior scientist in applied genomics under the global theme of biotechnology.

Robert M.D. Koebner, Ph.D., has been active in the world of wheat genetics and cytogenetics since embarking on his Ph.D. at the University of Adelaide, Australia, in 1981. This period coincided with the beginning of molecular mapping in plants, and thus Dr. Koebner was involved from an early stage in the development and application of markers in wheat. In 1986, he took up a postdoctoral post at the Plant Breeding Institute, Cambridge, United Kingdom. When the PBI was privatized in 1989, he transferred to the John Innes Institute (now the John Innes Centre) in Norwich, where he has worked in the Crop Genetics Department.

Reviewers

Kapil Bharti
National Institutes of Health
Bethesda, Maryland, U.S.A.

Luigi Cattivelli
Experimental Institute for Cereal Research
Foggia, Italy

Isabelle Debeaujon
Institut Jean-Pierre Bourgin, INRA
Versailles, France

John Draper
Institute of Biological Sciences
University of Wales
Aberystwyth, U.K.

Thomas Eulgem
University of California, Riverside
Riverside, California, U.S.A.

Robert Hasterok
University of Silesia
Katowice, Poland

Etienne-Pascal Journet
Laboratoire des Interactions Plantes Micro-organismes, INRA
Tolosan, France

Gábor Kocksy
Hungarian Academy of Sciences
Martonvasar, Hungary

Friedrich Kopisch-Obuch
Georg August Universität
Göttingen, Germany

Deepak Pental
University of Delhi, South Campus
New Delhi, India

Karam Singh
CSIRO Plant Industry
Canberra, Australia

Nagendra K. Singh
National Research Centre on Plant Biotechnology (NRCPB)
New Delhi, India

Imre E. Sommssich
Max Planck Institute for Plant Breeding
Cologne, Germany

Yunbi Xu
Cornell University
Ithaca, New York, U.S.A.

Shinjiro Yamaguchi
RIKEN Plant Science Center
Yokohama, Japan

Tong Zhu
Syngenta Biotechnology Inc.
Research Triangle Park, North Carolina, U.S.A.

Contributors

Markku K. Aalto
Department of Biological and Environmental Sciences
Division of Genetics
University of Helsinki
Helsinki, Finland

Baltazar A. Antonio
National Institute of Agrobiological Sciences
Tsukuba, Ibaraki, Japan

Tameka Bailey
Department of Plant Pathology and Program in Cell and Molecular Biology
University of Arkansas
Fayetteville, Arkansas, U.S.A.

Leónie Bentsink
Laboratory of Genetics
Wageningen University
Wageningen, the Netherlands

Wolfgang Busch
Department of Molecular Biology
Max Planck Institute for Plant Breeding Research
Tübingen, Germany

Emilio Fernández
Departamento de Bioquímica y Biología Molecular
Facultad de Ciencias
Universidad de Córdoba
Córdoba, Spain

Pierre R. Fobert
Protein Research Group
Plant Biotechnology Institute
National Research Council of Canada
Saskatoon, Saskatchewan, Canada

Henk Franssen
Department of Plant Science
Laboratory of Molecular Biology
Wageningen University
Wageningen, the Netherlands

Aurora Galván
Departamento de Bioquímica y Biología Molecular
Facultad de Ciencias
Universidad de Córdoba
Córdoba, Spain

David F. Garvin
USDA-ARS Plant Science Research Unit and
Department of Agronomy and Plant Genetics
University of Minnesota
St. Paul, Minnesota, U.S.A.

René Geurts
Department of Plant Science
Laboratory of Molecular Biology
Wageningen University
Wageningen, the Netherlands

David González-Ballester
Departamento de Bioquímica y Biología Molecular
Facultad de Ciencias
Universidad de Córdoba
Córdoba, Spain

Pekka Heino
Department of Biosciences
Division of Genetics and Institute of Biotechnology
University of Helsinki
Helsinki, Finland

Michael Holdsworth
Division of Agricultural and Environmental Sciences
University of Nottingham, Sutton Bonington Campus
Loughborough, U.K.

Graham J. King
Rothamsted Research
Harpenden, U.K.

Robert M.D. Koebner
Crop Genetics Department
John Innes Centre
Norwich, U.K.

Maarten Koornneef
Laboratory of Genetics
Wageningen University
Wageningen, the Netherlands
and
Max Planck Institute for Plant Breeding Research
Cologne, Germany

Mukesh Kumar
Zentrum für Molekularbiologie der Pflanzen
Allgemeine Genetik
Eberhard Karls Universität
Tübingen, Germany

Vicente Mariscal
Departamento de Bioquímica y Biología Molecular
Facultad de Ciencias
Universidad de Córdoba
Córdoba, Spain

E. Tapio Palva
Department of Biosciences
Division of Genetics and Institute of Biotechnology
University of Helsinki
Helsinki, Finland

Tressa J. Panikulangara
Zentrum für Molekularbiologie der Pflanzen
Allgemeine Genetik
Eberhard Karls Universität
Tübingen, Germany

Takuji Sasaki
National Institute of Agrobiological Sciences
Tsukuba, Ibaraki, Japan

Friedrich Schöffl
Zentrum für Molekularbiologie der Pflanzen
Allgemeine Genetik
Eberhard Karls Universität
Tübingen, Germany

Rajeev K. Varshney
International Crops Research Institute for the Semi-Arid Tropics
Patancheru, India

Yinong Yang
Department of Plant Pathology
and
Program in Cell and Molecular Biology
University of Arkansas
Fayetteville, Arkansas

Xiangjun Zhou
Department of Plant Pathology
and
Program in Cell and Molecular Biology
University of Arkansas
Fayetteville, Arkansas, U.S.A.

Contents

1 Development and Application of Genomic Models for Large-Crop Plant Genomes

Robert M.D. Koebner and Rajeev K. Varshney

CONTENTS

1.1 INTRODUCTION

Plant genomes vary enormously in size. A part of this variation is generated by polyploidy, which is ubiquitous in the plant kingdom; however, even between closely related, ostensibly diploid species, it can still vary by an order of magnitude. A notable, but not atypical example is the contrast between rice (1 C DNA content of 0.50 pg, equivalent to 450 Mbp) and barley (5.55 pg, 5300 Mbp). The gene content of these two species is thought to be rather similar, numbering something under 40,000, depending on the gene prediction program employed [1]. Thus, much of the difference in DNA content is made up of nongenic DNA—in particular, retrotransposons.

When large-scale genome sequencing became possible in the 1990s, the large size of the majority of the leading crop genomes was technically and financially prohibitive. This prompted the plant research community to identify species (in particular *Arabidopsis thaliana*) with more tractably sized genomes as genomic models. Technical improvements in the efficiency of sequencing achieved the finishing of the *Arabidopsis* genome by 2000 (4 years ahead of schedule) and the sequence was released with some fanfare in *Nature* [2].

At the time, *Arabidopsis* represented one of the first eukaryotes to be sequenced fully (along with *Saccharomyces cerevisiae*, human, and *Caenorhabditis elegans*). Its protein-encoding gene content has been estimated to be about 25,000 [2]. In the meantime, the genome sequence of *Arabidopsis* has been joined by those of a bewildering and ever-growing list of eukaryotic and prokaryotic organisms numbering over 300 as of December 2005 (http://www.genomesonline.org). Of the 40 fully sequenced eukaryotic genomes, 25 belong to simple organisms (protozoans and fungi), 7 are vertebrates, 3 are insects, 2 are nematodes, and 3 are plants (of which 2 are the *indica* and *japonica* subspecies of rice).

The divergence of the monocot from the dicot clade is an ancient event, currently dated using molecular clock methods applied to the chloroplast genome at 140 to 150 MYA during the late Jurassic to early Cretaceous periods [3]. Independent estimates based on mitochondrial sequences have placed it somewhat earlier, at 170 to 235 MYA [4]. Dating of the time of speciation within each clade has been attempted by applying molecular clock methodology to repetitive sequences such as retrotransposons, but sequence homology in this class of element between clades is insufficient to use this method to date the monocot–dicot divergence.

Thus, it was recognized at an early stage that the *Arabidopsis* genome sequence would probably be of only partial relevance to monocot genomes. With a genome size about three times larger than that of *Arabidopsis*, rice was rapidly identified as the donor of a suitable model monocot genome. Before completion of the rice genome sequence, it became apparent that only a poor level of commonality in gene order existed between *Arabidopsis* and rice [5], thereby justifying post hoc the need for a separate model for the two major plant clades.

Nevertheless, the two genomes do retain some similarity as a result of common descent. Although some 85% of predicted *Arabidopsis* proteins were found to share significant homology with those of rice, about a tenth of them show a strong level of conservation [6]; in addition, most monocot–dicot homologs maintain exon order as expected. Perhaps most surprisingly, in many homologs, intron number, position, and even relative size show a remarkable level of conservation [7]. Despite the apparent disparity in gene number between the two models (25,000 vs. 40,000), it has recently been claimed that only a few hundred, or at most a few thousand, rice genes appear to lack close homologs in *Arabidopsis* [1].

The infrastructure and efficiency of whole genome sequencing is now at a point at which it has become much more realistic to undertake on a large scale. Current crop species targets include oat, *Brassica* spp., orange, coffee, barley, soybean, cotton, ryegrass, alfalfa, tomato, banana, bean, poplar, castor oil, sorghum, and maize. A growing number of other species has been targeted for sequencing of the gene space (ESTs or similar). If these trends continue, it is likely that within 10 years, most of the major crop genomes will have been fully sequenced. In the meantime, species that are nodal in crop phylogenies may be chosen to serve to generate a network of submodels; a particular example of this lies behind the current proposal to sequence the grass *Brachypodium distachyon*.

This chapter attempts to take stock of model genomes' contribution to understanding of the genomes of crop species to date. Perhaps other contributors to this volume will show the lasting value that model species biology has made to crop improvement.

1.1.1 Dicot Models

1.1.1.1 *Arabidopsis thaliana* (Thale Cress)

Arabidopsis is by far the most well developed of the crop plant models. In addition to its completed genome sequence, it is easily transformable and enjoys a huge range of genetic (mutants, mapping populations, ecotypes) and genomic (cloned genes, libraries, arrays, markers, etc.) resources and an ever expanding database relating phenotype to genotype. The closest crop relatives to *Arabidopsis* are the three diploid *Brassica* species *rapa*, *nigra*, and *oleracea* that carry, respectively, the A, B, and C genomes as described in Reference 8. Although all of these represent rather minor crop species, the major contributor of *Brassica* spp. to agriculture is *B. napus* (oilseed rape or canola), which is an AC allotetraploid formed from the combination *B. oleracea* × *B. rapa*.

The lineages of *Arabidopsis* and *Brassica* are thought to have diverged from one another between 14 and 20 MYA [4]; this divergence has included a number of distinct polyploidization events because the present-day diploid *Brassica* spp. carry multiple paralogous copies of chromosomal segments collinear with the *Arabidopsis* genome. This copy number is most commonly three, so the inference is that the diploids must have evolved from a hexaploid ancestor [9,10]. Copy number is frequently less than three, varying in 4× *B. napus* from four to seven [10]. Within the triplicated paralogs, a common pattern of interspersed gene loss is emerging, with the result that each paralog typically carries a slightly different spectrum of the full gene set presumably present on the progenitor segment [11].

A further complication is that *Arabidopsis*, as revealed from its genome sequence, is a cryptic polyploid, carrying a sufficient number of large segmental duplications for an evolutionary history of at least four different large-scale duplication events to have been proposed [12]. Overall, an estimated 74 translocations, fusions, deletions, or inversions separate the genomes of *Arabidopsis* and *B. napus* [10], of which about one half are common to A and C genomes in present-day oilseed rape.

1.1.1.2 *Lotus japonicus* (Trefoil) and *Medicago truncatula* (Barrel Medic)

The Fabaceae, one of the largest families of flowering plants with 650 genera and over 18,000 species, is distinguished from other dicot families by its symbiotic relationship with nitrogen-fixing *Rhizobium*. The economic and nutritional importance of nitrogen fixation has been sufficient to justify targeting a model representative, and two competitive species are currently being pursued. *Medicago truncatula* has some importance in its own right as a forage crop in Australia. It has a small diploid genome (1 C DNA 0.48 pg) and a rapid generation time, is self-fertile, transformable, and is a prolific seed producer. *Lotus japonicus* is a short-life-cycle, perennial wild legume that also has a small genome size (1 C DNA 0.48 pg).

The genomes of both species are currently being sequenced (see, respectively, http://www.medicago.org and http://www.kazusa.or.jp/lotus/index.html). The two sequences show a high degree of similarity to one another [13]. Collinearity between *M. truncatula* and pea at the level of coarse genetic maps appears to be encouragingly high [14], although there is significant sequence divergence between those of *Lotus* and the major legume crop species soybean [13]. In a computational approach, *Lotus*, *Medicago*, and *Glycine* unigenes were BLASTed against non-legume unigene sets and the rice genome sequence to define legume-specific gene motifs; this delivered some 2500 such contigs, of which less than 3% showed any homology to any previously identified legume genes [15]. Such results underline the utility of a model legume to define sequences specific to this group of agriculturally important crop species.

1.1.1.3 *Populus trichocarpa* (Poplar or Black Cottonwood)

Conventional genetic approaches in trees are limited by the large size, long generation interval, and outcrossing mating system of most species. The need for a tree model reflects the importance of many traits that are not shared by an herbaceous annual plant such as *Arabidopsis*. Important among these are wood formation, longevity, seasonal growth, and hardiness. The genus *Populus* consists of 30 to 40 species, 4 of which have significant commercial importance. Selection and hybridization programs in poplars began in North America in the 1960s, and the most commonly exploited crosses have involved *P. trichocarpa*, *P. deltoides*, *P. nigra*, *P. grandidentata*, *P. alba*, *P. tremuloides*, and *P. tremula*.

Because the genomic resources of *P. trichocarpa* were the most developed at the time that genome sequencing was proposed, this species became the accepted tree model. It was chosen as the first tree for genome sequencing largely because of its modest genome size (0.6 pg)—about 40 times smaller than that of pine, the most important of all forestry species. It also has a number of other advantages over potential alternative tree species specifically related to its rapid juvenile growth, which allows for phenotypic assessments to be made relatively quickly; its well-established transformation and regeneration protocols; and the pre-existence of a body of genetic mapping, which includes placement and tagging of a number of quantitative trait loci (QTL). The final draft sequence was scheduled for release in early 2005, but is still awaited at the time of writing. Current status is updated on http://genome.jgi-psf.org/Poptr1/Poptr1.home.html.

1.1.2 Monocot Models

1.1.2.1 *Oryza sativa* (Rice)

Rice is the pre-eminent monocot model and is uniquely both a model and a crop in its own right. The particular importance of this duality lies in the much greater potential that this allows for transferring phenotype, as well as genotype, from model to crop. Rice is a tropical species and thus more likely to share pathogens and/or abiotic stresses with its tropical crop relatives such as the millets (and, to a lesser extent, maize and sorghum) than with its important temperate small-grain and pasture-grass relatives (wheat, barley, rye, oat, and ryegrass). Nevertheless, shared morphology and crop architecture among all cereal species do allow many phenotypic connections to be made. The dicot models, in contrast, are far removed in a crop morphology, making such transfers much less predictable.

The grasses belong to the Poaceae, which evolved from a common ancestor some 50 to 60 MYA; together, they provide an estimated 60% of global human calorific intake. The family includes at least 10,000 species, classified into 650 genera [16,17]. The crop species within the family fall into the three subfamilies: Pooideae (which includes the temperate cereals and ryegrass), Panicoideae (maize, sorghum, millets, sugar cane), and Bambusoideae (rice). Until the development of generic DNA technology, primarily in the 1990s, genetic research in each grass crop was conducted in isolation from that in the others. Before this time there was no secure way of verifying what had already been suspected for some time: that because these species were related by (albeit distant) descent, they were likely to share genetic content and, at least at a basic level, genetic mechanisms.

The first demonstration of what is now referred to as "comparative genetics" was carried out in the Solanaceae, where common RFLP linkage relationships in tomato and potato were uncovered using DNA probes developed from a tomato template [18]. The concept spread quickly to the Poaceae, and numerous cross-species comparisons began to appear in the literature during the early to mid-1990s [19–21]. These led to the construction of partial consensus maps linking maize with sorghum [22] and wheat with barley and rye [23]. A synthesis of these maps was generated by relating them all to that of the rice genome [24]. The concept of "synteny" elaborated by these cross-species comparisons of gene order reflects conservation over evolutionary time at the macroscale. Whether this was extendable to the microscale was questionable, given the large variation in genome size between individual Poaceae species.

The outcome of sequence-based comparisons in selective syntenic regions is that although gene structure and sequence are extremely well conserved between taxa, intergenic regions are highly divergent, even at the level of genotypes within a taxon [25]. Much of this intra- and interspecific divergence is generated by retroelement activity and, in particular, helitron-like transposons composed of multiple gene-derived fragments [26]. In addition, the increasing body of evidence generated from large-scale sequence comparisons between related taxa demonstrates how synteny is also disturbed by the presence of species-specific localized duplications and other forms of genome reorganization [27–29].

By the end of the 1990s, with the *Arabidopsis* genome project already well under way, rice became an increasingly attractive candidate for whole genome sequencing [30] in the private and public sectors. These efforts were combined to produce almost full genomic sequences of *japonica* and *indica* subspecies [6,31,32], along with a near-complete compendium of full-length cDNA sequence [33]. The finished sequence currently covers about 95% of the genome, including most euchromatic regions and

2 (out of 12) complete centromeres. Mirroring the situation in *Arabidopsis,* the genome sequence has revealed a history of polyploidization in the evolution of modern day rice, with about half of the gene content duplicated as paralogs.

1.1.2.2 *Brachypodium* spp. (False Bromes)

The false bromes are a group of non-cultivated grasses, mostly regarded in agriculture as weeds rather than as beneficial plants. The perennial *B. sylvaticum* (slender false brome) and the annual *B. distachyon* (purple false brome) have been suggested as intermediate models for the temperate cereals. *B. distachyon* has been claimed to have a genome size indistinguishable from that of *Arabidopsis* [34], but measurement of 1 C DNA content suggested that it is three times larger (0.36 pg; http://www.rbgkew.org.uk/cvalues/). The genome size of *B. sylvaticum* is slightly higher still (0.48 pg), but both genomes are smaller than that of rice.

The value of both as genomic models for the temperate grain cereals lies in their membership within the Pooideae clade and hence their much closer relationship to wheat, barley, rye, and oats than rice enjoys. The significance of this relationship has been confirmed in two recent positional cloning projects, one in wheat [35] and the other in barley [36] since both the quality of probe hybridization to and prediction of overall gene content in the target were superior in *Brachypodium* to that offered by rice [37]. Although *B. sylvaticum* has been proposed to date only as a genomic and not a biological model, *B. distachyon* does have a number of generic advantages as a functional genomic and biological model (self-fertility, in-breeding habit, short life cycle, small size [approximately 20 cm at maturity], lack of seed-head shatter, and undemanding growth requirements) [34]. At the time of writing, there is a concerted effort to develop *B. distachyon* as a fully functional genomic model, but this proposal remains controversial.

1.2 HARNESSING MODEL GENOMES FOR CROP GENETICS AND IMPROVEMENT

The impact of model genomes on crop species has been felt mainly in their delivery of a strategy for gene isolation in the large genome crop species. This strategy relies on the maintenance of synteny, assuming that gene content in the model in a specific genomic region is more or less conserved in the target crop genome. The model-to-crop paradigm follows a combination of:

- Mapping a trait to a defined genetic interval in the crop
- Identifying the corresponding genomic region in the model via the use of common genic markers (because it is substantially only the gene content, not the nongenic, largely retrotransposon-containing, repetitive content that is conserved across clades)
- Identifying a potential candidate sequence in the model on the basis of a relationship between predicted gene function (derived from the annotation of the model genome) and the target trait
- Validating the crop homolog of the candidate, demonstrated by allelic association and/or mutation complementation

The first major success of the model-to-crop genomic approach in the monocots came with the isolation of the "green revolution" wheat semidwarfing genes *Rht-B1* and *Rht-D1* [38]. Together, these two genes have been responsible for probably the most far-reaching and widespread change in the appearance of any crop worldwide. Their incorporation into the breeding pool has generated shorter plants that enjoy an enhanced grain yield potential, thanks to the consequent increase in harvest index, and are responsive to higher application rates of fertilizer without becoming liable to straw collapse.

The isolation of these genes predated the availability of the full rice or *Arabidopsis* genome sequence, but nevertheless relied heavily on genomic information from both model species. Critical to the success of their cloning was that the physiological nature of the semidwarf variants of wheat was similar to that of previously characterized mutants in maize and *Arabidopsis*. This allowed an approach whereby the rice ortholog of the *Arabidopsis gai* gene was identified from a rice EST collection. When this rice sequence hybridized to wheat DNA at the genomic locations of the *Rht-1* genes, the rice probe was exploited to extract the full genomic sequence of both of the wheat genes. Thereafter, the sequence and functional basis of these important semidwarf alleles were readily obtained.

Finishing the genome sequences of the models enabled the model-to-crop paradigm to be tested. A textbook illustration was provided by the recent successful cloning of the barley gene *Ppd-H1*, the major determinant of flowering time under long photoperiods [36]. Unlike the situation with *Rht*-1, the physiological model provided by *Arabidopsis* was not informative because the candidate genes provided by *Arabidopsis* did not map to the genomic location of the barley gene target. Thus, the initial step was to fine-map *Ppd-H1* in a conventional cross between parents carrying contrasting alleles, and the linked markers thereby derived then allowed for construction of a physical contig based on the presence of key marker loci on barley bacterial artificial chromosomes (BACs). The gene content of the homologous region in *Brachypodium sylvaticum* helped to define the matching region in rice, and the critical barley recombinants finally identified a region in the homologous rice segment that contained only a single candidate sequence.

This rice gene, *Os-PRR*, shares significant sequence homology with *Arabidopsis At-PRR7*, which, when mutated, leads to delayed flowering under long day conditions, just as the inactive form of *Ppd-H1* does in barley. *Ppd-H1* and *At-PRR* also share temporal patterns of expression. Finally, resequencing of the critical parts of *Hv-PRR* across varieties of known allelic status at *Ppd-H1* was able to demonstrate a correlation between a functional glycine to tryptophan change in a domain of the gene that is well conserved across taxa.

A more elaborate but essentially equivalent strategy was used to isolate the wheat gene responsible for determination of winter habit (vernalization requirement) [39]. Once again, a large mapping population, this time in the diploid wheat *Triticum monococcum*, was used to delineate a genetic interval of <0.1 cM containing the target. Sequencing of the 324 kb represented by this segment identified two genes, with no additional candidates present in the homologous segments of rice or sorghum. Both candidate genes had *Arabidopsis* homologs, but only one of them, *AP1*, is required for the transition between vegetative and reproductive phases in *Arabidopsis*; the other is a floral meristem identity gene. The association between sequence variation at *Tm-AP1* and phenotype was established by demonstration of three independent deletions distinguishing the promoter sequence of spring from winter accessions.

The most recent example of positional cloning in a monocot crop that has relied on the availability of model genomes is the isolation of the *Ph1* locus in wheat [35]. This "gene" is responsible for the diploid-like inheritance of hexaploid wheat, and its isolation was hampered at the outset by a lack of any verifiable allelic variation. Because of this, it was not possible to generate a fine-scale genetic map as a first step to defining the target genomic region. Instead, a series of overlapping deletion mutants was generated, and phenotype (loss of diploid-like chromosome pairing) was related with loss of genic markers in the *Ph1* region, which had been derived from synteny comparisons between wheat and rice and/or *Brachypodium.*

As a result, the number of genes present in the smallest genetic interval defined was over 30, and because the effect of *Ph1* is specific to polyploids, there were no sensible leads derived from the predicted function of any of these candidates. To progress beyond this point, it was necessary to sequence a substantial tract of wheat DNA directly; the identity of the locus was finally determined through an internal comparison among the individual A, B, and D genome segments.

A reasonable level of synteny between *Arabidopsis* and *Brassica* exists, the complications of segmental duplication notwithstanding [40], and the finished *Arabidopsis* sequence has been available for longer than that of rice; however, gene isolation in *Brassica* has relied more on functional homology than on positional cloning. Thus, having established function of a gene in *Arabidopsis*, primarily by mutation/complementation, homologs in *Brassica* have been extracted from genomic or cDNA libraries and function in *Brassica* established by associating variation in phenotype with polymorphism at the RFLP or sequence level. Beyond the *Brassica* spp., high rates of sequence divergence have greatly inhibited the success of orthologous cDNAs as hybridization probes against genomic DNA and restricted the applicability of the model to its immediate relatives.

1.3 PERSPECTIVE

The value of model plants in a strictly genomic context is probably ephemeral. This is primarily because large genomes are increasingly considered practical to sequence on cost or technical grounds. Within 10 years, it is likely that most of the major crops will have been sequenced, at least with respect to their gene space. At the same time, comparative genomics is showing that although gene order at the macroscale is well conserved over large taxonomic distances, the microsynteny necessary to predict sequence across species (and even, to a surprising extent, within species [25]) at the microscale is insufficient for a small number of models to be able to serve many diverse crop species. The cereals are exceptional in this respect, in that so many cereal crop species are clustered within a narrow taxonomic clade, but even for these, the models have their limitations.

The more lasting value of models will surely lie in the insights into plant biology that they will allow. Some of these will include the rapidly developing fields of epigenetic and micro-RNA-directed gene regulation, where *Arabidopsis* is already serving as a model organism for species well beyond the plant kingdom [41,42]. Many of the more specifically plant-orientated areas of biology informed by model species are covered by other contributions to this volume.

REFERENCES

1. Bennetzen, J.L. et al., Consistent overestimation of gene number in complex plant genomes, *Curr. Opin. Plant Biol.*, 7, 732, 2004.
2. *Arabidopsis* Genome Initiative, Analysis of the genome sequence of the flowering plant *Arabidopsis thaliana*, *Nature* 408, 796, 2000.
3. Chaw, S.M. et al., Dating the monocot–dicot divergence and the origin of core eudicots using whole chloroplast genomes, *J. Mol. Evol.*, 58, 424, 2004.
4. Yang, Y.W. et al., Rates of nucleotide substitution in angiosperm mitochondrial DNA sequences and dates of divergence between *Brassica* and other angiosperm lineages, *J. Mol. Evol.*, 48, 597, 1999.
5. Liu, H., Sachidanandam, R., and Stein, L., Comparative genomics between rice and *Arabidopsis* shows scant collinearity in gene order, *Genome Res.*, 11, 2020, 2001.
6. Goff, S.A. et al., A draft sequence of the rice genome (*Oryza sativa* L. ssp. *japonica*), *Science*, 296, 92, 2002.
7. Carels, N. and Bernardi, G., Two classes of genes in plants, *Genetics*, 154, 1819, 2000.
8. U, N., Genome analysis in *Brassica* with special reference to the experimental formation of *B. napus* and peculiar mode of fertilization, *Jpn. J. Bot.*, 7, 389, 1935.
9. Lysak, M.A. et al., Chromosome triplication found across the tribe Brassiceae, *Genome Res.*, 15, 516, 2005.
10. Parkin, I.A.P. et al., Segmental structure of the *Brassica napus* genome based on comparative analysis with *Arabidopsis thaliana*, *Genetics*, 171, 765, 2005.
11. Rana, D. et al., Conservation of the microstructure of genome segments in *Brassica napus* and its diploid relatives, *Plant J.*, 40, 725, 2004.
12. Vision, T.J., Brown, D.G., and Tanksley, S.D., The origins of genomic duplications in *Arabidopsis*, *Science*, 290, 2114, 2000.
13. Choi, H.K. et al., Estimating genome conservation between crop and model legume species, *Proc. Natl. Acad. Sci. USA*, 101, 15289, 2004.
14. Kalo, P. et al., Comparative mapping between *Medicago sativa* and *Pisum sativum*, *Mol. Genet. Genomics*, 272, 235, 2004.
15. Graham, M.A. et al., Computational identification and characterization of novel genes from legumes, *Plant Physiol.*, 135, 1179, 2004.
16. Bennetzen, J.L. and Freeling, M., Grasses as a single genetic system—genome composition, collinearity and compatibility, *Trends Genet.*, 9, 259, 1993.
17. Kellogg, E.A., Relationships of cereal crops and other grasses, *Proc. Natl. Acad. Sci. USA*, 95, 2005, 1998.
18. Bonierbale, M.W., Plaisted, R.L., and Tanksley, S.D., RFLP maps based on a common set of clones reveal modes of chromosomal evolution in potato and tomato, *Genetics*, 120, 1095, 1988.
19. Ahn, S. and Tanksley, S.D., Comparative linkage maps of the rice and maize genomes, *Proc. Natl. Acad. Sci. USA*, 90, 7980, 1993.
20. Kurata, N. et al., Conservation of genome structure between rice and wheat, *Bio-Technology*, 12, 276, 1994.

21. Devos, K.M. et al., Chromosomal rearrangements in the rye genome relative to that of wheat, *Theor. Appl. Genet.*, 85, 673, 1993.
22. Dufour, P. et al., Comparative genetic mapping between duplicated segments on maize chromosomes 3 and 8 and homoeologous regions in sorghum and sugarcane, *Theor. Appl. Genet.*, 92, 1024, 1996.
23. VanDeynze, A.E. et al., Molecular-genetic maps for group-1 chromosomes of Triticeae species and their relation to chromosomes in rice and oat, *Genome*, 38, 45, 1995.
24. Moore, G. et al., Cereal genome evolution—grasses, line up and form a circle, *Curr. Biol.*, 5, 737, 1995.
25. Brunner, S. et al., Evolution of DNA sequence nonhomologies among maize inbreds, *Plant Cell*, 17, 343, 2005.
26. Morgante, M. et al., Gene duplication and exon shuffling by helitron-like transposons generate intraspecies diversity in maize, *Nat. Genet.*, 37, 997, 2005.
27. Salse, J. et al. New *in silico* insight into the synteny between rice (*Oryza sativa* L.) and maize (*Zea mays* L.) highlights reshuffling and identifies new duplications in the rice genome, *Plant J.*, 38, 396, 2004.
28. Sorrells, M.E. et al., Comparative DNA sequence analysis of wheat and rice genomes, *Genome Res.*, 13, 1818, 2003.
29. Klein, P.E. et al., Sequence-based alignment of sorghum chromosome 3 and rice chromosome 1 reveals extensive conservation of gene order and one major chromosomal rearrangement, *Plant J.*, 34, 605, 2003.
30. Goff, S.A., Rice as a model for cereal genomics, *Curr. Opin. Plant Biol.*, 2, 86, 1999.
31. Yu, J. et al., A draft sequence of the rice genome (*Oryza sativa* L. ssp *indica*), *Science*, 296, 79, 2002.
32. International Rice Genome Sequencing Project, The map-based sequence of the rice genome, *Nature*, 436, 793, 2005.
33. Rice Full-Length cDNA Consortium, Collection, mapping, and annotation of over 28,000 cDNA clones from *japonica* rice, *Science*, 301, 376, 2003.
34. Draper, J. et al., *Brachypodium distachyon*. A new model system for functional genomics in grasses, *Plant Physiol.*, 127, 1539, 2001.
35. Griffiths, S. et al., Molecular characterization of *Ph1* as a major chromosome pairing locus in polyploid wheat, *Nature* (London), 439(7077), 749, 2006.
36. Turner, A. et al., The pseudo-response regulator *Ppd-H1* provides adaptation to photoperiod in barley, *Science*, 310, 1031, 2005.
37. Foote, T.N. et al., Construction and analysis of a BAC library in the grass *Brachypodium sylvaticum*: its use as a tool to bridge the gap between rice and wheat in elucidating gene content, *Funct. Integr. Genomics*, 4, 26, 2004.
38. Peng, J.R. et al., "Green revolution" genes encode mutant gibberellin response modulators, *Nature*, 400, 256, 1999.
39. Yan, L. et al., Positional cloning of the wheat vernalization gene *VRN1*, *Proc. Natl. Acad. Sci. USA*, 100, 6263, 2003.
40. Lukens, L. et al., Comparison of a *Brassica oleracea* genetic map with the genome of *Arabidopsis thaliana*, *Genetics*, 164, 359, 2003.
41. Martienssen, R.A., Doerge, R.W., and Colot, V., Epigenomic mapping in *Arabidopsis* using tiling microarrays, *Chromosome Res.*, 13, 299, 2005.
42. Lippman, Z. et al., Role of transposable elements in heterochromatin and epigenetic control, *Nature*, 430, 471, 2004.

2 Conserved Mechanisms of Dormancy and Germination as Targets for Manipulation of Agricultural Problems

Michael Holdsworth, Leónie Bentsink, and Maarten Koornneef

CONTENTS

2.1 DORMANCY AND GERMINATION OF SEEDS—DEFINITIONS AND ECOLOGICAL SIGNIFICANCE

The seed is the structure in which a usually fully developed plant embryo is dispersed and that enables it to survive the period between seed maturation and establishment of the next generation as a seedling after it has germinated. The dry, quiescent seed is well equipped to sustain extended periods of unfavorable conditions. Dormancy, defined as the failure of an intact viable seed to complete germination under conditions favorable for germination, is an adaptive trait optimizing germination to the best suitable time that enables the species to complete its life cycle. The environmental conditions required for germination are not defined specifically, but in practice refer to conditions that allow germination of a nondormant seed batch of the species under investigation [1].

Dormancy varies in a quantitative way often described by deep and non-deep or strong and weak dormancy. Dormancy is the property of the seed, so the degree of dormancy defines which conditions should be met to make the seeds germinate. Therefore, a more precise method of defining the dormancy status of a seed batch is to describe the environmental requirements for germination (temperature range or time of after-ripening required to overcome dormancy) [2].

A complication in seed dormancy research is that the germination assays used are a measure of the integration of many events that happened in the history of the seed (dormancy) and the various environmental factors acting during germination. For a better understanding of seed dormancy and germination, it is important to distinguish between these two processes. Germination (or germination *sensu stricto* or visible germination) is defined as embryo protrusion, which depends on embryo expansion (which is a growth, mainly cell expansion, process) driven by water uptake. After radicle protrusion, seedling establishment takes place, which requires mobilization of reserves and growth of the seedling. Seedling establishment is often included in the seed germination process (Figure 2.1).

Different types of dormancy, including primary, secondary, seasonal, and coat-imposed dormancy, have been defined. Primary dormancy is induced during seed development. Dormancy is most likely induced or initiated during the later phases of seed development, as can be concluded by the absence of seed dormancy in mutants that have a strongly disturbed seed maturation, such as ABA-*insensitive 3* (*abi3*), *fusca 3* (*fus3*), and *leafy cotyledon* (*lec1* and *lec2*). Because virtually all of the cellular and metabolic events known to occur before the completion of germination in nondormant seeds also occur in imbibed dormant seeds, failure of the embryo axis to elongate seems to represent the mechanism of dormancy [3]. Secondary dormancy can be induced when imbibed seeds cannot germinate because of an unfavorable environmental factor; this indicates that dormancy induction mechanisms continue to operate even after loss of primary dormancy.

Dormancy and germination are determined by balance of the growth potential of the embryo and the constraints imposed by the tissues surrounding it. The balance between these forces and their relative contribution as well as differences in the response to environmental conditions results in the fact that dormancy can be very different between species. In many species, the seed envelope imposes a strong physical constraint to radicle protrusion. This explains why envelope characters affect the dormancy status of the seed and also why weakening these envelopes (which can be the testa, the endosperm layer, or both) leads to germination.

Dormancy and germination are strongly influenced by environmental factors, which are mainly light and temperature, as well as soil factors, among which nitrate is best known. These environmental factors can act during formation of the seed (maternal factors) and during the imbibed stage of the mature seed. According to Vleeshouwers et al. [2], changes in dormancy levels of imbibed seeds depend only on temperature. In addition, some factors may act during conditions of low metabolic activity due to low water content and explain the after-ripening effect, a strong factor influencing dormancy release. How endogenous and environmental factors interact is largely unknown, with the exception of the induction of gibberellin (GA) synthesis during imbibition.

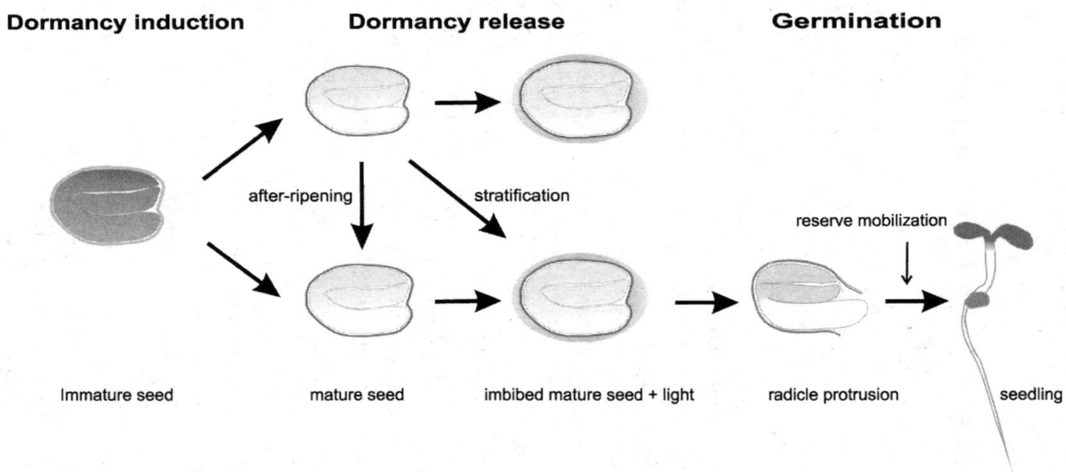

FIGURE 2.1 **(Color Figure 2.1 follows p. 144.)** A general outline of dormancy and germination in *Arabidopsis* seeds.

The ecological significance and large variation in dormancy characteristics between species are described in reviews by Baskin and Baskin [1,4] and references there. Seasonal dormancy can delay germination until conditions are appropriate for growth and consequently may influence fitness. In temperate climates, delaying germination until spring (when the cold winter changes the dormancy status) can prevent winter mortality. However, autumn germination may have a selective advantage when the risk of winter mortality is low by enabling the plants to flower earlier or at a larger size [5].

2.2 DORMANCY AS AN AGRICULTURAL PROBLEM

Dormancy has been an agricultural problem since early farmers first started to domesticate wild plant species [3]. Those features of dormancy that provide ecological advantages present agronomic disadvantages within a farmed system. At one extreme, many weed species show very high levels of dormancy when shed from the plant, thereby infesting land that subsequently requires long-term treatment to remove succeeding generations [6]. At the other extreme, lack of dormancy in crops is considered part of the "domestication syndrome" that provides the benefit of early seedling establishment, but the disadvantage of possible germination before harvest and reduced quality of the seed [7].

Within crop species, competence for dormancy is associated with several agronomic problems. The capacity of oil seed rape (*Brassica napus*) to display secondary dormancy means that in successive harvests from fields originally sown with rape, "volunteer" plants can appear as germination is triggered following ploughing and light-activated germination [8]. Wild rice types (including red rice) can also present problems as weeds, in part because of dormancy characteristics. Within the Triticeae tribe of the grass family, the important agricultural species barley, rye, and wheat (bread and pasta) can show extreme reductions in dormancy during grain development, leading to a complex set of traits together known as preharvest sprouting (PHS) [9]. Alternatively, seeds can display high levels of primary dormancy, thus requiring heating and storage treatments before use, for example, in malting (where uniform germination is essential) [10]. Storage and heating requirements to remove moisture (and the potential for sprouting) from wheat seeds harvested under damp conditions are major environmental costs because of the energy used during storage.

PHS is a particularly important problem associated with seed quality, and quality issues for downstream processors can result from different phenomenology. For example, rape seed oil content is reduced in harvests containing seeds showing PHS; typically, the oil contains free fatty acids associated with increased cloudiness that can be green due to the presence of chlorophyll [11]. Typically, industries associated with crushing to extract oil will not buy seeds containing greater than 2% free fatty acid. Seed lots containing high levels of hydrolytic enzyme activity associated with sprouted wheat and rye grains (principally due to enzymatic activity of alpha-amylase) produce flour that, if used for baking, provides low-quality loaves of bread, with a typically sticky crumb structure and poor loaf volume.

Triticeae PHS is one component of a complex interrelated set of phenotypes characterized by inappropriate preharvest alpha-amylase production by grains [9].

It is measured commercially using the Hagberg falling number test (HFN) that provides an indication of starch breakdown by endogenous enzymes [12]. In the United Kingdom, an HFN score above 250 is usually required for bread-making wheat, and above 225 for grower contracts [13]. Sprouting damage is caused by interaction between development within the maturing seed and prevailing environmental conditions. Typically, it occurs when cool, damp conditions are encountered before harvest through excessive moisture remaining in intact ears or as a side effect of increased exposure of ears to ground moisture following lodging. Under these conditions, seeds do not contain enough dormancy to prevent the onset of germination; this phenomenon has also been termed postmaturity sprouting (PoMS; [9]).

In addition to physical environmental damage, biological damage of wheat seeds by larvae of orange wheat blossom midge (OBM; *Sitodiplosis mosellana*) can also lead to sprouting, although this mechanism is of far less economic significance. This occurs at an earlier stage of seed development than PoMS, when grains still have a high moisture content, and sprouting may be the result of damage to embryo surrounding structures, leading to a reduction in dormancy capacity [9]. PHS is associated with production of low- and high-pI alpha-amylase (*amy2* and *amy1*, respectively), normally expressed in mature grains of harvested seeds as part of the germination process [14]. From an economical perspective, overexpression of PHS in particular years can cause financial problems for farmers.

Throughout the world, where the capacity for cool, damp conditions during the later stages of maturation of harvests exists, PHS has been an important problem [15,16]. In Germany, at least 5 of the last 15 years have seen PHS damage to rye crops; in Poland 6.4% of wheat production and 8% of rye could not be sold for consumption due to sprouting in 1998. In 2000, up to 45% of wheat production in northern France was used as feed due to low quality as a result of sprouting. In the United Kingdom, where it is estimated that the average yearly loss due to PHS damage of the wheat crop is ca. GBP 18M (J. Flintham, personal communication), 2004 was a particularly bad year due to prevailing U.K. weather conditions. In this year almost three quarters of group 1 (high quality used for bread making) varieties failed to meet quality standards, with a national average HFN of 231—well below that required for bread making [17].

The unpredictable nature of the environmental input to PHS has meant that plant breeders have found it very difficult to provide effective strategies for increased genetic resistance. Field testing of breeding lines and analysis of HFN are not very useful tools because they monitor a highly complex set of interacting subtraits. Although PHS is unpredictable at the microscale, evidence suggests that long-term global weather patterns can predict PHS occurrence. There is a high degree of correlation between average HFN of the U.K. wheat crop and fluctuation of the North Atlantic oscillation (NAO), a measurement of the difference in air pressures between the Azores Iberian peninsula region and Arctic Iceland region of the North Atlantic [13]. The NAO has a cycle of 8 years, mirroring the cycling of average U.K. HFN. Although not helpful in preventing damage to crops, this correlation provides evidence that continued breeding for resistance is important, even during sustained periods of low environmentally induced damage.

2.3 ROLE OF THE MODEL ORGANISM *ARABIDOPSIS* IN DEFINING GENETIC CONTROL OF DORMANCY

Seed dormancy has been studied in many plant species [18]. This happened partly because researchers were interested in comparison of species and ecological significance of differences found between them. Weed scientists studied species with strong dormancy where this had an impact on the weedy character [19]. Model species have also been used for study where this has provided important insights to particular mechanisms. This has been based on suitability for the analysis of specific phenomena, such as light-induced germination in the case of lettuce (because this species shows a strong response to light).

In recent decades, as in other fields of biology, model species have been used to provide genetic variants. The study of plant hormone mutants in *Arabidopsis* and tomato showed convincingly the importance of abscisic acid (ABA) and GA for seed germination. In addition, the availability of other types of genetic variants and access to the full genome sequence of *Arabidopsis* has made genetics and molecular biology indispensable to the study of biological processes, including seed dormancy, because these resources allow identification of the genes and also processes that control the trait. In addition, they allow the application of "omic" approaches including full genome gene expression analysis and efficient proteomics.

Because the required resources for these studies are presently only available in *Arabidopsis* and rice, it is no surprise that biological research is focusing on these species. Genetic research in other species, such as barley and *Avena fatua*, where genetic variation for dormancy is available will certainly benefit from studies on these models. Sequence similarity and synteny between gene order in different species will allow the transfer of knowledge from model species to other species. This requires that mechanisms involved in a process are similar across species. Although such similarities are seen (discussed later), one should not overlook the possibility that some processes or subprocesses can be specific to species (or more likely species group). This is considered likely for seed dormancy in which qualitatively and quantitatively large differences in dormancy phenotypes (and also possible mechanisms) exist [1].

Although earlier studies on the role of light quality during seed development and during germination used *Arabidopsis*, the use of plant hormone mutants has stimulated research on seed biology in *Arabidopsis*. Much of the work on seed dormancy in *Arabidopsis* up to 2002 has been reviewed previously [20,21].

Several plant hormones affect seed dormancy. The importance of GA in seed dormancy in *Arabidopsis* was confirmed by the identification of nongerminating mutants that could be restored by the application of GA to the imbibition medium [22]. That application of GA biosynthesis inhibitors such as paclobutrazol (PAC) and uniconazol during imbibition inhibited germination indicated that de novo synthesis of GA is needed; this has recently been confirmed by measuring GA levels [23]. The way in which light and cold signals promote the transcription of GA biosynthesis genes (especially GA 3-oxidases) and how this signal is transduced to activate genes that affect cell expansion are presently some of the best-known parts of hormone signaling related to germination.

A dormant genotype resembles a GA-deficient mutant because neither germinates in light on water. However, because differences between imbibed nongerminating GA mutants and wild-type at the protein level are very limited and only become more obvious during radicle protrusion [24], it seems plausible that GA acts (only) at the stage of radicle protrusion when a growth potential should be developed that allows protrusion of the radicle through the surrounding envelopes. GA is not absolutely necessary for germination, which is indicated by the capacity of GA-deficient mutants to germinate when the seed coat is mechanically removed or genetically weakened [25]. The late rise in GA levels after imbibition and the fact that germination can be inhibited rather late by uniconazol [26] suggest that the role of GA is relatively late. That dormancy is different from lack of germination is suggested by the fact that dormancy-related genes are expressed during seed development and that some nondormant mutants such as *fus3* and *delay of germination 1* (*dog1*) still require GA for germination [27; Bentsink et al., unpublished]. This would be in agreement with the hypothesis that primary dormancy is induced during seed maturation.

Dormancy can only be assessed after imbibition during which much metabolic activity and also gene expression changes take place. Prolonged imbibition in conditions where germination does not occur can lead to secondary dormancy, which indicates that dormancy reinduction can take place or that the developmental phase of the seed can be reversed to pregermination (maturation phase) as has been suggested [28]. This developmental process might be partially under epigenetic control as indicated by the *pickle* mutant in which maturation is prevented and that encodes a chromatin-modeling protein [29].

Another plant hormone that affects seed dormancy is ABA. The ABA biosynthesis mutants were identified based on the fact that nondormant ABA-deficient mutants do not require GA for germination [30]. ABA signal transduction mutants also show a dormancy phenotype similar to that of ABA biosynthesis mutants [31]. ABA signal transduction mutants that also were characterized by lack of dormancy could be selected directly by their resistance to germination-inhibiting concentration of ABA [32,33]. ABA-deficient mutants in all species studied thus far showed an absence of dormancy. The role of ABA may be twofold. On one hand, it seems to be required for induction of dormancy during seed development, where *ABI3* is an important downstream component.

Karssen et al. [34] concluded that ABA induces a dormancy state during seed maturation; they observed that ABA levels are high halfway during seed development. This ABA is partly from maternal origin. Important for dormancy induction was a transient peak of ABA produced at a late stage during seed maturation by the embryo proper [34]. The conclusion that ABA was present during imbibition at levels so low that it could not inhibit germination did not take into account the possibility that ABA could be resynthesized during imbibition. This was observed later for dormant seeds of the dormant accession Cape Verde Islands (Cvi) [35] and was also suggested by the fact that application of inhibitors of ABA biosynthesis promotes germination [25,35,36].

ABA levels in mature seeds also contribute to the inhibition of germination and need to be metabolized before germination takes place. The essential gene for this breakdown appears to be ABA 8′-hydroxylase (*CYP707A*). Null mutants of this gene show a strongly reduced germination [37]. The way in which these effects on hormone metabolism are determined by the dormancy status of seeds set during maturation and the way in which GA and ABA levels depend on the levels of each other is still not clear.

Among the other hormones, clear, although not decisive, roles were found for brassinosteroids (BR) and ethylene. Ethylene seems to be required for normal fast germination of seeds [38], which may also be related to the sensitivity of seeds to ABA [39,40]. BR can promote germination of GA mutants by bypassing the GA requirement [41], which is also an effect of applied ethylene [42]. However, BR mutants germinate normally and BRs do not enhance germination of wild-type seeds [41]. Phytochrome mutants have allowed dissection of the role of different phytochrome species in seed germination [43]; phytochrome B was especially shown to induce GA biosynthesis after exposure to red light [44].

The importance of the seed coat or testa could also be demonstrated using genetics because almost all mutants with an altered testa color or structure showed reduced germination. Wild-type *Arabidopsis* seeds are brown because of the presence of brown proanthocyanindins (condensed tannins) that are flavonoid end products in the inner layer of the inner integument. These compounds affect the structure of the cells in that layer and confer additional resistance to the protruding radicle [45]. It is not clear whether the thick-walled single layer of endosperm (aleurone) in mature seeds [46] plays a significant role in preventing germination because no mutants affected specifically in this layer have been identified. However, the relevance of this layer, which needs to be weakened to allow germination in species such as tomato and tobacco, is suggested by the expression of cell wall-weakening enzymes in the aleurone layer at the onset of germination [23,47].

Mutants affected in storage reserve mobilization indicate that this process is not required for germination but is essential for seedling establishment [48]. However, the *comatose* mutant isolated as a nongerminating mutant has a defect in an ABC transporter affecting lipid breakdown [49]. This suggests that, differently from other lipid mobilization genes, this gene affects germination in *stricto senso*.

The application of microarray technology and proteomics added another dimension to our understanding of seed germination. The relevance of transcription initiation for germination may be limited because the transcriptional inhibitor α-amanitin does not inhibit germination, as does cycloheximide [28], and hardly affects the levels of major proteins during imbibition. However, for storage mobilization and hexose metabolism during establishment, transcription certainly plays a role. Apparently, many changes in transcription observed during imbibition in microarray experiments before radicle protrusion could be related to seedling establishment and not to germination per se, which seems mainly driven by mRNAs already present in the dry seed [23,50]. However, a role of newly synthesized transcripts shortly after imbibition cannot be excluded because, at this time point, α-amanitin may not be fully effective.

Although progress has been made in understanding of dormancy and germination, especially by using the tools available for *Arabidopsis*, many questions about both processes remain unanswered. Because progress in the understanding of dormancy and germination has been focused on the role of GA and ABA, this emphasis might neglect the importance of other factors. These may be those controlled by genes represented by mutants (e.g., *reduced dormancy, rdo1* to *rdo4* [51]) that have not yet been cloned or for which only quantitative trait loci (QTL) positions are known [36,52].

The importance of maternal factors that are not directly related to the structure of the testa is not well understood. Indications about the importance of such factors come from the fact that two DOF zinc finger genes, *DAG1* and *DAG2* (*DOF affected in germination*), influence germination. These genes (*DAG1* inhibits and *DAG2* promotes germination) are expressed in the vascular tissue of developing seeds but not in mature seeds or during imbibition [53]. Furthermore, the mechanism of after-ripening and moist chilling (stratification) is not well understood. Although the effect of the latter treatment on GA biosynthesis during imbibition is convincingly shown [50], it seems unlikely that this is the only role of cold because this treatment is far more effective than applied GA in breaking dormancy in strongly dormant genotypes [35,36].

Genetic variation for seed dormancy within species is present between accessions of wild plants and among varieties of cultivated plants. The large environmental effects on the expression of germination characteristics and the involvement of many genes make dormancy genetically a typically quantitative trait. Such traits are now more amenable to genetic analysis because the position of individual QTL and the relative contribution of these loci can be determined. QTL analysis for seed dormancy requires permanent or immortal mapping populations such as recombinant inbred lines (RILs) because it allows testing a large number of genetically identical seeds per genotype in different environmental conditions.

QTL analysis for seed dormancy has been reported for *Arabidopsis* [36,52,54]. This analysis can be followed by study of the individual genes (or chromosome regions containing specific dormancy QTL) by fine mapping and subsequent cloning. In this way genes can be identified that control seed dormancy; furthermore, genes that control adaptation to specific environments can also be identified. The genes identified in the study of natural variation can be the same as those identified in mutant screens. However, there are several reasons why this is not always the case. First the parent lines used for mutation experiments can be mutated for specific genes. In the case of *Arabidopsis*, many natural accessions show much stronger seed dormancy than the commonly used laboratory accessions Landsberg *erecta* (L*er*) and Columbia (Col). Also, mutants that show strong pleiotropic effects such as most ABA mutants will not survive in nature. Therefore, genes identified by analyzing natural variation are expected to be ecologically relevant.

The analysis of different RIL populations thus far identified more than twelve regions on the *Arabidopsis* genome where QTLs associated with dormancy are located [36; Bentsink et al., unpublished]. Many of these do not colocate with known dormancy loci. Depending on the parents of the progeny analyzed, the same or different regions are identified [Bentsink et al., unpublished].

The feasibility to clone such genes has been demonstrated for the *DOG1* locus, of which L*er* contains a weak allele and the dormant accession Cvi a strong allele [36]. This gene, which is expressed during seed development and down-regulated during imbibition, encodes a gene of unknown function that is assumed to control genes responsible for dormancy induction (Bentsink, unpublished data). The cloning of additional QTL and the study of more dormant lines in which regions containing strongly dormant alleles have been introgressed from other accessions into an *Ler* background is ongoing and should identify additional genes involved in dormancy induction, dormancy maintenance, and dormancy breakage.

2.4 MECHANISM OF DORMANCY AND GERMINATION IN CEREALS

Much research into dormancy has been spurred by the need to define methods of control for highly dormant weed species. The possible usefulness of different weed species as models for investigation has been reviewed recently [55]. Wild oat (*Avena fatua*), which is a highly pernicious weed, has been used as the major model for physiological investigations of grass dormancy for over 100 years (Graham Simpson's dormancy publications database is available at http://library.usask.ca/dbs/seed.html). This species is a good physiological model, demonstrating deep dormancy imposed by embryo and surrounding maternal structures. Many studies have been carried out using wild oat to define biochemical and physiological changes controlling dormancy. For example, an increase in glycolysis and/or the Krebs cycle has been shown to be an important determinant of dormancy breakage in this species [56].

Genetic studies have been attempted using wild oat. Many distinct genetic lines exist with differing dormancy characters [57–59] and these have provided the material for QTL analyses that have indicated interactions between loci that promote germination or dormancy. However, because the wild oat genome is hexaploid, with a large genome size (approx. 11,000 Mbp) and very little physical genome information, it is a poor model to use at this level of investigation [55]. Several molecular studies have also used this genetically defined material to study changes in gene expression associated with dormancy and germination phases [60–62]. In one study, the expression of the wild oat ortholog of maize Vp1/*Arabidopsis* ABI3 was shown to be highly correlated with seed dormancy status [60].

The genetic components contributing to PHS have been investigated using a variety of methodologies from whole plant to single seed. Physiological studies of developing wheat caryopses have analyzed the relationship between ABA content of seeds and dormancy capacity. It has been argued that it is important to define the timing of dormancy induction during seed development in order to assess the influence of applied hormones and changes in environment [63]. King [63] analyzed the environmental influence on in-ear sprouting and excised grain germination, comparing cold humid conditions with warm dry. These analyses established a window of dormancy induction in wheat associated with grain drying and suggest that development favors germination if grain desiccation is prevented at maturity.

In addition, studies including this one have shown that dormancy imposition was not related to embryo ABA content [63–65]. Other studies comparing dormant and nondormant seeds (of varieties or induced mutants) demonstrated large differences in responsiveness of embryos to ABA [65,66]. Therefore, it is likely that sensitivity to ABA at specific developmental periods, in association with the induction of desiccation, are key processes related to dormancy induction and sensitivity to environmental conditions that may induce premature germination. The chromosomal locations of wheat embryo sensitivity to ABA and dormancy have been investigated [67]. A cross between wheat cv. Chinese spring (nondormant and ABA insensitive) and line Kitakei-1354 (dormant, ABA sensitive) was used to demonstrate the importance of 4AL and 2D for these characters.

At present, the major genetic components used to increase resistance to sprouting in wheat are the red grain color (*R*) loci located on group 5 chromosomes [68]. Dominant alleles promote expression of red pigments (phlobaphenes) within the maternally derived pericarp that tightly surrounds the embryo. It is not known whether the *R* genes provide increased dormancy or are very closely linked to other dormancy-promoting loci (for example, *Vp-1*, discussed later).

Recently, it has been suggested that the *R* genes encode an myb-like transcription factor that controls expression of genes of the flavonoid pathway within the pericarp during grain development [69]. Although the *R* loci can be used to provide some resistance to PHS, they also provide characters that reduce the perceived quality of the flour for certain markets. These include color (red wheat seeds produce discolored flour that is perceived negatively by the noodle-making industry), production energy costs, and taste perception. White wheat types contain no active *R* loci and in general therefore are more prone to PHS. Breeders, particularly in North America and Australia, have concentrated on producing white wheats with increased dormancy levels that should have increased market potential [70,71].

2.5 POTENTIAL FOR INFORMATION TRANSFER FROM MODEL SYSTEMS TO AGRONOMICALLY IMPORTANT SYSTEMS

Seed dormancy of weeds and sprouting characteristics of crops provide ongoing problems for plant breeders, agrochemical companies, farmers, and downstream processors. Long-term durable solutions are an important target. Genetic alteration of crops to increase resistance to sprouting would provide benefits of quality assurance and sustainability. Because these phases of development are complex, many confounding factors need to be addressed to achieve useful alterations in traits associated with seed performance.

From the breeding perspective, PHS is phenotypically difficult to manipulate because it results from complex physiological and environmental inputs and the environmental inputs are difficult or impossible to control and influence. Multiple genetic loci input into the trait at different levels (morphologically) and stages of seed growth; loci unassociated with seed development per se can influence susceptibility. In addition, mechanisms can be species specific, making it difficult to transfer physiological information from one example to another. Candidate genes that influence dormancy and germination, defined through genetic and molecular approaches, offer important potential tools that could be utilized as highly informative molecular markers for selection and/or genes for manipulating development through transgenesis.

Genetic approaches to understand PHS in cereals have used QTL and mutant studies to define important regions of the genome [10,72–79], as well as introgression studies to add new genetic material with improved trait characteristics. The D-genome diploid progenitor *Triticum tauschii* has been used as a donor of embryo and maternal mechanisms for sprouting resistance, via construction of synthetic hexaploid wheat with *T. turgidum* [80]. Similarly, *T. monococcum* types with high sprouting resistance have been used to introgress this character into triticale [81]. Several QTL studies in wheat and rice have indicated regions of the genome that influence traits of dormancy and sprouting resistance or susceptibility. In addition, a comparative genetic approach has been used to indicate QTLs from wheat/rice/maize that have conserved syntenic relationships (i.e., that lie at the same chromosomal positions in the genome) [82]. These represent important potential targets for candidate loci having conserved functions associated with dormancy or germination traits.

Genetic studies in cereal crops suffer the disadvantage of complexity of gene isolation. This is caused by the extreme difficulty of positional cloning of major gene loci and QTLs—for the most part, because of genome size and polyploidy. Recently, several rice and wheat loci have been cloned through positional methodologies [83,84], but in general this remains a long-term and problematic approach, especially for QTLs of small effect. Therefore, the definition of candidate genes defined in simpler model systems provides a complementary approach to identify and characterize regulatory molecules. Models can be utilized at different levels; in the study of mechanisms controlling dormancy these include physiological or biochemical approaches relating biochemical changes to physiological characteristics, molecular or biochemical approaches providing mechanistic understanding of gene expression regulation and subsequent function, and the use of "simpler" genetic systems.

Studies in tomato, *Arabidopsis*, and maize have identified processes and regulator molecules controlling the initiation of germination and genetic loci regulating dormancy initiation and the transition to germination [85]. The alpha-amylase promoter has been used as a model for transcriptional regulation in studies of aleurone function and hormone responsiveness in association with postgermination events in cereals [86]. Results obtained from all these approaches have been integrated into models for regulation from promoter–transcription factor interactions through developmental changes. These models have the potential to define function and usefulness of identified candidate genes in agriculturally relevant species and environments. Several examples of candidate genes identified using model systems have been studied and reveal some highly conserved aspects of cell signaling and control of gene expression in flowering plants in relation to post seed-shed development. However, it is likely that because this phase of development is so complex and environmental interactions species specific, many aspects of control may not be shared by all species.

An indication of the importance of specific candidate genes can be understood by analysis of conserved function and structure. Here, the focus is on factors associated with ABA and GA signal transduction and synthesis and control of expression of alpha-amylase, a key marker for germination and sprouting in cereals. A simple model (based on information obtained from monocot and dicot species) showing

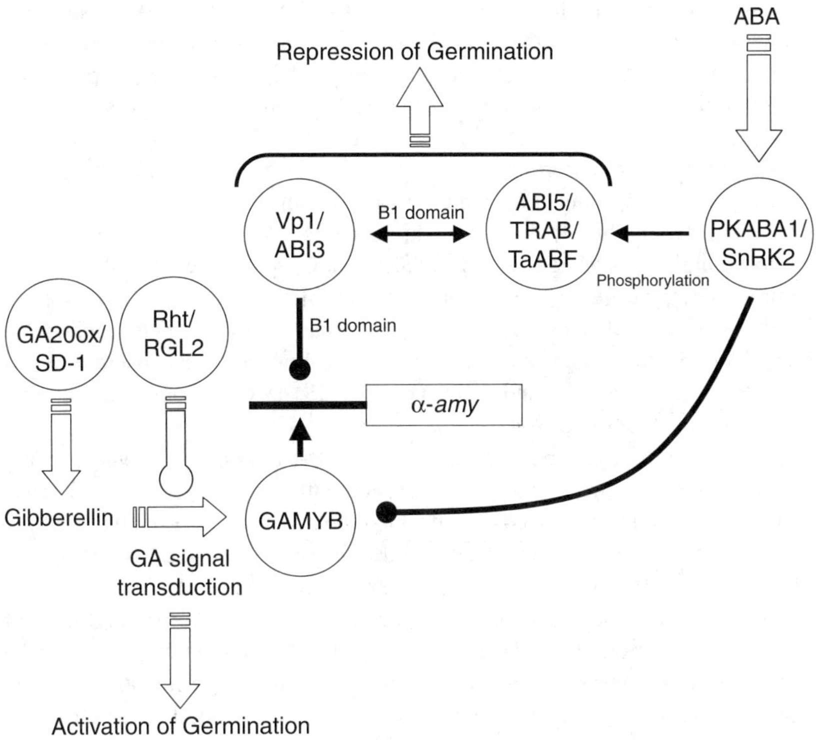

FIGURE 2.2 Conceptual framework of possible interactions between regulatory molecules involved in ABA and GA control of alpha-amylase synthesis, dormancy, and germination. Solid lines indicate physical interactions, open arrows point to functional consequences. Activation is indicated by arrows, repression by blocked ends.

candidate genes and possible interactions between key regulator molecules and pathways associated with alpha-amylase production is shown in Figure 2.2, and components of the model are discussed next.

Mutants have been isolated in *Arabidopsis*, maize, and rice that exhibit a viviparous phenotype similar to that of PHS [87,88]. The phenotypes of several mutants in *Arabidopsis*, including severe alleles of *abi3*, *fus3*, and *lec2*, are outwardly similar to those of mutation at the maize *Vp-1* locus. In each case, dormancy is completely lost and seeds display a viviparous developmental pattern, including the absence of seed maturation programs and the premature activation of germination- and post-germination-associated development [87]. These loci have all been cloned and encode highly similar proteins with DNA-binding and transcriptional activation functions, suggesting that they act in one part by repressing gene expression programs associated with germination [14]. Conserved structure and function also suggest a central and conserved role in the transition to germination. The phenotypes of these mutants are superficially similar to that of seeds within ears displaying PHS in wheat, although it is also possible that vivipary in these mutants results from disruptions in other developmental pathways.

The role of Vp-1 in controlling resistance to PHS and in ABA sensitivity has been investigated [89]. In this case, it was shown that wheat and closely related progenitors exhibit significant alternative splicing of homolog transcripts (derived from each of the single locus homologous genome positions) and it was not possible to observe full-length Vp-1 protein in wheat embryo nuclei. The suggestion that one reason that wheat shows a propensity to sprout is caused by a lack of effective Vp-1 activity was tested using transgenic wheat containing a correctly spliced *Vp-1* cDNA derived from a highly dormant wild oat ecotype (where *Vp-1* expression was shown to be highly correlated with dormancy status) under the control of a constitutive promoter. Transgenic wheat plants showed increased resistance to sprouting in the ear, and isolated seeds also displayed increased sensitivity to applied ABA. Both results indicate, as suggested, that wheat embryos have the capacity to express a higher level of dormancy (and resistance to PHS) through increased activity of "correctly" expressed Vp-1 protein.

Others have reported genetic differences in ABA sensitivity of wheat embryos [65]. One study has shown that a QTL associated with sprouting resistance is present over the chromosomal region containing *Vp-1* on the long arm of 3DL [70]. The *Vp-1* gene is therefore a good candidate for development of markers associated with reduced splicing of transcripts or increased expression of specific homologs.

The Vp-1 protein has been shown to repress expression from a high-pI alpha-amylase promoter through the B1 domain located in the middle of the protein [90], providing one molecular explanation for Vp-1 repression of germination-associated pathways. This region has also been shown to interact with another conserved transcription factor from *Arabidopsis* and rice (ABI5/TRAB1, respectively) [91,92] that is highly similar to the wheat protein TaABF [93]. ABI3 and ABI5 have been studied in detail in *Arabidopsis*, where their roles in ABA-related control of seed development and germination are well established [94]. The TaABF protein interacts with a wheat ABA-induced Ser/Thr SnRK2 protein kinase PKABA1 [93], suggesting that phosphorylation of this transcription factor is an important component of function.

Interestingly, ABI5 has also been shown to be phosphorylated in an ABA-dependant manner in imbibed *Arabidopsis* seeds, mirroring observations in wheat embryos [95], and TRAB1 is phosphorylated in response to ABA treatment in rice [96]. Similar kinases exist in *Arabidopsis* (that contains 10 SnRK2 genes)—two of which exhibit expression related to ABA responses—and may be good candidates for functional analysis during seed development and germination [97]. In tomato, expression of the regulatory subunit of the SnRK1 complex (LeSNF4) is associated with lack of germination in ABA-treated imbibed seeds [98]. These observations provide evidence of a conserved pathway regulating the activation of ABA responses, induction of ABA-regulated molecular interactions, and repression of germination.

PKABA1 is induced by ABA in wheat and barley [99,100] and suppresses GA-inducible alpha-amylase gene expression in barley [100]. PKABA1 has been shown to down-regulate GAMYB—a transcription factor that is part of GA-regulated responses—and is required for activation of expression through a GA response element present in all GA-inducible alpha-amylase promoters [101–103]. Rice orthologs of PKABA1 appear to exist (SAPK 1 and 2) and expression of SAPK1 was induced by ABA, although the protein is apparently not activated by ABA [104].

GAMYB was originally identified in barley aleurone cells, but has subsequently been shown to play a role in rice aleurone by conferring GA responsiveness of amylase production [105]. (These experiments used TOS17 transposon insertion lines to remove expression of the gene.) GAMYB is induced by GA [102], indicating that this transcription factor is used to integrate signals from GA and ABA signal transduction. Three *Arabidopsis* MYB proteins have been identified with similarity to GAMYB that can substitute for barley GAMYB in transactivating the barley alpha-amylase promoter [106], although their role in controlling germination is not known.

A paradigm for demonstrating the conservation of function of agriculturally important genes is that of the relationship between *Arabidopsis GIBBERELLIN INSENSITIVE* (*GAI*) and wheat *Reduced height* (*Rht*) [107]. Both encoded proteins belong to the DELLA subfamily of the GRAS family of plant regulatory proteins. In both cases, dominant mutant alleles produce plants of shortened stature and reduced fertility. Dominant alleles exert their effect by reducing the capacity for GA-induced degradation of the mutant protein; interestingly, alleles in *Arabidopsis* wheat and maize contain deletions of amino acids around a highly conserved section of the protein (the DELLA domain) [107]. These proteins are selected for degradation through the ubiquitin–proteasome pathway via interaction with the F-box protein SLEEPY (SLY); this degradation does not occur with dominant mutant alleles [108,109].

Arabidopsis contains five DELLA protein genes (RGA, GAI, RGL1, RGL2, and RGL3) that are expressed at different levels throughout development [110], whereas wheat contains only one homolog group on chromosomes 4B and 4D [111]. Of the five *Arabidopsis* genes, *RGL2* has been shown to be the major determinant of repression of the initiation of germination; mutant seeds show reduced sensitivity to PAC and enhanced germination potential [112]. Early work with the *Rht3* allele showed that this reduces wheat aleurone amylase expression during germination and exogenous GA responsiveness of amylase production by the aleurone [113]. In addition, this allele has been shown to reduce susceptibility to prematurity alpha-amylase production [114] that occurs in the absence of sprouting [9].

Genes associated with many of the steps of GA metabolism have been isolated and analyzed from a variety of species [115]. Individual enzymes may provide useful candidates for manipulation of grain GA biosynthesis capacity and hence sensitivity to sprouting. For example, one dwarfing locus in rice (analogous in importance to the *Rht* locus in wheat) (*sd-1*) encodes a GA 20-oxidase gene that catalyses several of the later stages of GA biosynthesis [116–118]. Recently, genes representing enzymes from this pathway have been mapped using comparative approaches in wheat, rice, and barley [119]. This approach provides useful information that should allow analysis of the extent to which GA metabolism loci account for chromosomal regions regulating GA-associated phenotypes (including germination) and an indication of the relationship of QTLs associated with GA phenotypes and GA metabolism loci.

As the function of candidate genes in *Arabidopsis* and other model systems is revealed, their usefulness for the manipulation of agronomically important characters associated with dormancy and germination can be tested. The importance of dormancy induction mechanisms and ABA sensitivity in wheat embryos as determinants of susceptibility to PHS means that candidate loci affecting these responses in *Arabidopsis* should be priority targets for further investigation. The recent identification in *Arabidopsis* of the key ABA catabolic enzyme ABA 8′-hydroxylase [37] offers one avenue for analysis of this process in wheat embryos, as mutants display hyperdormancy. Several other novel regulators influencing ABA sensitivity have recently been described, including *SAD1* and *ABH1* [120,121].

Analysis of crop orthologs of *Arabidopsis* genes associated with QTLs that enhance dormancy potential will also offer new possibilities for manipulation of PHS resistance and some indication of pathways important for dormancy in weed species. Large-scale bioinformatic comparisons of *Arabidopsis* gene sequences associated with dormancy and germination with rice genome and wheat/barley EST information should provide candidate orthologous sequences for further analysis.

Conserved function allows easy integration of information from model studies to practical application. However, there are several caveats to the use of this approach. It is notable that QTL studies in *Arabidopsis* have revealed loci previously undetected in mutational screens [36]. Candidate gene studies use conservation as a major tool, and it may be that little variation will be present at loci with highly conserved functions, thereby reducing the usefulness of such genes for marker-assisted breeding. Lastly, it is of course important to identify variation within the species and trait. Candidate genes and pathways can provide a framework to detect species-specific components; however, in some cases, pathways may be taxon specific and not held in common with those present in distantly related model species. This latter point is exemplified by the identification of vernalization proteins in wheat that do not appear to correspond to those used in vernalization pathways in *Arabidopsis* [83].

Several agronomic problems are associated with germination and dormancy. These can be caused by deterioration of quality of seeds in crops or survival of weed seeds in the soil contaminating and competing with subsequent crops. Understanding underlying genetic and molecular mechanisms of the different processes that input into dormancy of the embryo and surrounding structures is an important component of strategies designed for improving seed quality or controlling seed survival. Whether via marker-assisted selection or production of transgenic plants, basic research in plant science should provide compelling tools for plant breeders to manipulate traits associated with seed development and subsequent dormancy and germination, as well as for agrochemical companies to devise novel chemistries to disrupt the dormancy mechanism of weed seeds.

REFERENCES

1. Baskin, J.M. and Baskin, C.C. A classification system for seed dormancy. *Seed Sci. Res.* 14, 1–16, 2004.
2. Vleeshouwers, L.M. et al. Redefining seed dormancy: an attempt to integrate physiology and ecology. *J. Ecol.* 83, 1031–1037, 1995.
3. Bewley, J.D. Seed germination and dormancy. *Plant Cell* 9, 1055–1066, 1997.
4. Baskin, C.C. and Baskin, J.M. *Seeds: Ecological, Biogeography, and Evolution of Dormancy and Germination.* Academic Press, London, 1998.

5. Donohue, K. Germination timing influences natural selection on life-history characters in *Arabidopsis thaliana*. *Ecology* 83, 1006–1016, 2002.
6. Dyer, W.E. Exploiting weed seed dormancy and germination requirements through agronomic practices. *Weed Sci.* 43, 498–503, 1995.
7. Harlan, J.R. *Crops and Man*, 2nd ed. American Society of Agronomy, Madison, WI, 1992.
8. Lutman, P.J.W. et al. The long-term persistence of seeds of oilseed rape (*Brassica napus*) in arable fields. *J. Agric. Sci.* 141, 231–240, 2003.
9. Lunn, G.D. et al. Mechanisms leading to excess alpha-amylase activity in wheat (*Triticum aestivum*, L.) grain in the U.K. *J. Cereal Sci.* 33, 313–329, 2001.
10. Li, C.D. et al. A major QTL controlling seed dormancy and preharvest sprouting/grain alpha-amylase in two-rowed barley (*Hordeum vulgare* L.). *Aust. J. Agric. Res.* 54, 1303–1313, 2003.
11. Staff. Arable news: high chitting hits rape. *Farmers Wkly. Interactive*, 17 August 2004.
12. Hagberg, S. A rapid method for determining alpha-amylase activity. *Cereal Chem.* 37, 218–222, 1960.
13. Kettlewell, P. et al. Development of a scheme for Pre-Harvest Sprouting prediction of Hagberg Falling Number in wheat. In *Eighth International Symposium on Pre Harvest Sprouting*. (D. Weipert, Ed.). Association of Cereal Research, Detmold, Germany, 1999, 9–14.
14. Holdsworth, M. et al. Genetic control mechanisms regulating the initiation of germination. *J. Plant Physiol.* 158, 439–445, 2001.
15. Derera, N.F. A perspective of sprouting research. In *Fifth International Symposium on Preharvest Sprouting in Cereals*. (K. Ringlund, E. Mosleth, and D.J. Mares, Eds.). Westview Press, Boulder, CO, 1990, 3–11.
16. Williams, P. European campaign approaching the last leg. *Farmers Wkly.*, 25 August 2000.
17. Staff. Arable news: majority of wheat fails the grade. *Farmers Wkly. Interactive*, 28 September 2004.
18. Bewley, J.D. and Black, M. *Seeds: Physiology of Development and Germination*. Plenum Press, New York, 1994.
19. Basu, C. et al. Weed genomics: new tools to understand weed biology. *Trends Plant Sci.* 9, 391–398, 2004.
20. Bentsink, L. and Koornneef, M. Seed dormancy and germination. In *The Arabidopsis Book*. (C.R. Somerville and E.M. Meyerowitz, Eds.). American Society of Plant Biologists, Rockville, MD, 2002, DOI 10.1199 tab.0050.
21. Koornneef, M. and Karssen, C.M. Seed dormancy and germination. In *Arabidopsis*. (E.M. Meyerowitz and C.R. Somerville, Eds.). Cold Spring Harbor Laboratory Press, Cold Spring Harbor, NY, 1994, 313–334.
22. Koornneef, M. and Van der Veen, J.H. Induction and analysis of gibberellin-sensitive mutants in *Arabidopsis thaliana* (L.) Heynh. *Theor. Appl. Genet.* 58, 257–263, 1980.
23. Ogawa, M. et al. Gibberellin biosynthesis and response during *Arabidopsis* seed germination. *Plant Cell* 15, 1591–1604, 2003.
24. Gallardo, K. et al. Proteomics of *Arabidopsis* seed germination. A comparative study of wild-type and gibberellin-deficient seeds. *Plant Physiol.* 129, 823–837, 2002.
25. Debeaujon, I. and Koornneef, M. Gibberellin requirement for *Arabidopsis* seed germination is determined both by testa characteristics and embryonic abscisic acid. *Plant Physiol.* 122, 415–424, 2000.
26. Nambara, E. et al. Effects of the gibberellin biosynthetic inhibitor uniconazol on mutants of *Arabidopsis*. *Plant Physiol.* 97, 736–738, 1991.

27. Raz, V. et al. Sequential steps for developmental arrest in *Arabidopsis* seeds. *Development* 128, 243–252, 2001.
28. Rajjou, L. et al. The effect of alpha-amanitin on the *Arabidopsis* seed proteome highlights the distinct roles of stored and neosynthesized mRNAs during germination. *Plant Physiol.* 134, 1598–1613, 2004.
29. Ogas, J. et al. PICKLE is a CHD3 chromatin-remodeling factor that regulates the transition from embryonic to vegetative development in *Arabidopsis. Proc. Natl. Acad. Sci. USA* 96, 13839–13844, 1999.
30. Léon-Kloosterziel, K.M. et al. Isolation and characterization of abscisic acid-deficient *Arabidopsis* mutants at two new loci. *Plant J.* 10, 655–661, 1996.
31. Nambara, E. et al. A regulatory role for the ABI3 gene in the establishment of embryo maturation in *Arabidopsis thaliana. Development* 121, 629–636, 1995.
32. Koornneef, M. et al. The isolation and characterization of abscisic acid insensitive mutants of *Arabidopsis thaliana. Physiol. Plant.* 61, 377–383, 1984.
33. Nambara, E. et al. A screen for genes that function in abscisic acid signaling in *Arabidapsis thaliana. Genetics* 161, 1247–1255, 2002.
34. Karssen, C.M. et al. Induction of dormancy during seed development by endogenous abscisic acid—studies on abscisic-acid deficient genotypes of *Arabidopsis thaliana* (L.) Heynh. *Planta* 157, 158–165, 1983.
35. Ali-Rachedi, S. et al. Changes in endogenous abscisic acid levels during dormancy release and maintenance of mature seeds: studies with the Cape Verde Islands ecotype, the dormant model of *Arabidopsis thaliana. Planta* 219, 479–488, 2004.
36. Alonso-Blanco, C. et al. Analysis of natural allelic variation at seed dormancy loci of *Arabidopsis thaliana. Genetics* 164, 711–729, 2003.
37. Kushiro, T. et al. The *Arabidopsis* cytochrome P450CYP707A encodes ABA 8′-hydroxylases: key enzymes in ABA catabolism. *EMBO J.* 23, 1647–1656, 2004.
38. Siriwitayawan, G. et al. Seed germination of ethylene perception mutants of tomato and *Arabidopsis. Seed Sci. Res.* 13, 303–314, 2003.
39. Beaudoin, N. et al. Interactions between abscisic acid and ethylene signaling cascades. *Plant Cell* 12, 1103–1115, 2000.
40. Ghassemian, M. et al. Regulation of abscisic acid signaling by the ethylene response pathway in *Arabidopsis. Plant Cell* 12, 1117–1126, 2000.
41. Steber, C.M. and McCourt, P. A role for brassinosteroids in germination in *Arabidopsis. Plant Physiol.* 125, 763–769, 2001.
42. Karssen, C.M. et al. Key role for endogenous gibberellins in the control of seed germination. *Ann. Bot.* 63, 71–80, 1989.
43. Shinomura, T. et al. Mode of phytochrome B action in the photoregulation of seed germination in *Arabidopsis thaliana. Plant J.* 13, 583–590, 1998.
44. Yamaguchi, S. et al. Phytochrome regulation and differential expression of gibberellin 3 beta-hydroxylase genes in germinating *Arabidopsis* seeds. *Plant Cell* 10, 2115–2126, 1998.
45. Debeaujon, I. et al. Influence of the testa on seed dormancy, germination, and longevity in *Arabidopsis. Plant Physiol.* 122, 403–413, 2000.
46. Penfield, S. et al. Reserve mobilization in the *Arabidopsis* endosperm fuels hypocotyl elongation in the dark, is independent of abscisic acid, and requires PHOSPHOENOLPYRUVATE CARBOXYKINASE1. *Plant Cell* 16, 2705–2718, 2004.
47. Dubreucq, B. et al. The *Arabidopsis* AtEPR1 extensin-like gene is specifically expressed in endosperm during seed germination. *Plant J.* 23, 643–652, 2000.
48. Eastmond, P.J. et al. Postgerminative growth and lipid catabolism in oilseeds lacking the glyoxylate cycle. *Proc. Natl. Acad. Sci. USA* 97, 5669–5674, 2000.

49. Footitt, S. and Cohn, M.A. Developmental arrest: from sea urchins to seeds. *Seed Sci. Res.* 11, 3–16, 2001.
50. Yamauchi, Y. et al. Activation of Gibberellin biosynthesis and response pathways by low temperature during imbibition of *Arabidopsis thaliana* seeds. *Plant Cell* 16, 367–378, 2004.
51. Peeters, A.J.M. et al. Characterization of mutants with reduced seed dormancy at two novel rdo loci and a further characterization of rdo1 and rdo2 in *Arabidopsis*. *Physiol. Plant.* 115, 604–612, 2002.
52. Van Der Schaar, W. et al. QTL analysis of seed dormancy in *Arabidopsis* using recombinant inbred lines and MQM mapping. *Heredity* 79, 190–200, 1997.
53. Gualberti, G. et al. Mutations in the Dof zinc finger genes DAG2 and DAG1 influence with opposite effects the germination of *Arabidopsis* seeds. *Plant Cell* 14, 1253–1263, 2002.
54. Clerkx, E.J.M. et al. Analysis of natural allelic variation of *Arabidopsis* seed germination and seed longevity traits between the accessions Landsberg erecta and Shakdara, using a new recombinant inbred line population. *Plant Physiol.* 135, 432–443, 2004.
55. Foley, M.E. Weeds, seeds, and buds—opportunities and systems for dormancy investigations. *Weed Sci.* 50, 267–272, 2002.
56. Adkins, S.W. and Simpson, G.M. The physiological-basis of seed dormancy in *Avena fatua*. 9. Characterization of 2 dormancy states. *Physiol. Plant.* 73, 15–20, 1988.
57. Jana, S. et al. Genetic basis of dormancy and differential response to sodium azide in *Avena fatua* seeds. *Can. J. Bot.* 66, 635–641, 1988.
58. Symons, S.J. et al. Secondary dormancy in avena fatua—induction and characteristics in genetically pure dormant lines. *Physiol. Plant.* 68, 27–33, 1986.
59. Foley, M.E. and Fennimore, S.A. Genetic basis for seed dormancy. *Seed Sci. Res.* 8, 173–182, 1998.
60. Jones, H.D. et al. Genotype and environment interact to control dormancy and differential expression of the VIVIPAROUS 1 homologue in embryos of *Avena fatua*. *Plant J.* 12, 911–920, 1997.
61. Johnson, R.R. et al. Characterization of cDNA clones for differentially expressed genes in embryos of dormant and nondormant *Avena fatua* L. caryopses. *Plant Mol. Biol.* 28, 113–122, 1995.
62. Li, B.L. and Foley, M.E. Cloning and characterization of differentially expressed genes in imbibed dormant and after-ripened *Avena fatua* embryos. *Plant Mol. Biol.* 29, 823–831, 1995.
63. King, R.W. Manipulation of grain dormancy in wheat. *J. Ex. Bot.* 44, 1059–1066, 1993.
64. Black, M. et al. Water content, raffinose, and dehydrins in the induction of desiccation tolerance in immature wheat embryos. *Plant Physiol.* 120, 463–471, 1999.
65. Walker-Simmons, M.K. ABA levels and sensitivity in developing wheat embryos of sprouting resistant and susceptible cultivars. *Plant Physiol.* 84, 61–66, 1987.
66. Kawakami, N. et al. ABA insensitivity and low ABA levels during seed development of nondormant wheat mutants. *J. Ex. Bot.* 48, 1415–1421, 1997.
67. Noda, K. et al. Chromosomes responsible for sensitivity of embryo to abscisic acid and dormancy in wheat. *Euphytica* 123, 203–209, 2002.
68. Flintham, J.E. Different genetic components control coat-imposed and embryo-imposed dormancy in wheat. *Seed Sci. Res.* 10, 43–50, 2000.
69. Himi, E.N., and Noda, K. R gene for wheat grain colour might be a Myb-type transcription factor. In *Proceedings of the Tenth International Wheat Genetics Symposium*. (N.E.R. Pogna, A. Pogna, and A. Galterio, Eds.). Instituto Sperimentale per la Cerealicoltura, Rome, Italy, 2003, 958–960.

70. Mares, D. et al. Dormancy in white-grained wheat: progress towards identification of genes and molecular markers. *Euphytica* 126, 47–53, 2002.
71. McCaig, T.N. and Depauw, R.M. Breeding for preharvest sprouting tolerance in white-seed-coat spring wheat. *Crop Sci.* 32, 19–23, 1992.
72. Swanston, J.S. et al. Germination and malting properties of mutants derived from malting barley cv. triumph. *Cereal Chem.* 79, 392–396, 2002.
73. Warner, R.L. et al. Dormancy in white-grain mutants of Chinese spring wheat (*Triticum aestivum* L.). *Seed Sci. Res.* 10, 51–60, 2000.
74. Cai, H.W. and Morishima, H. Genomic regions affecting seed shattering and seed dormancy in rice. *Theor. Appl. Genet.* 100, 840–846, 2000.
75. Lin, S.Y. et al. Mapping quantitative trait loci controlling seed dormancy and heading date in rice, *Oryza sativa* L., using backcross inbred lines. *Theor. Appl. Genet.* 96, 997–1003, 1998.
76. Gu, X.Y. et al. Multiple loci and epistases control genetic variation for seed dormancy in weedy rice (*Oryza sativa*). *Genetics* 166, 1503–1516, 2004.
77. Anderson, J.A. et al. RFLP analysis of genomic regions associated with resistance to preharvest sprouting. *Crop Sci.* 33, 453–459, 1993.
78. Zanetti, S. et al. Genetic analysis of preharvest sprouting resistance in a wheat × spelt cross. *Crop Sci.* 40, 1406–1417, 2000.
79. Mares, D.J. and Mrva, K. Mapping quantitative trait loci associated with variation in grain dormancy in Australian wheat. *Aust. J. Agric. Res.* 52, 1257–1265, 2001.
80. Gatford, K.T. et al. Novel resistance to pre-harvest sprouting in Australian wheat from the wild relative *Triticum tauschii*. *Euphytica* 126, 67–76, 2002.
81. Sodkiewicz, W. Diploid wheat—*Triticum monococcum* as a source of resistance genes to preharvest sprouting of triticale. *Cereal Res. Commun.* 30, 323–328, 2002.
82. Gale, M.D. et al. Cereal comparative genetics and preharvest sprouting. *Euphytica* 126, 21–25, 2002.
83. Yan, L.L. et al. The wheat VRN2 gene is a flowering repressor down-regulated by vernalization. *Science* 303, 1640–1644, 2004.
84. Yano, M. et al. Hd1, a major photoperiod sensitivity quantitative trait locus in rice, is closely related to the *Arabidopsis* flowering time gene CONSTANS. *Plant Cell* 12, 2473–2483, 2000.
85. Koornneef, M. et al. Seed dormancy and germination. *Curr. Opin. Plant Biol.* 5, 33–36, 2002.
86. Lovegrove, A. and Hooley, R. Gibberellin and abscisic acid signalling in aleurone. *Trends Plant Sci.* 5, 102–110, 2000.
87. Holdsworth, M. et al. Molecular and genetic mechanisms regulating the transition from embryo development to germination. *Trends Plant Sci.* 4, 275–280, 1999.
88. Miyoshi, K. et al. Characterization of viviparous mutants in rice. *Breed. Sci.* 50, 207–213, 2000.
89. McKibbin, R.S. et al. Transcripts of Vp-1 homeologues are misspliced in modern wheat and ancestral species. *Proc. Natl. Acad. Sci. USA* 99, 10203–10208, 2002.
90. Hoecker, U. et al. Integrated control of seed maturation and germination programs by activator and repressor functions of viviparous-1 of maize. *Genes Devel.* 9, 2459–2469, 1995.
91. Hobo, T. et al. A bZIP factor, TRAB1, interacts with VP1 and mediates abscisic acid-induced transcription. *Proc. Natl. Acad. Sci. USA* 96, 15348–15353, 1999.
92. Nakamura, S. et al. Physical interactions between ABA response loci of *Arabidopsis*. *Plant J.* 26, 627–635, 2001.

93. Johnson, R.R. et al. The abscisic acid-responsive kinase PKABA1 interacts with a seed-specific abscisic acid response element-binding factor, TaABF, and phosphorylates TaABF peptide sequences. *Plant Physiol.* 130, 837–846, 2002.
94. Brocard-Gifford, I.M. et al. Regulatory networks in seeds integrating developmental, abscisic acid, sugar, and light signaling. *Plant Physiol.* 131, 78–92, 2003.
95. Lopez-Molina, L. et al. ABI5 acts downstream of ABI3 to execute an ABA-dependent growth arrest during germination. *Plant J.* 32, 317–328, 2002.
96. Kagaya, Y. et al. Abscisic acid-induced transcription is mediated by phosphorylation of an abscisic acid response element binding factor, TRAB1. *Plant Cell* 14, 3177–3189, 2002.
97. Hrabak, E.M. et al. The *Arabidopsis* CDPK-SnRK superfamily of protein kinases. *Plant Physiol.* 132, 666–680, 2003.
98. Bradford, K.J. et al. Abscisic acid and gibberellin differentially regulate expression of genes of the SNF1-related kinase complex in tomato seeds. *Plant Physiol.* 132, 1560–1576, 2003.
99. Anderberg, R.J. and Walker-Simmons, M.K. Isolation of a wheat cDNA clone for an abscisic acid-inducible transcript with homology to protein-kinases. *Proc. Natl. Acad. Sci. USA* 89, 10183–10187, 1992.
100. Yamauchi, D. et al. Molecular analysis of the barley (*Hordeum vulgare* L.) gene encoding the protein kinase PKABA1 capable of suppressing gibberellin action in aleurone layers. *Planta* 215, 319–326, 2002.
101. Gubler, F. et al. Gibberellin-regulated expression of an Myb gene in barley aleurone cells—evidence for Myb Transactivation of a high-pl alpha-amylase gene promoter. *Plant Cell* 7, 1879–1891, 1995.
102. Gomez-Cadenas, A. et al. Gibberellin/abscisic acid antagonism in barley aleurone cells: site of action of the protein kinase PKABA1 in relation to gibberellin signaling molecules. *Plant Cell* 13, 667–679, 2001.
103. Lanahan, M.B. et al. A gibberellin response complex in cereal alpha-amylase gene promoters. *Plant Cell* 4, 203–211, 1992.
104. Kobayashi, Y. et al. Differential activation of the rice sucrose nonfermenting1-related protein kinase2 family by hyperosmotic stress and abscisic acid. *Plant Cell* 16, 1163–1177, 2004.
105. Kaneko, M. et al. Loss-of-function mutations of the rice GAMYB gene impair alpha-amylase expression in aleurone and flower development. *Plant Cell* 16, 33–44, 2004.
106. Gocal, G.F.W. et al. GAMYB-like genes, flowering, and gibberellin signaling in *Arabidopsis*. *Plant Physiol.* 127, 1682–1693, 2001.
107. Peng, J.R. et al. "Green revolution" genes encode mutant gibberellin response modulators. *Nature* 400, 256–261, 1999.
108. Dill, A. et al. The DELLA motif is essential for gibberellin-induced degradation of RGA. *Proc. Natl. Acad. Sci. USA* 98, 14162–14167, 2001.
109. Dill, A. et al. The *Arabidopsis* F-box protein SLEEPY1 targets gibberellin signaling repressors for gibberellin-induced degradation. *Plant Cell* 16, 1392–1405, 2004.
110. Tyler, L. et al. DELLA proteins and gibberellin-regulated seed germination and floral development in *Arabidopsis*. *Plant Physiol.* 135, 1008–1019, 2004.
111. Borner, A. et al. The relationships between the dwarfing genes of wheat and rye. *Euphytica* 89, 69–75, 1996.
112. Lee, S.C. et al. Gibberellin regulates *Arabidopsis* seed germination via RGL2, a GAI/RGA-like gene whose expression is up-regulated following imbibition. *Genes Devel.* 16, 646–658, 2002.

113. Fick, G.N. and Qualset, C.O. Genetic control of endosperm amylase activity and gibberellic-acid responses in standard-height and short-statured wheats. *Proc. Natl. Acad. Sci. USA* 72, 892–895, 1975.
114. Gale, M. The genetics of preharvest sprouting in cereals, particularly in wheat. In *Preharvest Field Sprouting in Cereals*. N.F. Derera, ed., CRC Press Inc., Boca Raton, FL, 1989, 85–110.
115. Hedden, P. and Kamiya, Y. Gibberellin biosynthesis: Enzymes, genes and their regulation. *Ann. Rev. Plant Physiol. Plant Mol. Biol.* 48, 431–460, 1997.
116. Ashikari, M. et al. Loss of function of a rice gibberellin biosynthetic gene, GA20 oxidase (GA20ox-2) led to the rice "green revolution." *Breed. Sci.* 52, 143–150, 2002.
117. Monna, L. et al. Positional cloning of rice semidwarfing gene, sd-1: rice "green revolution gene" encodes a mutant enzyme involved in gibberellin synthesis. *DNA Res.* 9, 11–17, 2002.
118. Spielmeyer, W. et al. Semidwarf (sd-1) "green revolution" rice contains a defective gibberellin 20-oxidase gene. *Proc. Natl. Acad. Sci. USA* 99, 9043–9048, 2002.
119. Spielmeyer, W. et al. Isolation of gibberellin and metabolic pathway genes from barley and comparative mapping in barley, wheat and rice. *Theor. Appl. Genet.* 109, 847–855, 2004.
120. Hugouvieux, V. et al. An mRNA cap binding protein, ABH1, modulates early abscisic acid signal transduction in *Arabidopsis*. *Cell* 106, 477–487, 2001.
121. Xiong, L.M. et al. Modulation of abscisic acid signal transduction and biosynthesis by an Sm-like protein in *Arabidopsis*. *Dev. Cell* 1, 771–781, 2001.

3 Utilization of *Arabidopsis* and *Brassica* Genomic Resources to Underpin Genetic Analysis and Improvement of *Brassica* Crops

Graham J. King

CONTENTS

3.1 INTRODUCTION

Brassica species are the closest crop relatives to the reference dicotyledon plant species, *Arabidopsis thaliana*; they contribute to a diverse range of agricultural and horticultural crops worldwide, encompassing vegetable, salad, oil, mustard, fodder and nonfood uses. Throughout the genus, taxa are characterized by the wide range of developmental adaptations, many of which have been domesticated into crops that include oilseed rape/canola and swede (*B. napus*); cabbage, cauliflower, broccoli, Brussels sprout (*B. oleracea*); Chinese cabbage, pak choi, turnip, and oil (*B. rapa*); and the mustards and associated oils crops (*B. nigra, B. juncea, B. carinata*). These crops contribute basic food energy, nutrients, and secondary metabolites to human and livestock diets, as well as providing a potentially increasing number of nonfood uses. In addition, brassicas can be beneficial as break crops with soil remediation properties and contribute to rotational cropping systems. *Brassica* species are naturally mostly outbreeding, with a strong and well described sporophytic self-incompatibility system.

Among the major challenges facing crop improvement in changing economic, market, and climate conditions is the ability to harness genetic diversity through an information-led approach that maximizes information from model species. For brassicas, as with other crops, there is a wide range of valuable traits for which genetic variation exists, but where understanding is required. These include improving harvest index and yields in the context of changing climate and reduced inputs, optimizing harvestable and processed product quality fit for purpose, and identifying the scope for increased added value through nutritional, prophylactic health, or nonfood use.

A wide range of genetic resources are available for *Brassica*. These may be used in genetic analysis or as a source of alleles for introgression into new crop varieties. There is considerable scope to accelerate introgression by use of gene- or locus-specific genetic markers. An increasing amount of genomic information is accumulating, together with ready access to information derived from the related Brassicaceae model plant *Arabidopsis*.

There has been considerable progress in the genetic analysis of agronomic and related plant traits in *Brassica*. However, compared with *Arabidopsis*, there has been relatively slow progress in identifying and characterizing the behavior of the underlying genes, genomic regulatory networks, and associated metabolism. This has partly resulted from the complexities of genome organization, based on segmental chromosome duplication and divergence, existing within the diploid brassicas. These are compounded in the amphidiploid species such as *B. napus* and *B. juncea*.

For effective crop-based research, it is essential to be able to navigate between trait and gene and thus integrate information from agronomy, breeding, genetics, and genomics. Genomic information is the key to exploiting knowledge gained at the level of gene expression, biochemistry, metabolism, and physiology. To manipulate crop traits based on genomic knowledge, genes need to be located in their relevant genomic context in order to understand their regulation and any evolutionary selection pressures. In practical terms, this knowledge may then be used to develop locus-specific molecular genetic markers for use in marker-assisted selection or to introduce novel alleles via site-selected mutagenesis or transgenic modification.

At the time of writing, developments within the international research communities have led to establishment of the Multinational *Brassica* Genome Project, which was initiated in 2002. [1] This is a long-term project, an early output of which is the Multinational *Brassica rapa* Genome Sequencing Project, which is generating contiguous sequence of gene-coding regions over the *B. rapa* "A" genome [137]. This builds on previous initiatives to develop public-access genomic resources, which have included expressed sequence tags (ESTs), bacterial artificial chromosome (BAC) genomic libraries, BAC-based physical contigs, saturated sequence-tagged genetic maps, and associated populations.

This chapter will concentrate on providing examples of current developments in *Brassica* research that have benefited from genetic and genomic approaches, highlighting areas well positioned to capitalize on data, information, and knowledge from the model *Arabidopsis* for the benefit of crop improvement. As such, it does not aim to provide a comprehensive review of *Brassica* research, which has a long and rich history based on diverse genetic resources and breeding systems amenable to genetic analysis [2,3]. The central role of comparative genomic approaches will be highlighted, together with an assessment of current gaps in knowledge.

3.2 COMPARATIVE GENOMICS

The ability to carry out comprehensive comparative genomics depends upon the adoption of common standards and nomenclature for describing various constituent entities or objects such as linkage group, locus, gene identity, and trait. It has taken some time for the linkage group nomenclature for *Brassica* species to converge because different researchers have developed linkage maps based upon diverse sets of arbitrary markers. The availability of sequence-tagged markers and subsequent development of maps anchored to the *Arabidopsis* genome is currently accelerating convergence in this area. The linkage group nomenclature established by Parkin et al. [4] and Sharpe et al. [5] for *B. napus*, by Bohuon et al. [6] for *B. oleracea*, and by Lagercrantz et al. [7] for *B. nigra* has now been adopted to describe other linkage maps. This nomenclature scheme allows alignment of linkage groups for the amphidiploid genomes with their constituent diploid linkage groups.

Thus, *B. rapa* (R1-R10) corresponds to the *B. napus* (N1-N10) A genome, and *B. oleracea* (O1-O9) corresponds to the *B. napus* (N11-N19) C genome. For *B. oleracea*, the linkage groups have now been aligned and oriented with respect to the karyotype [8]. Recently, the pattern of chromosome segments conserved within the A and C genomes of *B. napus* has been aligned with collinear regions of the *Arabidopsis* genome [129].

3.2.1 Genome Organization

The relationship between the canonical *Brassica* genomes has been characterized in the schema commonly referred to as the "triangle of U." [9] Three distinct diploid genomes or "cytodemes" are recognized and each is represented by a type species (Table 3.1). The genome sizes vary and have been estimated to range from 470 Mbp (*B. nigra*) to 1540 Mbp (*B. carinata*) [10,11]. In contrast, *Arabidopsis* species have five chromosomes, with a genome size for *A. thaliana* of ~120 Mbp [12].

3.2.1.1 Collinearity and Chromosome Segmental Duplications

Different hypotheses have been proposed to account for the origin of and relationships between contemporary crucifer genomes. Based on comparison of linkage maps, Truco et al. [13] have proposed a possible chromosome phylogenetic pathway based on an ancestral genome of at least five, and no more than seven, chromosomes. A number of studies have focused on specific regions of *Arabidopsis* and compared them with the genome organization in *Brassica* species. For example, a comparison

TABLE 3.1
Relationships among the Species of *Brassica*, with Chromosome Numbers (*n*) and Indicative Genome Sizes

Genome	Species	*n* =	~Genome size
A	*B. rapa*	10	500–550 Mbp
B	*B. nigra*	8	470
C	*B. oleracea*	9	600–650
AB	*B. juncea*	18	1100–1500
AC	*B. napus*	19	1130–1240
BC	*B. carinata*	17	1540

of *B. napus* with *Arabidopsis* Chr 5 based on sequence-tagged (RFLP) markers [14] revealed six highly conserved copies in *B. napus* that corresponded to an 8-Mb segment from Chr 5. This included a single inversion that appeared to be the primary rearrangement that accounted for two lineages since divergence of the genera.

These results were used to suggest that the constituent genomes of *B. napus* were generated from a fusion of three ancestral genomes that together had strong similarities to the contemporary *Arabidopsis* genome. This is consistent with the hypothesis that diploid *Brassica* genomes evolved from a common hexaploid ancestor. For specific regions, a genetic distance of 1 cM in *B. napus* was found to be equivalent to 285 kb in *Arabidopsis* [14], although this figure may vary considerably across the genomes.

Within different regions of the genome, comparisons between *Arabidopsis* and *Brassica* suggest differential patterns of divergence [15,16], with reciprocal translocations described at a genetic level in natural and resynthesized amphidiploid *B. napus* [4,5]. The first evidence for this came from RFLP probes, which allowed detection of a reciprocal chromosomal transposition found in several oilseed *B. napus* genotypes, and it involved exchange of interstitial homologous regions on N7 and N16 [17]. This was confirmed by cytological analysis on synaptonemal complexes. Up to a third of the physical length of the N7 and N16 chromosomes appeared to be involved, although only a few recombination events were detected in the region.

It is interesting that this region corresponds to an inverted segmental duplication that has been described in *B. oleracea* O6 [18], with the self-incompatibility locus (S) located at or near the junction of the duplication. Higher seed yields were associated with parental configurations of the *B. napus* rearrangement in segregating progenies [17], with complete complements of homologous chromosomes from the diploid A and C genome progenitors of *B. napus*.

A more comprehensive survey of mapped RFLP probes has shown that 73% of genomic clones detect two or more sequences in the *Brassica* A and C diploid genomes [19]. Most duplicate loci appear to be in distinct linkage groups as collinear blocks of linked loci and display a variety of rearrangements, including inversions and translocations, following duplication. The presence of some identical rearrangements can be taken as evidence that these occurred before divergence of the two species. For some of the linkage groups, their current organization appears to be consistent with earlier centric fusion and/or fission processes that may have played an important role in the evolution of *Brassica* genomes [19].

Overall, it appears that at least 16 gross chromosomal rearrangements can account for differences between these two diploid genomes since divergence from a common ancestor [19]. It also appears that there are homologous loci in the C genome for almost every mapped locus in the "A" genome. Given current efforts to sequence the complete *B. rapa* genome, this homologous conservation should facilitate the inference of gene location in other *Brassica* genomes through comparison of saturated physical maps and the complete genome sequence of *B. rapa*.

The ability to infer information about the organization of diploid *Brassica* genomes has been complicated by the recognition that a large amount of internal duplication has taken place within the *Arabidopsis* genome. Up to 80% of this genome appears to comprise duplicated sequences with about 20% of genes tandemly duplicated [20]. Thus, the overall pattern of rearrangements and genome organization may not represent a simple one-to-one correspondence, as recognized by Lan et al. [21]. At a more detailed level, O'Neill and Bancroft [15] have demonstrated that the collinearity between a neighboring set of 19 predicted gene sequences in *Arabidopsis* and the corresponding replicated loci in *Brassica* is characterized by local rearrangements and deletions.

The *Brassica* genome can be regarded as a mosaic of segments sharing a common ancestry with *Arabidopsis*. Parkin et al. [129] have recently compiled a comprehensive data set that allows the pattern of duplicated chromosome segments in *B. napus* to be compared with the *Arabidopsis* genome. They used evidence from 359 sequenced *Brassica* RFLP probes to detect 1232 loci in *B. napus*. Comparative sequence analysis with the *Arabidopsis* genome revealed 550 homologous sequences from which they could infer relative chromosomal position. They were able to identify 21 blocks conserved within *Arabidopsis* that, following replication and rearrangement, can account for almost 90% of the genetic map of *B. napus*. They estimate that a minimum of 74 gross rearrangements (38 and 36 in the A and C genomes, respectively) may have separated the two lineages since their divergence 14 to 24 MYA [130].

Pairs or sets of genes are said to be orthologous when they have diverged following a speciation event; paralogous genes arise as a result of duplication events. The organization of the diploid *Brassica* genomes can be represented as a mosaic of segments that align in varying degrees of collinearity with corresponding orthologous regions of the *Arabidopsis* genome. The extent of collinearity between paralogous segments may differ significantly as a result of the time elapsed since their divergence and the degree to which they may have been subject to infection by transposable elements.

3.2.1.2 Genetic Markers

The use of molecular markers in *Brassica* has recently been reviewed in detail [22]. The emphasis here will be on developments in genetic markers that have the potential to provide additional information about genome evolution between *Brassica* and other taxa, and *Arabidopsis* in particular. The increased availability of markers amenable to reproducible high-throughput screening, combined with locus-specific linkage to genes underlying crop traits, provides a key technology for crop improvement involving marker-assisted selection [131].

3.2.1.2.1 Marker Systems

In general, simple sequence repeats (SSRs, or microsatellites) have the advantage that they are relatively often polymorphic, locus specific, sequence tagged, and transferable between laboratories. They have been isolated and characterized from *Arabidopsis* and a range of *Brassica* species [23–26] and primer sets used to develop informative and locus-specific sequence-tagged genetic markers. The process of marker development has conventionally been based upon sequencing of clones randomly selected from SSR-enriched genomic libraries. When the original clone sequences are of sufficient quality and made available, the sequence flanking the repeat can be used for comparative genomic analysis.

For *B. rapa,* up to 90% of 228 microsatellites identified in one study were found to amplify corresponding regions in other *Brassica* species, and 40% of primer sets amplified *Arabidopsis* loci [24]. A related technology, inter simple sequence repeats (ISSRs), involves using one locus-specific primer flanking the SSR together with a nonanchored primer. This has been used to amplify and sequence 44 fragments from cauliflower [27], where the majority of the internal regions of the ISSRs had homologies with known sequences (e.g., from *Arabidopsis*)—primarily with protein-coding genes implicated in DNA interaction and gene expression. This is one line of evidence to suggest that there are long and numerous regions conserved with the *Arabidopsis* genome. More recently, the availability of genome survey sequence (GSS) and other data sets for the *B. oleracea* genome has allowed informatic detection of SSRs [28,29].

A comparison of different sequence-tagged molecular markers has been carried out in order to establish their relative ease of development [30]. This involved using EST sequences from *Brassica* and *Arabidopsis* to design primer pairs to amplify gene sequences using sequence characterized amplified region (SCAR), cleaved amplified polymorphic sequence (CAPS), and PCR-RF-SSCP (a combination of CAPS and single-strand conformation polymorphism) markers. The level of polymorphism detected was assessed in two sets of *B. oleracea* breeding lines. More single genes were amplified using primer pairs from *B. oleracea* data than from *B. rapa* or *Arabidopsis*. The PCR-RF-SSCP method was the most efficient, yielding informative assays between even closely related parent lines in situations where CAPS assays were uninformative.

3.2.1.2.2 Gene-Specific Markers and Comparative Genomics

Considerably more information about genome evolution and organization can be obtained by developing genetic linkage maps based upon markers directly associated with coding regions. For *Brassica* species, a number of technologies have been developed that make use of genic information from *Brassica* or *Arabidopsis*.

The sequence-related amplified polymorphism (SRAP) technique [31] was developed to provide a relatively high level of information per assay. SRAP is based on a two-primer system that involves core as well as filler sequences and provides a similar level of efficiency to AFLPs, while being technically simpler. This approach was used to develop markers from *B. oleracea* cDNAs [32]. Of these, 169 had similarity to genes in the *Arabidopsis* genome. Orthologous and paralogous genes were identified by the clear differences in their similarity score values. In common with other comparative studies, the resulting genetic map [32] revealed extensive collinearity between the two genomes over chromosomal segments, including many inversions and segmental indels, as well as an uneven distribution of large-scale duplications.

However, based on the particular markers used, it was found that most of the duplicated segments corresponded to *Arabidopsis* Chr 1 and Chr 5, whereas representation in *Brassica* of segments from *Arabidopsis* Chr 2 and Chr 4 was lower. SRAP has also been used to determine the level of diversity among oilseed rape (*B. napus*) cytoplasmic male sterility maintainer and restorer lines [33]. In this case, 118 polymorphic loci were used to calculate similarity indices from between 0.46

and 0.97. A subsequent cluster analysis was successful in grouping the lines and was in agreement with existing pedigree data.

A more directed approach for gene marker development has involved development of amplified consensus genetic markers (ACGM) [34]. Based on sequence analysis and design of PCR primers, this technique allows rapid sequencing of homologous genes from species within the same phylogenetic family, as well as detection of intragenomic polymorphism. As a demonstration, a set of 32 ACGMs has been used to amplify genes from *Arabidopsis* and *B. napus* [35]. The polymorphism detected with ACGMs is primarily associated with intron sequences and, in this study, allowed mapping of 43 genes, together with attribution of homologs to A or C genomes within *B. napus*. A similar approach has been used with primers that amplified regions of 22 putatively orthologous functional loci in *Arabidopsis* species and *B. oleracea* [36].

3.2.1.3 Karyotype and Physical Map Analysis

Chromosome fluorescent *in situ* hybridization (FISH) has been successfully applied to *Brassica* and *Arabidopsis* to locate genes, markers, and marker loci in the context of existing karyotypes and chromosomal features. The linkage maps of *B. oleracea* have been aligned and oriented to the karyotype using a combination of RFLP-, cosmid-, and BAC-labeled probes [8]. Comparative fiber-FISH mapping has been demonstrated between *B. rapa* and *Arabidopsis* [37]. This technique allows direct measurements to be made on DNA fibers isolated from chromatin and provides a valuable level of resolution between that achieved with chromosomal FISH and currently available genomic sequence or physical maps of *Brassica*. These results supported the hypothesis that chromosomal duplications played a major role in expansion and evolution of *B. rapa*, which contrasts with the alternative of regional expansion due to accumulation of repetitive sequences in intergenic regions, such as has been found in grass genomes [38].

Chromosome FISH mapping of low-copy sequences from *Arabidopsis* BACs onto *B. oleracea* allows investigation of patterns of chromosomal duplication and relative physical distances. This has been used to demonstrate conserved organization of two BACs on two *B. oleracea* chromosomes [39]. A combination of *Arabidopsis* and *Brassica* BACs has also enabled confirmation of conservation of the order of genomic DNA between a chromosomal segment of *Arabidopsis* Chr I and two duplicated segments on *B. oleracea* linkage group O6 [40]. The FISH analysis was able to resolve the inverted duplication on O6, together with a short inversion within one duplicated copy. This study also demonstrated that although genetic distance (frequency of recombination) between the two segments diverged, the physical distance appeared relatively conserved.

Lysak et al. [132] recently used a multicolor FISH approach to investigate the pattern of rearrangements within the Brassicaceae. They fluorescently labeled adjacent segments within an *Arabidopsis thaliana* BAC contig of ~8.7 Mb from chromosome 4 and used the visualization of three colors to trace homologous chromosome regions in 21 species. Their data were consistent with the Brassiceae tribe being a monophyletic group, with all species analyzed descending from a common hexaploid ancestor. They confirmed the presence of three copies of the segment in the three diploid *Brassica* species and six in the amphidiploids. Detailed analysis of the amphidiploid *B. juncea* (AB genome) and its two component species, *B. rapa* (A) and *B. nigra* (B), indicated that two of the three A genome homologs in *B. juncea* had a structure that deviated from that of *B. rapa*. This most likely occurred due to translocation and inversion event.

3.2.2 Genome Rearrangements

3.2.2.1 Mechanisms

One of the primary mechanisms for genomic change among plants is whole genome or segmental polyploidy. Superimposed on such variation are the ongoing processes and effects of transposition. Transposable elements (TEs) are a major component of plant genomes and, together with chromosomal segmental duplications, are likely to account for most of the differences in genome size between *Brassica* species.

It has been possible to determine some of the patterns of TE amplification, diversification and loss since divergence of TEs between *Arabidopsis* and *B. oleracea* by making use of the shotgun genomic sequence data available for *B. oleracea* [41] and subjecting these to comparative analysis with the *Arabidopsis* genome [42,133]. From this analysis, it appears that nearly all TE lineages are shared between the genera, with the number of elements in each lineage larger for *B. oleracea* [42]. In both species, Class I retroelements are the most abundant, of which LTR and non-LTR elements are the most prevalent. Within *B. oleracea* several families of class II (DNA) elements are present in very high copy numbers and can account for much of the observed genome expansion. The TIGR plant repeat database contains a collation of repeat sequences for *Arabidopsis* and *Brassica* [43], and these are coded into superclasses, classes, and subclasses based on sequence and structure similarity.

The distribution of long terminal repeat (LTR) transposons has also been investigated in the *B. oleracea* genome by probing specific elements onto gridded BAC libraries [133]. Analysis of these data provided estimates of between 90 and 320 copies of individual Ty1 (*copia*-like) and Ty3 (*gypsy*-like) retrotransposons per haploid genome. This was consistent with sequence analysis of the same elements in available shotgun genome survey sequence, which indicated between 60 and 570 elements. There was minimal evidence for clustering of the two retrotransposon groups, which was also substantiated by FISH analysis that showed that each had a characteristic chromosomal distribution. Taken together, these results suggest that preferential sites, and perhaps control mechanisms, may exist for the insertion or excision of the different retrotransposon groups. There was only evidence for a single LINE element in the BAC analysis and none from the sequence analysis.

3.2.2.2 Timing

Recent estimates of the timing of the whole-genome duplication that has occurred in *Arabidopsis* have been based on two alternative hypotheses. These are based either on the assumption that duplicated segments diverged from an autotetraploid form (38 MYA) before divergence from *Brassica* [44] or that the ancestor was allotetraploid and the duplication occurred less than 38 MYA, thus contributing to the *Arabidopsis–Brassica* divergence.

Duplicated blocks within the *Arabidopsis* genome [45] have been detected using protein sequence similarity, together with estimates of the level of synonymous substitution present between duplicated genes. This allows estimates of the relative age of segments and suggests that *Arabidopsis* underwent two distinct episodes of duplication. One of these would have occurred as a polyploidy prior to the *Arabidopsis–Brassica* divergence (14 to 40 MYA), and another, older set of duplicated blocks would have been formed following the monocotyledon–dicotyledon divergence [45]. A comparative analysis involving the *Arabidopsis* and *Capsella rubella* genomes [46] found that collinearity is more pronounced than that for *Arabidopsis* and different *Brassica* species.

Brassica genomes are remarkably plastic. In a comparison of the *B. oleracea* linkage map and the *Arabidopsis* genome sequence, Lukens et al. [47] found strong evidence for genome duplication and rearrangement within the diploid *Brassica* species, but less evidence for triplication. It appears that large-scale translocations combined with tetrasomic inheritance can account for some but not all genomic changes observed in the more recent amphidiploid species formed from the base diploids. Superimposed on these large-scale events at the level of chromosomal segment, transpositions and other small-scale sequence changes contribute to continuing genomic novelty [47].

3.2.2.3 Recent Genome Rearrangements and Breeding Introgression

The increased and widespread production of canola and rapeseed over the past 30 years has resulted from the development of modern low-glucosinolate cultivars that produce high-protein meal for animal feed. The low-glucosinolate trait was initially introduced from 'Bronowski,' and residual segments of this genotype are present in modern cultivars and may still contribute to reduced yield, poorer winter hardiness, and lower oil content. It has been found that at least 15 segments are still present in the 'Tapidor' genotype [48], representing about 30% of the *B. napus* genome. Sharpe and Lydiate [48] have shown that just three 'Bronowski' donor segments contain loci that can explain more than 90% of the variation for total seed glucosinolates. This level of linkage drag is common in breeding programs, reflecting a relatively low or discontinuous level of recombination. With knowledge of the location of such introgressed loci, there is considerable scope for using high-resolution marker-assisted selection (MAS) to eliminate any associated deleterious alleles.

3.3 GENE HOMOLOGY AND DIVERGENCE OF FUNCTION

3.3.1 Detecting Orthologous and Paralogous Gene Coding Sequences

The DNA sequences in coding regions of the *Arabidopsis* and *Brassica* genomes are highly conserved and have, on average, about 85% similarity [49]. As already described, collinearity of gene order exists between these two genomes over regions covering as much as 30 cM in *Arabidopsis* [49]. Many studies indicate that a high proportion of loci in *Arabidopsis* are present in at least three copies in haploid *Brassica* genomes [50,51,129] For example, Lan et al. [160] presented a detailed comparative map of *B. oleracea* and *A. thaliana* based primarily on RFLP mapping of *Arabidopsis* expressed sequence tags (ESTs). They found a one-to-one correspondence between linkage groups accounting for 57% of comparative loci.

A similar study by Babula et al. [52] used 110 *Arabidopsis* ESTs as molecular markers in *B. oleracea* and found that 95 were informative and suitable to use for construction of an RFLP genetic linkage map. These EST-based markers yielded 212 new loci that covered all nine linkage groups and confirmed previous patterns of collinear organization, albeit with varying levels of sequence conservation. By taking into account the extensive duplication within the *Arabidopsis* genome, they were able to identify long conserved regions covering entire chromosome arms in both genomes, suggesting that these are probably shared by descent. This was consistent with other studies [129], as was the presence of extensive rearrangements in many chromosome regions.

The difficulty associated with identifying the most closely related segments between *Arabidopsis* and *Brassica* genomes is exacerbated by the duplications present within both species. This can occasionally lead to ambiguous criteria being used to identify orthologous regions. To address this, Lukens et al. [53] compared the positions of genetically mapped and sequenced loci in *B. oleracea* to positions of putative orthologs present within the *Arabidopsis* genome. They defined explicit criteria to distinguish orthologous from paralogous loci and developed a conservative algorithm to identify collinear loci between the genomes, as well as using a permutation test to evaluate the significance of the regions. This analysis enabled 34 significant *Arabidopsis* regions to be identified that were collinear with up to 30% of the *B. oleracea* genetic map. The findings were consistent with a high level of rearrangement in *B. oleracea* genome since divergence from *Arabidopsis*, probably as a result of polyploidization.

Coding sequence divergence can be assessed through comparison of nonsynonymous (Ka) to synonymous (Ks) changes between coding regions [54]. This has been carried out for a small sample of *B. rapa* ESTs and orthologous *Arabidopsis* coding regions [55]. Among the 218 sequences sampled, the distribution of Ka:Ks ratio was unimodal; substitution rates were more variable at nonsynonymous sites and no evidence suggested that Ka and Ks were positively correlated. Therefore, it would appear from this sample that there was no evidence for any of the genes having evolved in response to positive selection. As more complete data sets become available and are compared, it should become possible to relate any patterns and selective constraints to the evolutionary history of different chromosomal segments.

3.3.2 Consequences of Genome Duplication

Osborn et al. [56] have addressed some of the issues relating to understanding the mechanisms associated with novel genetic variation and gene expression in polyploids. When genetic variation is achieved via the processes of gene duplication and intergenomic heterozygosity, this may increase variation in dosage-regulated gene expression and changes in regulatory interactions, as well as rapid genetic and epigenetic changes. For crop plants, the consequences of gene duplication are likely to be manifested in greater developmental plasticity and the ability to respond to changing environments. Genetic crop improvement, therefore, may involve selection of optimal combinations of alleles at duplicated loci and the avoidance of any associated linkage drag effects.

3.3.2.1 Redundancy of Gene Function

In *Arabidopsis*, many insertion mutants have no obvious phenotypic effect [57], which in part may be due to redundancy of function among duplicated genes. An example of this is found with the shatterproof genes SHP1 and SHP2, which are MADS box transcriptional regulators that must both be simultaneously down-regulated or removed to generate a silique nondehiscence phenotype [58]. These two genes are located within a chromosomal block that appears to have duplicated about 100 MYA [57].

A comparative study involving another related MADS box gene in *Arabidopsis* and *Brassica* has also indicated possible redundancy of gene function. The *ap1/cal* double mutant in *Arabidopsis* displays an arrest at inflorescence initiation and forms a cauliflower-like curd. Such lines are also apetalous. In contrast, *B. oleracea* cauliflowers that have recessive loci that contain the orthologous genes *Bocal-a/Boap1-a* arrest at the same stage, but have wild-type flowers. Because *B. oleracea* contains multiple loci of each of these genes, it appears this could represent differential function during ontogeny [59] consistent with the duplication–degeneration–complementation (DDC) model [60].

This model predicts that degenerative mutations in regulatory elements tend to increase the probability of duplicate gene preservation rather than the converse. It also predicts that the normal mechanism associated with preservation of duplicate genes involves partitioning of ancestral functions rather than the evolution of new functions.

3.3.2.2 Divergence of Gene Function

Studies of genes in isolation can provide useful insights into the processes driving functional divergence and specificity. However, it is important to understand such changes within their genomic context because this may provide information about mechanisms underlying phenomena such as heterosis and epistasis.

For some classes of genes, especially those involved in recognition pathways, there is often a requirement for rapid divergence of specificity, as is the case with pathogen resistance recognition genes. Glycine-rich pollen surface proteins (GRPs) have diverged substantially between *Arabidopsis* and *B. oleracea*, and this makes identification of homologous genes impossible, although they are only separated by ~20 million years [61]. Fiebig et al. [61] have sequenced eight members of the GRP cluster in four related crucifer species, as well as 11 flanking genes. They found that GRP genes change more rapidly than neighboring genes, are more repetitive, and have undergone more insertion and deletion events. However, this variation occurs concurrently with conservation of the repeat amino acid composition. The sequence analysis provided evidence that genes flanking the cluster had undergone strong purifying selection, with relaxed selective constraints in the first exon of the GRP. As a result, these researchers conclude that rapid GRP evolution was due primarily to the processes of duplication and deletion, together with divergence of repetitive sequences.

A comparison of sequence and gene content has been carried out in a targeted region (*ABI1-Rps2-CK1*) in *Arabidopsis* and *B. oleracea* [62]. *BoRps2* is present in single copy, with a segment containing an additional *N-myr* gene between *Rps2* and *CK1*. The *Arabidopsis* homologs for this gene are on different chromosomes, where they are linked with additional homologs for *CK1*. There are high levels of sequence identity for coding sequences of all genes. However, in most cases, *Brassica* has larger intra- and intergenic noncoding spacers than does *Arabidopsis*. As has been found in some other studies, the promoters of these genes were poorly conserved, except for several sequences of a few nucleotides.

Comparison of duplicate copies of caseine kinase-like (*CK1*) genes in *Arabidopsis* and *B. oleracea* enabled separation into two major groups based on exon number and sequence identity [63] and thus assignment to orthology and parology. A more thorough examination has been carried out for recent (10,000 years for *B. napus*) and ancient (~20 MYA from *Arabidopsis*) divergence of blocks of genes with an intermediate set (A genome vs. C genome: 4 million years). This was based on a comparison of *Arabidopsis* genome sequence and clones from BAC libraries [64].

There appears to have been extensive divergence of gene content between *B. rapa* paralogous segments and homologous segments in *Arabidopsis*. It also appears that the pattern of gene loss in *B. rapa* and *B. oleracea* is similar, with a small number of species-specific rearrangements. From such comparisons, it can be concluded that the evolution of genome microstructure is an ongoing process [64], although there has been little or no change in microstructure as a consequence of the hybridization event forming modern *B. napus* amphidiploids.

3.4 ACCOUNTING FOR TRAIT VARIATION IN *BRASSICA* CROPS

Brassica species display a notable phenotypic plasticity in terms of morphology and developmental adaptation, which is apparent in the diversity of crop morphotypes. To understand the genetic and genomic basis underlying phenotypic variation and its relevance to crop improvement fully, it is important to be able to assign variance among different components of environmental effects, as well as components of genetic variation. Comparative studies between *Arabidopsis* and *Brassica* can also be particularly informative, especially when a similar range of phenotypes may be observed.

Forward genetic approaches involving analysis of natural or chemically induced allelic variation have been widely used both in model and crop. Much of the current understanding of genes underlying plant development initially arose from mutational and segregation analysis in *Arabidopsis*. Within *Brassica*, progress in identifying major and quantitative gene effects has primarily relied on access to a range of different segregating populations and associated linkage maps based on molecular markers. Beyond this, progress has been made through candidate gene and some map-based cloning efforts facilitated by development of large-insert BAC libraries and associated physical contigs anchored to the *Arabidopsis* genome [64].

The ease of transformation and establishment of the complete genomic sequence in *Arabidopsis* have led to an array of reverse genetic approaches, which have provided deep insights into gene function. There are large collections of T-DNA insertion lines and transposon knock-out lines, as well as constitutive and inducible RNAi lines that saturate the *Arabidopsis* genome. When combined with the ready ability to up- and down-regulate specific genes and complement mutations, as well as more sophisticated resources such as enhancer traps and cell lineage markers, a previously unforeseen level of detail is being accumulated. Although many of these resources and much of the information can be transferred to the related *Brassica* crop species, there is a need to understand the most appropriate use of comparative analyses. This is particularly relevant in the context of the extent of genome-wide duplication events that characterize the *Brassica* species.

3.4.1 QTL Analysis

Among the *Brassica* crops, many agronomic traits are comprised of components having a polygenic inheritance, with varying degrees of genotype times environment ($G \times E$) interaction. The availability of segregating populations with associated linkage maps provides the opportunity to detect quantitative trait loci (QTL). QTL may be detected in F2 or derivative populations, although the estimates of variance components associated with $G \times E$ interaction will not be as reliable as those obtained from replicated populations of recombinant inbred or doubled haploid populations of homozygous lines.

In *Arabidopsis*, QTL studies using recombinant inbred (RI) populations have increasingly been used to detect the genetic basis of more complex traits [65]. Although the resolving power of QTL analysis is limited by the number of recombinants available in a population or set of populations [66], for *Arabidopsis*, any QTL detected can usually be readily resolved further through the screening of additional near isogenic lines (NILs) or STepped Aligned Inbred Recombinant (STAIR) lines [67]. This then allows resolution to a relatively small number of candidate genes, which may then be functionally analyzed using knock-out mutants, RNAi, or other resources.

For *Brassica* species, the availability of reference doubled haploid mapping populations with well-covered linkage maps has allowed initial detection and comparison of many QTL effects. However, the lack of resolving power, in terms of number of recombinants, presents a major limitation to widespread use of genomic information in *Brassica* crops. Further resolution can be achieved through use of substitution lines [68,92], which provide similar resolving power to NILs. To date, little use has been made of association genetic approaches and linkage disequilibrium to detect or resolve trait loci in *Brassica*, although such approaches [134] could readily be applied to the wide range of genetic resources that represent the domesticated gene pool. This would provide a sound basis for extending the role of MAS in crop breeding by enabling rapid screening of germplasm for novel variation, followed by introgression of specific alleles into breeding lines.

3.4.2 Mutational Analysis

TILLING (targeted induced local lesions in genomes) allows screening pools of PCR products from plants that have been chemically mutagenized, and it has successfully been applied to *Arabidopsis* [69]. The methodology allows identification and isolation of mis-sense and non-sense mutant alleles within target (candidate) genes and is often more suitable than transgenic reverse genetic approaches for crop plants. In particular, genomes containing multiple duplicated copies of genes may be able to withstand a high mutant load and thus require relatively smaller numbers of lines to be screened. In addition, TILLING allows identification of an allelic or paralogous series of mutations with which to study gene dosage effects and epistatic interactions. As well as providing a tool for inducing variation and thereby identifying gene function in model and crop plants, there is considerable potential for using this approach in generating and identifying natural variation.

EcoTILLING [136] has been adopted to identify natural genetic variants that can provide considerable information about gene function. The technique can also be useful for association mapping and linkage disequilibrium analysis [135] and is likely to become an important tool for crop improvement by allowing rapid identification of allelic variation that can then be introgressed into new varieties using MAS. In *Arabidopsis*, Henikoff and Comai [136] were able to detect small deletions, insertions, and microsatellite polymorphisms, as well as single nucleotide polymorphisms (SNPs). EcoTILLING can also be used to establish the level of heterozygosity within a gene [135], which could be important in screening crop lines to assess uniformity or the basis of hybrid vigor.

Earlier studies of EMS-induced mutation in *B. napus* [70] generated plants with increased and decreased flowering times. In these experiments, there were no changes in seedling emergence with EMS concentrations between 0 to 1%. To date, no reports of TILLING applied to *Brassica* species have been published, although a number of screening programs are under way.

Other mutagens have been used effectively in *Brassica* to induce genetic or epigenetic variation. Dunnemann and Grunewaldt [71] used *N*-nitroso-*N*-methyl-urea (NMU), which is an alkylating agent that acts on DNA to affect base pairing and thus induces SNPs. They observed a range of developmental phenotypes at a rate of 14 to 28% with 20 n*M* NMU. Epigenetic mutations may be induced by treatment with 5-azacytidine, which incorporates into DNA during replication in place of cytidine and appears to inhibit the subsequent action of methyltransferase, thus effecting a reduction of 5-methylcytosine (^{5m}C), the major methylated nucleotide in eukaryote genomes. By treating imbibing seeds of *B. oleracea*, a range of developmental variants can be generated at high frequency [72]. Although mostly transmitted through mitosis and not meiosis, the phenotypes observed were similar in type to those observed as a result of somaclonal variation in tissue culture or with mutation by NMU.

3.5 CHARACTERIZING SPECIFIC GENE FAMILIES

Comparative genomic approaches based on candidate genes from *Arabidopsis* have been applied to characterizing a range of traits in *Brassica* crop species. A number of examples relating to different areas of biology and crop traits are sufficient to illustrate the progress that has been made and to highlight complexities that can emerge as a result of paralogous genes arising from historical segmental duplications. The relevance of such information for crop improvement lies in the ability to determine the underlying genetic complexity and inter-relationships of particular traits and to understand the scope for selecting allelic combinations likely to provide predictable crop phenotypes in particular growing conditions.

3.5.1 Development

3.5.1.1 Flowering

QTLs controlling flowering time have been identified in *Brassica* species [68,73,74], some of which have been associated with genomic regions that contain orthologs of *CONSTANS*, a regulator that plays a key role in the photoperiodic flowering pathway. *CONSTANS* is controlled by the circadian clock and in *Arabidopsis* promotes flowering in long days. Four orthologs from homologous loci have been isolated from *B. napus* lines that had different flowering times [75]. These *BnCO* genes all appear to be expressed in *B. napus,* and the functional conservation of one allele (*BnCOa1*) has been confirmed by complementation of the *Arabidopsis co-2* mutation in a dosage-dependent manner [75].

Two different alleles of *CONSTANS* that possess identical DNA coding sequence have also been obtained from *B. nigra* (*BniCOa*), although they were isolated from early flowering and late flowering lines [76]. Because these did not show any differential effect on flowering time, Osterberg et al. [76] deduced that the variation influencing flowering time must be outside the coding region (*cis*-regulation or another gene). Further investigation showed that a *B. nigra* ortholog of the *Arabidopsis CONSTANS-like1* gene (*BniCOL1*) was located 3.5 kb upstream of *BniCOa,* and this did display sequence divergence among alleles of early and late flowering lines. It was also found that a single indel polymorphism in the *BniCOL1* coding region was present in several natural populations of *B. nigra* and, in most cases, had a significant association with flowering time.

More detailed analysis showed that the intergenic sequence between *BniCOL1* and *BniCOa* had a prominent peak of divergence 1 kb downstream of the *BniCOL1* coding region and may in fact contain regulatory elements for the downstream *BnCOa* gene. Further comparison of the indel among 41 sequences of complete *BniCOL1* revealed a moderate rate of within-population recombination, with no evidence for selection [77]. This is an exemplary example of the care and attention to detail required to interpret candidate- and map-based cloning studies, especially in complex crop genomes.

3.5.1.2 Vernalization

Several QTLs have been detected that account for vernalization in *Brassica* species. VFR2 is a major QTL for vernalization-responsive flowering time [78] in *B. rapa.*

The chromosome region in which VFR2 is located is syntenous with a region of *B. napus* that controls the same trait, as well as a region of *Arabidopsis* Chr 5 that contains several flowering time loci. Kole et al. [78] have backcrossed the late allele into an early flowering line and detected an additive effect attributable to the late allele. FLC is a type II MADS-box repressor of flowering and is down-regulated in response to exposure to cold temperatures. This stable epigenetic switch is required for the winter-annual habit of late flowering ecotypes of *Arabidopsis* [79]. It also appears to provide an explanation for cold vernalization in biennial brassicas and other species.

The expression patterns in *B. rapa* are consistent with those in *Arabidopsis*, and an RFLP detected by the *Arabidopsis FLC* sequence was found to cosegregate exactly with the VFR2 QTL in 414 gametes [78]. In the *B. rapa* biennial parent, *BrFLC* RNA is up-regulated, whereas under cold treatment it is down-regulated. Four *BrFLC* orthologs have been cloned [80] and sequenced. There appears to be no evidence for differential rates of evolution, with the Ka:Ks ratios of nonsynonymous to synonymous substitutions suggesting they are not under strong purifying selection. The *BrFLC1*-3 gene has been mapped to regions that are collinear with the top of *Arabidopsis* Chr 5, which is consistent with a polyploid origin.

Another paralog, *BrFLC5,* maps near a junction of two collinear regions of *Arabidopsis*. One of these contains an *FLG*-like gene (*AGL31*). However, all *BrFLC* sequences appear more closely related to *FLC* than *AGL31*. Kole et al. [161] have concluded that the duplicated *BrFLC* genes appear to have a similar function and to interact in an additive manner to modulate flowering time. Thus, one of the consequences of segmental chromosomal duplication appears to have been to increase the sensitivity and adaptive range with respect to changes in location and environment.

Environmental and endogenous flowering signals in *Arabidopsis* are integrated by a range of transcriptional regulators. These include the MADS-box gene AGL20, which appears to be a flowering activator downstream of FLC. Knockouts of *AGL20* have a late flowering phenotype, whereas when activated, it promotes early flowering even in the presence of strong expression of FLC [81]. The role of AGL20 appears to have been conserved in the Brassicaceae, with the orthologs from *B. rapa* (*BrAGL20* genes) at least 94% identical [82]. When the *BrAGL20* genes were constitutively expressed in *Cardamine felxuosa* (a long-day Brassicaceae that does not respond to vernalization), it was found that although some transgenic plants flowered very early, other antisense plants had delayed flowering.

3.5.1.3 Floral Development

It has been suggested that oilseed *B. napus* genotypes with reduced or no petals would possess greater photosynthetic efficiency and activity. To manipulate this trait, hairpin (hnRNA) gene silencing has been used to silence the B-type MADS-box floral organ identity genes *APETALA3* and *PISTILLATA* in *Arabidopsis* and in *B. napus* [83]. This engineering approach made use of an *AP1* promoter that regulates transcription in a second whorl-specific manner. The transgenic *Arabidopsis* plants had male fertile flowers in which the petals were converted into sepals. The corresponding transgenic

B. napus plants also had male fertile flowers and sepaloid petals. In both cases, the phenotypes were stable and heritable, underlying the conservation of function. This study also underlines the experimental value of focusing on key transcriptional regulators to understand and manipulate crop plant development.

It should be noted that natural variation present within the *Brassica* gene pool can also result in a reduction or absence of petal tissue. For example, Fray et al. [84] have identified cosegregation of two loci (*STAP*) that control the production of stamenoid petals in homologous positions in *B. napus*, and have isolated orthologs of the *Arabidopsis CURLY LEAF* (*CLF*) gene from the same genetic loci. They considered *CLF*, which pleiotropically affects leaf and flower development, to be a candidate gene for *STAP*. More recently, Jiang et al. [85] have established that the apetalous phenotype of a mutant *B. napus* line (ap-Tengbe) was regulated by cytoplasmic genes interacting with two pairs of nuclear genes [85].

3.5.1.4 Gene Identification in Model and Crop

Such investigations highlight the need for parallel experimental approaches that make use of the functional genomic resources for a model plant, as well as detailed and exhaustive study of natural variation present at multiple loci within a crop plant. Simple knockout experiments often cannot provide full insight into the functioning of key regulatory genes that may exist in two or more copies in the genome. The use of gene-silencing approaches such as RNAi may be effective in such studies, but will not always reveal the subtlety of regulation or interactions among multiple loci.

A valuable demonstration of the suitability of crop plants for isolating key regulatory genes that provide additional information about model systems is illustrated by the characterization of the *BABY BOOM* (*BBM*) gene [86]. This was isolated following subtractive hybridization of RNA from embryogenesis-induced microspores of *B. napus,* against a nonembryogenic sample. Such an approach would have been challenging in *Arabidopsis* because no optimized procedure is currently available for production of microspore-generated embryos.

BBM is preferentially expressed in developing embryos and seeds and is similar to the AP2/ERF family of transcription factors. The *Brassica* (*BnBBM*) and *Arabidopsis* (*AtBBM*) sequences have a high level of similarity (85%) and conserved intron–exon boundaries. When the *BBM* gene was ectopically expressed in *Arabidopsis* and *Brassica*, it gave rise to spontaneous formation of somatic embryos and cotyledon-like structures on seedlings, as well as similar ranges of pleiotropic effects.

The preceding examples highlight the importance of understanding the function of paralogous loci in their genomic context, as well as the need to identify and sequence multiple copies of gene-coding sequences from *Brassica* based on candidate genes in *Arabidopsis*. Testing the conservation of gene action over a wider taxonomic range can also be informative. For example, the *OsMADS1* gene from rice is functional across angiosprem subclasses and has been successfully introduced into *B. rapa* under a constitutive promoter [87]. The transgenes appeared to be expressed, with one line notably involving homeotic replacement of a carpel with another flower.

3.5.2 Oil and Fatty Acid Pathways

Fatty acid synthesis and metabolism is of primary importance for production and genetic improvement of *Brassica* oil crops. Understanding the basis of harvestable yield (oil content) and quality (fatty acid profile) has considerable potential to enable these traits to be manipulated by transgenic or conventional means. Triacylglyerols and proteins are used as the major seed storage reserves in the developing embryos of the Brassicaceae. The synthetic and modification pathways and products are well conserved, and considerable information is now available about the genes coding for key enzymatic steps in synthesis, elongation, and modification steps.

The accumulation of knowledge about plant lipid metabolism, including fatty acid synthesis and modification, has benefited from availability of information and functional genomic resources in *Arabidopsis*, as well as genetic variation in *Brassica* species. Considerable progress has been made by mining the *Arabidopsis* genome sequence. This has already demonstrated that most of the genes encoding enzymes involved in lipid biosynthesis are represented by gene families that include a diverse array of isoform functions [88].

Quantitative genetic studies of oil synthesis in *Arabidopsis* have so far revealed a QTL that can account for some of the variation in seed oil content and fatty acid composition [89]. This study was based on a recombinant inbred population between Landsberg *erecta* and Cape Verdi Islands ecotypes, where the QTLs included two major and two minor loci accounting for 42% of variation in oil content. As with other areas of metabolism, there is interest in knowing whether some or all of the genetic variation observed can be accounted for by specific candidate enzyme-coding genes or whether regulatory genes make major contributions. In this study, significant QTLs for linoleic acid and linolenic acid appeared to colocate with the fatty acid desaturase 3 (FAD3) locus in *Arabidopsis*, and one for oleic acid with FAD2 [89].

Within the plastid, synthesis of fatty acids requires carboxylation of acetyl-CoA, which is catalyzed by acetyl-CoA carboxylase (ACCase). The same enzyme is also used in a number of biosynthetic pathways within the cytosol for fatty acid elongation. Two genes located in a tandem 25-kbp duplicated region near the centromere of *Arabidopsis* Chr1 code for two multifunctional ACCase isoforms [90].

Fatty acid desaturases important in plant lipid metabolism are located in the endoplasmic reticulum (ER) and other subcell compartments. *B. juncea* plants have been transformed with ADS1, which is an *Arabidopsis* homolog of yeast and mammalian acyl-CoA Delta9 desaturases [91]. These had a significant decrease in the level of seed saturated fatty acids. These preliminary data suggested that *Arabidopsis* ADS1 encoded a Delta9 desaturase. However, as well as altering the level of saturated fatty acids in *Brassica*, it also affected the levels of monounsaturated fatty acids. This highlights some of the complexities associated with regulation of fatty acid metabolism that may be compounded in amphidiploid *Brassica* species.

Substitution lines developed between two varieties of *B. napus* have been used to identify a range of QTLs for oil content and fatty acid profile [92]. Of 13 QTLs detected that affected fatty acid composition, 7 also affected total seed oil content. The approach of using substitution lines appeared to substantiate and provide additional resolution over previous results from segregating DH populations [93].

Erucic acid is the main component of storage fatty acids in oilseed *Brassica*, with the levels controlled by activity of the fatty acid elongation 1 (FAE1) gene. Fatty acid elongases have been cloned from *Arabidopsis* and *Brassica*, including *B. rapa* and *B. juncea* [94], and the gene sequences compared with corresponding loci in high- and low-erucic acid lines of *B. rapa* as well as from *B. oleracea* and *B. napus*. FAE1 sequences from *Brassica* and *Arabidopsis* have a high level of nucleotide and amino acid sequence conservation. This study concluded that differences in 13 amino acid positions in the central part of the protein were responsible for differences in erucic acid levels between low- and high-erucic acid lines. QTL analysis in *B. juncea* identified loci on two linkage groups; SNP markers to FAE1.1 and FAE1.3 cosegregated with the QTLs that accounted for 60 and 38% of the total phenotypic variance, respectively.

Quantitative transcriptional analysis has shown that mRNAs of various components involved in lipid biosynthesis are expressed in a coordinated manner during *B. napus* embryogenesis and are present in constant molar stoichometric ratios [95]. For biotin carboxylase, similar amounts of RNA were found in *Brassica* embryos [95] and *Arabidopsis* siliques [96]. Although Girke et al. [97] found similar levels of mRNAs for fatty acid synthase components between leaves and seeds using a cDNA microarray, this was not substantiated by O'Hara et al. [95]. Using quantitative northern analysis and RT-PCR they found that embryos accumulated between 3- and 15-fold more transcripts per unit total RNA than young leaf tissue. In cases where there appears to be divergence of expression pattern, it is necessary to distinguish between inherent behavior of the promoter and ancillary effects of transgene insertion site or transitory silencing.

Oil within the developing embryo is accumulated in oil bodies that are small droplets containing mostly triacylglycerol and are surrounded by a phospholipid/oleosin annulus. As the major protein component of developing embryos, oleosins have been suggested to play a structural role in stabilizing the lipid body during dessication of the seed by preventing coalescence of the oil [98]. In *Arabidopsis*, pollen-specific oleosin-like proteins (olleopollenin) genes are located in a tandemly repeated cluster.

Comparative analysis between *Arabidopsis* and *Brassica* [98] of the complete set of oleosin genes confirmed that they were subject to rapid evolution, including whole gene duplication and loss events, as well as a high rate of nonsynonymous mutations and indels in coding sequence. Evidence suggested that lineages leading to *Arabidopsis* and *Brassica* arose from independent duplications, consistent with the overall pattern of variation deduced from collinearity studies. Based on the Ka:Ks ratios of nonsynonymous to synonymous divergence, this class of gene appears to be among the most rapidly evolving.

A survey of the *Arabidopsis* gene pool for seed oil content, very long chain fatty acids, and polyunsaturated fatty acids has demonstrated extensive natural allelic variation [99]. A core collection derived from the original 360 accessions has been selected and should be valuable for gene identification, as well as more detailed dissection of the genetic regulation of seed lipid traits. Comparative studies using similar approaches are likely to be very effective, especially when applied to the wider gene pools beyond relatively modern mono- or oligophyletic oilseed crops such as *B. napus*.

Interest in production of crop plants that possess enhanced or optimized nutritional capacity is increasing. One trait of particular interest is the accumulation of omega-3 very long chain polyunsaturated fatty acids (VLC-PUFAs). These have been shown to help prevent cardiovascular disease, metabolic syndrome, and progression towards other prevalent Western pathologies such as type-2 diabetes and obesity. Unfortunately, VLC-PUFAs are not normally present in the oils of higher plants. However, it has been demonstrated that, by introducing genes encoding key enzymes originating from marine microalgal, algal, and fungal species, heterologous reconstitution of VLC-PUFA biosynthesis can be achieved in transgenic plants [138,139]. This work was carried out in *Arabidopsis* and tobacco, as well as the oilseed crop linseed, demonstrating proof-of-concept accumulation (to low levels) of VLC-PUFAs.

Contrary to earlier expectation, recent reports have demonstrated that some genotypes of *Brassica juncea* have the capacity to accumulate higher significant amounts of these valuable fatty acids [140]. This highly significant finding indicates that there may be considerable scope to engineer *Brassica* germplasm metabolically to produce economically viable amounts of VLC-PUFAs. At present, there is no knowledge of the inferred intrinsic species-specific variation in terms of combinations of naturally occurring alleles that may contribute to expression and regulation of the relevant pathways. This would provide information to guide genetic improvement through prebreeding of lines amenable to wide-scale production.

3.5.3 Glucosinolate Pathways

The genes in the Brassicaceae uniquely enable production of glucosinolates (GLS), which break down to isothiocyanates such as sulphoraphane that are known to provide some protection against a range of human cancers [100–102]. This has been ascribed to the ability of isothiocyanates to induce phase 1 and 2 detoxification enzymes in mammalian cells, which can then lead to reduction in the rate of tumor development. Glucosinolates also have a potentially significant role in some aspects of herbivore defense and signaling mechanisms, as well as in phytoremediation through interactions with soil microorganisms.

The role of comparative studies with *Arabidopsis* in elucidating the various genes involved in glucosinolate synthetic, modification, and breakdown pathways has been well covered in recent reviews (e.g., Wittstock and Hailkier [103]). Different glucosinolate products vary considerably as a result of variation in activity of different steps in the synthetic and side-chain modification pathways. This variation occurs throughout the *Brassica* genepool and arises from mutations in loci controlling such steps. Genes involved in synthesis of glucosinolates have been isolated from *Arabidopsis* and *Brassica*. Those that regulate side-chain length of *GSL-PRO* result in three carbon glucosinolates, whereas *GSL-ELONG* results in four-carbon glucosinolates. Segregation of *GSL-PRO* and *GSL-ELONG* has been shown to be independent in *B. oleracea* [104], and double recessive plants produce only trace amounts of aliphatic glucosinolate. Candidate genes from the *Arabidopsis* genome sequence have been used to clone several *GLS* genes, including *BoGSL-ELONG* [105].

Comparative sequence analysis has provided a view of the conservation of gene order between *Arabidopsis* and *B. oleracea* in a region containing several GLS genes.

By comparing a region from *Arabidopsis* Chr 4 [106], Gao et al. demonstrated that a high level of collinearity existed, with 23 out of 37 genes present and in the same orientation in *B. oleracea*. However, whereas three 2-oxoglutarate-dependent dioxygenase (*AOP*) genes are present in the *Arabidopsis* region, together with an additional *AOP* pseudogene, in *B. oleracea* two of these genes are locally duplicated and a third (*AOP3*) was not present. This arrangement was also conserved between different *B. oleracea* crop types (collard and broccoli). A more extensive survey did indicate that the copy number and sequence of the *Brassica* AOP2 gene varied across the gene pool.

QTL studies in *B. napus* have accounted for much of the phenotypic variation for seed glucosinolate [107], with three QTLs detected in common between two populations were located in homologous regions. This suggested that seed glucosinolate accumulation is controlled by duplicated genes. In *B. juncea* [108], QTLs have been identified associated with 3-butenyl that were consistent across years and experimental sites. Other QTLs for 2-propenyl and total GLS were detected, although some of these were highly inconsistent in different environments.

GSL-ALK affects desaturation of side-chains. In *Arabidopsis* and *B. oleracea* [104], it cosegregates with the *GSL-OH* responsible for side-chain hydroxylation. An ortholog cloned from *B. oleracea* has been transformed into *Arabidopsis* [109]. This resulted in detectable transcriptional activity and associated changes in the glucosinolate profiles of leaf and seed tissues.

Myrosinase is the only known S-glycosidase in plants and contributes to degradation of glucosinolates to isothiocyanates and nitriles [110]. Thangstad et al. [110] have carried out promoter-fusion experiments to determine tissue specificity of myrosinase expression in *Brassica*, *Arabidopsis*, and *Nicotiana*. By fusing myrosinase promoters from *B. napus* and *A. thaliana* to a GUS reporter gene and transforming into *A. thaliana* and *B. napus* as well as *Nicotiana*, Thangstad and colleagues were able to determine the cell types in which they were expressed.

They found that the *Arabidopsis TGG1* promoter directed expression within guard cells and phloem myrosin cell idioblasts of the transgenic *A. thaliana* plants, whereas the *B. napus* Myr1.Bn1 promoter resulted in cell-specific expression in idioblast myrosin cells of immature and mature seeds, as well as the myrosin cells of phloem in *B. napus*. The *B. napus* promoter resulted in an expression pattern similar to *TGG1* in the guard cells. This differential pattern of expression may result from locus-specific divergence of *cis*-acting factors in the amphidiploid *B. napus* because only one promoter sequence was tested in this study.

In practical terms, advances have been made in manipulation of glucosinolate pathways and contents in *Brassica* harvestable products. For example, Faulkener et al. [111] describe a broccoli hybrid with a tenfold increase in the level of 4-methylsulphinylbutyl glucosinolate. Tissue from this was able to induce more than a 100-fold increase in quinone reductase within Hepa 1c1c7 cell lines.

3.5.4 Pollen–Stigma Interactions and Self-Incompatibility

Brassica species have for many decades provided the model for understanding sporophytic pollen self-incompatibility (SI) systems. In diploid brassicas, self-incompatibility is controlled by genes at a single locus (S), with the transmembrane receptor kinase (*SRK*) gene expressed in the stigma and the S-locus cysteine-rich *SCR* ligand gene is expressed in the pollen coat. A considerable body of work has been carried out to characterize the allelic variation and function of the S-locus, particularly in *B. oleracea* and *B. rapa* [112]. In the case of self-incompatibility, relatively few functional insights have been achieved through experimental work within *Arabidopsis*. This is primarily because *Arabidopsis thaliana* is self-compatible and has no functional S-locus, although a functioning and orthologous S-locus system with similar levels of allelic variability as *Brassica* is present within the perennial species *A. lyrata* 113].

The lack of a functional S-locus in *A. thaliana* has enabled the genetic basis of the Brassicaceae SI system to be confirmed through transfer of the relevant components between *Arabidopsis* species. Nasrallah et al. [114] isolated the *SRK* and *SCR* genes from one *S*-locus haplotype of *A. lyrata* and were able to demonstrate by complementation that these genes alone are sufficient to confer a self-incompatible phenotype upon self-fertile *A. thaliana*. From this, they concluded that all the other components of the relevant signaling cascade had been conserved within *A. thaliana*. This key experiment provided the impetus to analyze other aspects of self-incompatibility with the relevant forward and reverse functional genomics tools available for *A. thaliana*.

It is apparent that the transition from inbreeding to outbreeding can occur rapidly during evolution and crop domestication. This has been demonstrated from studies in *Brassica* [115] and *Arabidopsis* [114]. Ekuere et al. [115] analyzed the genetic control of self-incompatibility in populations derived from crosses between resynthesized lines of *B. napus* and oilseed rape cultivars. They were able to detect evidence for latent S-alleles in at least two *B. napus* rapeseed cultivars and also demonstrated that the S-phenotype was masked by an unlinked suppressor system common to oilseed rape. Based on this analysis, they suggested that similar latent S-alleles may be widespread throughout the domesticated rapeseed gene pool and that, moreover, they may be associated with the highly conserved C-genome S-locus of these crop types.

3.5.5 Host Resistance to Pathogens

Brassica crops are subject to attack by a range of parasitic organisms, including viruses, bacterial, fungi, oomycetes, and various insect pests. This section will focus on a small number of examples in which knowledge of resistance mechanisms determined in *Arabidopsis* is likely to provide insights into crop-based resistance. There are, however, several caveats to a comparative approach that will lead to genetically determined durable field resistance. These include the different life history and ecological context of *Arabidopsis* compared with *Brassica* crops, as well as global issues relating to rate of change and spread of pathogen populations as a result of international trade in seed and crop products.

The major *Brassica* pathogens include viruses such as turnip (TuMV) and cauliflower (CaMV) mosaics; bacteria such as *Xanthomonas campestris* pv. *campestris* (Xcc) and *Pseudomonas* spp.; fungi such as *Leptosphaeria maculans*, *Pyrenopeziza brassicae*, *Altenaria brassicae*, and *Fusarium oxysporum*; and the oomycetes

Hyaloperonospora peronospora (formerly *Peronospora parasitica*) and *Albugo candidans*, as well as the protozoan clubroot pathogen *Plasmodiophora brassicae*. For a number of these pathogens, a large research effort has focused on the same species and their interaction with the model plant *Arabidopsis* [141,142]. This has enabled considerable progress to be made in elucidating the detailed mechanisms of host–parasite recognition, signal transduction, and response.

However, as Hammond-Kosack and Parker [143] have pointed out, most attempts to harness this knowledge to engineer improved disease resistance in crops have so far failed. In terms of transgenic approaches that introduce components of resistance mechanisms, commercial exploitation has not been possible because of detrimental effects on plant growth, development, and crop yield [143]. Crop improvement that makes use of biotechnological approaches is increasingly focused on marker-assisted breeding, as well as a more targeted use of transgenes that involves vectors containing highly regulated transgenes able to confer resistance in several distinct ways.

To develop comprehensive and effective marker-assisted strategies for crop improvement, it is necessary to understand the underlying genetic and functional mechanisms of host–parasite interaction. In terms of host-specific resistance, it has become apparent that plants are able to resist pathogen attack by eliciting an active defense response that mostly leads to cell death or hypersensitive response. This involves dramatic cellular reprogramming [143] with activation of a signal transduction cascade mediated by plant disease resistance recognition (R) genes [144]. These have been well characterized and classified in *Arabidopsis* and other model species [145].

In a compatible response, avirulence genes encode parasite elicitor molecules that interact directly or indirectly with the corresponding plant host R gene product through R gene recognition of elicitors in a ligand–receptor interaction. Depending upon the class of R gene, there is then an interaction with different component pathways in the signal transduction cascade [146]. To date, very little research has been published that focuses on isolating the different components of resistance directly from *Brassica* crop species. Although there was initial progress in using rapid cycling *Brassica*s as a tool for characterizing resistance genes [147,148], in recent years the potential of these valuable resources has unfortunately not been fully realized.

Major pathotype-specific resistance gene loci have been mapped in *Brassica* for a number of pathogens. These include TuMV [149,150] and *Xanthomonas* [151], as well as markers to *Albugo* resistance [152,153]. Kole et al. [154] made use of comparative mapping in their characterization of *Albugo* white rust resistance loci in *B. rapa* and *B. napus*. By comparing map positions of resistance genes in these two species, they were able to identify loci where additional resistance loci may be located. Alignment of the *Brassica* maps to the physical map of the *Arabidopsis* genome identified regions to target for comparative fine mapping.

As in other crops, a number of studies have focused on identifying resistance gene analogs (RGAs) in *Brassica* species. Such information can assist in understanding the level of conservation between *Arabidopsis* and *Brassica* genomes either in terms of genome location or gene sequence. Most studies to date have focused on the NBS-LRR class of recognition genes rather than genes further down the signal transduction pathway.

A combination of 103 *Arabidopsis thaliana* ESTs homologous to cloned plant R genes and 36 *Brassica* R-gene homologs has been mapped in *Brassica napus* to identify candidate R-gene loci and explore collinearity with *Arabidopsis* [155]. These results indicated no apparent rapid divergence of the R-gene containing loci between the two genomes. As with many recognition genes, NBS-LRR genes are highly variable, although some conserved motifs are commonly used to amplify RGAs from genomic DNA.

Vicente and King [156] used this approach to isolate RGAs from *B. oleracea*, the sequences of which were highly variable, although most of them showed similarity to known disease-resistant genes. *Brassica*-specific primers were then used to amplify and map five groups of RGAs, and four locus-specific sequences were confirmed as being expressed. Probing the specific gene sequences onto a BAC library indicated that these specific genes may only be present in low copy number. In another study, 44 *B. napus* RGAs were identified, of which a third were expressed from a subset of 29 [162]. The sequence specificity allowed discrimination of each genotype within a *B. napus* collection.

Although examples of conservation of specific resistance recognition genes within the Brassicaceae exist, the pattern of changes in local organization and relative rate of evolution of resistance loci is likely to be highly complex. Grant et al. [157] investigated the conservation associated with homologs of the *Pseudomonas syringae* pathovar *maculicola* (*RPM1*) bacterial resistance gene, which is completely absent in *Arabidopsis thaliana* accessions that lack RPM1 function. Collinearity of genes flanking *RPM1* is conserved between *B. napus* and *Arabidopsis*, with four additional *B. napus* loci in which the flanking marker synteny is maintained, but *RPM1* is absent. The *B. napus rpm1*-null loci have no detectable nucleotide similarity to the *Arabidopsis rpm1*-null allele; thus, it appears that RPM1 evolved before the divergence of the Brassicaceae and has been deleted independently in the *Brassica* and *Arabidopsis* lineages. The general conclusion from such results is that functional polymorphism at R gene loci can arise from gene deletions.

Although much is now known of the diversity and interactions between plant host and parasites from work on *Arabidopsis* and genomic information about resistance recognition genes is being accumulated, at present little progress has been made on matching the molecular diversity to known functional resistances in crop brassicas. For example, Malvas et al. [158] investigated homologs of the *RPS2* disease resistance gene and found that 2.5-kb fragments were conserved at a level of 95 to 98% homology among *Brassica* species and that the homolog was constitutively expressed in *Brassica oleracea.* However, they found no linkage between the gene and resistance to blackrot caused by *Xanthomonas campestris* pv. *campestris*. This underlines the need for careful genetic analysis and gene isolation in well-characterized resistant lines.

Given the progress being made in *Brassica* comparative genomics and the ability to map-base clone genes rapidly, a large number of major resistance gene alleles should soon be isolated. This will provide information to understand the relative range and rate of resistance recognition gene variation in the domesticated *Brassica* species compared to *Arabidopsis*, as well as "in-gene" molecular markers to assist in breeding introgression and resistance gene pyrimiding strategies. However, although major gene resistance is valuable in some situations, the understanding of more durable "field" resistance is one of the ultimate goals in terms of integrated crop disease management. In this area, relatively little progress has been made through use of the model species *Arabidopsis*, primarily because of differences in terms of its ephemeral life history and unsuitability to the study of crop field pathogenesis and epidemiology.

3.6 UNDERSTANDING GENE REGULATION

3.6.1 Gene Expression and Use of Transcriptional Arrays

Transcriptional analysis (transcriptomics) based on gridded arrays of gene sequences has provided a powerful approach to understand the coordinated expression of genes and deduce the regulatory networks associated with different cell types, stages of development, and responses to changing environments. For *Arabidopsis*, a range of technological platforms exists, including those based on ESTs, amplified gene probes, and short or long oligonucleotides. This has been made possible due to the comprehensive whole-genome and EST data sets available for the model species.

For *Brassica*, transcriptomics has to date been less widely used, although the increasing availability of sequence data and validation of *Arabidopsis* resources will enable more detailed analysis. There have been concerns that the presence of multiple copies of closely related paralogous gene sequences may provide equivocal results. However, in any transcriptomic analysis, the initial identification of up- or down-regulated transcripts does require detailed quantitative verification, which requires development of locus-specific PCR assays. Amagai et al. [116] have used an *Arabidopsis* cDNA macroarray to detect signals from hybridization with cDNA isolated from anthers and pistils of *B. oleracea*. This resulted in identification of 53 putative anther-specific genes, including a number that had already been characterized. A third of the clones had RT-PCR expression patterns consistent with that detected in the *Arabidopsis* macroarray.

More recently, Lee et al. [117] carried out a sensitivity analysis by comparing cDNAs and corresponding 60- to 70-mer oligonucleotides for 192 *Arabidopsis* genes. In addition to demonstrating that the sources of variation were similar for *Arabidopsis* and *B. oleracea* RNA, they showed that cDNA and oligonucleotides were similar in their ability to detect changes in expression, with a common subset of significant genes.

3.6.2 Characterization of *cis*-Regulatory Regions

The behavior of promoters exchanged between *Arabidopsis* and *Brassica* provides a useful tool to investigate functional motifs and properties. In many cases, a very similar range of tissue or temporal expression is observed. For example, the promoter of an *Arabidopsis* late embryogenesis abundant protein (AtEm1), when fused to a GUS reporter, was found to be highly active in vascular tissues of *B. napus* embryo and pollen grains [118] and was also active in other late developing floral organs.

In other cases, the function of *cis*-acting regions across species can provide considerable insights into the relationships between sequence divergence and specificity of action. The promoter of *AGAMOUS* (*AG*) from *Arabidopsis* has been introduced into *B. napus* and shown to have an expression pattern limited primarily to the reproductive organs and nectarium [119]. However, the tissue-specific pattern in this case was not conserved between species. For example, the AG *cis* elements did not express in the ovules of *B. napus*, although the pattern was temporally similar to that observed during early development of *Arabidopsis*. Pylatuik et al. [119] conclude from these experiments that the regulatory factors controlling the generalized local expression of *AG* were conserved between these species, although those that control the temporal and tissue-specific expression have not been conserved. Future enhancer traps lines may be able to provide the relevant level of information on the subtlety of cell-specific control associated with *cis*-regulatory regions from paralogous copies within *Brassica* genomes.

To date, a number of different approaches have been used to isolate *cis*-regulatory regions from orthologous copies of candidate genes. Based on the *Apetala3* gene, an ortholog was isolated from *B. napus* (*BnAP3*) and the 5′ region of the cDNA, then retrieved by rapid amplification of cDNA ends (RACE) [120]. Expression analysis of the multiple alleles from *B. napus* demonstrated that they are expressed in floral as well as nonfloral tissues. Further information was obtained by transforming *BnAP3* into wild-type *Arabidopsis* and *B. napus* under the control of a reproductive organ specific promoter. Consistent with the behavior of the *Arabidopsis* ortholog, the transgenic plants had carpels converted to stamens. As well as complementing the *ap3 Arabidopsis* mutant, the *BnAP3* gene also directed the conversion of carpels to stamens in the fourth whorl.

A plant-wide survey searching for conserved regulatory sequence elements has been carried out by Hulzink et al. [121]. This was based on the relative importance of 5′ untranslated regions (UTR) of several pollen transcripts and the conservation of genetic programs in pollen. By analyzing 5′-UTR sequences of pollen and sporophytic expressed genes, they identified several pollen-specific elements containing various consensus sequences, several of which were preferentially associated with genes from dicotyledons, wet-type stigma plants, or plants with bicellular pollen. The analysis included three sequence elements that were preferentially found in the 5′-UTR of pollen-expressed genes in *Arabidopsis* and *B. napus*. This approach has generic application for identifying functional motifs in *cis*-regulatory regions through identification of consensus sequence or by bioinformatic comparison of sequence-determined physical properties that might affect binding of *trans*-acting regulators.

Margulies et al. [122] have outlined a rationale for identifying multispecies conserved sequences (MCSs) based on analyses between vertebrate genomes. Although they found that about 70% of the bases within MCSs are located within noncoding regions, they deduced that most of the sequences had no known function. Many MCSs corresponded to clusters of transcription factor-binding sites, noncoding RNA transcripts, and other candidate functional elements in vertebrates.

Although it should be noted that vertebrate genomes have a different distribution of coding and noncoding sequence than dicotyledonous plants, Colinas et al. [123] used a similar comparative bioinformatic approach to analyze the degree of conservation between upstream noncoding regions of *B. oleracea* and *Arabidopsis*. This demonstrated that there is likely to be significant conservation of promoter regions between *Arabidopsis* and *Brassica* and that such sequence comparisons between two species at this level of divergence could reveal functional *cis*-regulatory elements.

Another approach has been suggested by Gilchrist and Haughn [135], who speculated that by identifying blocks of conserved sequence within relatively unconserved noncoding regions, TILLING could help identify regulatory domains. This would have considerable application in the comparative analysis of *Arabidopsis* and *Brassica*.

3.7 FUTURE DEVELOPMENTS AND APPLICATIONS

As outlined in the examples given in the previous sections, there has been considerable progress in transferring information and experimental approaches from the model *Arabidopsis* to the crop *Brassica* species. A number of experimental areas still need additional focus of effort to exploit current knowledge fully. These include the ability to identify and resolve gene effects, to distinguish between locus-specific gene sequences, and to understand fully the interactions that may occur in complex genomes comprising multiple duplicated chromosomal segments. Crop improvement has additional requirements to identify and characterize sources of genetic diversity and to understand the constraints and control of recombination. Gene identification and haplotype reconstruction will benefit from developments in highly parallel locus-specific SNP detection, which should also allow high-throughput MAS for breeding.

Recombinational resolution currently limits inferences that can be made from conventional genetic analysis. This may be addressed through concerted development of overlapping sets of near-isogenic, substitution, or STAIR lines. A resolution gap then remains between trait loci (major gene or QTL) and existing BAC-based physical maps or the emerging genomic sequences. As more gene sequences are anchored in the context of *Brassica* genetic maps, so we will understand and be able to reconstruct the detailed relationships with the *Arabidopsis* genome. For practical purposes this will allow more rapid interpolation and identification of additional candidates within any given region. Understanding the adaptive significance of segmental and whole-chromosome polyploidy, especially with regard to interactions with the environment, is a further challenge. These may be mediated via variation in *cis*-acting regulatory sequences, with epigenetic mechanisms superimposed that confer locus or cell-specific variation in local or global patterns of gene regulation.

3.7.1 Assessing and Exploiting Genetic Diversity for Crop Improvement

The ability of genetic improvements in *Brassica* and other crops to meet complex and changing requirements of production and market conditions will be constrained by the ability to generate and select breeding lines that contain optimal combinations of alleles. As an increasing amount of information becomes available that contributes to understanding the interactions between genotype to phenotype, so will the ability to utilize a wider range of allelic variation within the gene pool. This places an emphasis on conserving, accessing, and understanding genetic diversity at the genomic level.

There is considerable scope to use combinations of locus-specific markers to explore the extensive genetic diversity within the *Brassica* gene pool. Although development of informative SNP markers can be problematic in some *Brassica* material because of the requirement to distinguish paralogs from homologous loci, such approaches are likely to be increasingly valuable in addressing the relationships and limitations of natural and domesticated variation. The success of association genetics and linkage disequilibrium studies in other species indicates that this should be a powerful approach for identifying alleles relevant for crop improvement. Development of "graphical genotypes" that display DNA marker data to create a graphical image for the genomic constitution of an individual [124] will provide an important link for breeding pedigrees, allelic variation, and crop phenotypes.

In general, modern crops have a relatively narrow genetic base that does not reflect the existence of extensive allelic variation within the wider gene pool or germplasm collections. For *Brassica* crops, existing natural variation has been exploited for particular target traits such as resistances, low glucosinolates, or modified oil profiles. Despite this, a considerable degree of genetic erosion has occurred compared with traditional land-race germplasm. *Ex situ* genetic resource collections for *Brassica* exist throughout the world, and core collections have been developed from these to screen for novel sources of fungal or other resistance traits [125].

Although they are valuable in identifying hotspots of variation within the relevant gene pools, such collections consist of heterogeneous and heterozygous material. This limits their long-term use for correlating detailed genetic studies. As a result, the concept of diversity fixed foundation sets (DFFSs) has been devised [126]. These are defined as *an informative set of genetically fixed lines representing a structured sampling of diversity across a gene pool;* they provide the advantage of being suitable for replicated, coordinated, or distributed analysis at molecular and trait levels. DNA and seed of fixed lines are being made available in the public domain to provide a common reference resource for *B. oleracea*, related C-genome species, and *B. napus*, with additional sets being prepared for *B. rapa*.

3.7.2 Information Management and Bioinformatics

The continuing development of crop genetics and genomics is increasing the requirement for access to a wide range of relevant information and data. The establishment of international community resources and generation of large amounts of interconnected and persistent data provide great opportunities for interdisciplinary research and involvement of stakeholders (see, for example, http://www.oregin.info). A range of databases and other information resources exists for *Brassica* [1], but require integration. These include genetic resources, genetic mapping, genomic and functional genomics, as well as disparate sources of legacy trait and pedigree data.

In addition, the close relationship with *Arabidopsis* provides a large amount of functional and reference information of direct relevance to the *Brassica* community. This is starting to be collated in a number of public-domain systems [29,159]. In all situations, it is important to be able to assess and determine provenance, status, and information content of data sets. The need for stable nomenclature for objects and entities such as chromosomes and linkage groups is becoming increasingly important as data integration progresses. For crop genetics in general, there are opportunities to establish community-wide acceptance of trait and other ontologies [128].

Throughout this chapter, the emphasis has been on the complexity of genome organization that underlies the plasticity of *Brassica* crop phenotypes. Developing a deeper understanding of the ramifications and constraints arising from this will be important in the context of identifying and characterizing relevant genetic variation that can then be exploited in future breeding programs.

ACKNOWLEDGMENT

The author is funded by the U.K. Biotechnology & Biological Sciences Research Council.

REFERENCES

1. http://www.brassica.info.
2. Gómez-Campo, C., Ed., *Biology of Brassica Coenospecies,* Elsevier, Amsterdam, 1999.
3. Paterson, A.H. et al., *Brassica* genomics: a complement to, and early beneficiary of, the *Arabidopsis* sequence, *Genome Biol.*, 2, 1011.1–1011.4, 2001.
4. Parkin, I.A.P. et al., Identification of the A and C genomes of amphidiploid *Brassica napus* (oilseed rape), *Genome*, 38, 1122–1131, 1995.
5. Sharpe, A.G. et al., Frequent nonreciprocal translocations in the amphidiploid genome of oilseed rape (*Brassica napus*), *Genome*, 38, 1112–1121, 1995.
6. Bohuon, E.J.R. et al., Alignment of the conserved C genomes of *Brassica oleracea* and *Brassica napus*, *Theor. Appl. Genet.*, 93, 883–839, 1996.
7. Lagercrantz, U. and Lydiate, D.J., RFLP mapping in *Brassica nigra* indicates differing recombination rates in male and female meioses, *Genome*, 38, 255–264, 1995.
8. Howell, E.C. et al., Integration of the cytogenetic and genetic linkage maps of *Brassica oleracea*, *Genetics*, 161, 1225–1234, 2002.
9. U, N., Genomic analysis in *Brassica* with special reference to the experimental formation of *B. napus* and peculiar mode of fertilization, *Jpn. J. Bot.*, 7, 389–452, 1935.
10. Arumuganathan, K. and Earle, E.D., Nuclear DNA content of some important plant species, *Plant Molec. Biol. Repr.*, 9, 211–215, 1991.
11. Bennett, M. and Smith, J.B., Nuclear DNA amounts in angiosperms, *Phil. Trans. Royal Soc. Lond. B*, 274, 227–274, 1976.
12. The *Arabidopsis* Genome Initiative (AGI), Analysis of the genome of the flowering plant *Arabidopsis thaliana*, *Nature*, 408, 816–820, 2000.
13. Truco, M.J. et al., Inter and intragenomic homology of the Brassica genomes: implications on their origin and evolution, *Theor. Appl. Genet.*, 93, 1225–1233, 1996.
14. Parkin, I.A.P., Lydiate, D.J., and Trick, M., Assessing the level of collinearity between *Arabidopsis thaliana* and *Brassica napus* for *A. thaliana* chromosome 5, *Genome*, 45, 356, 2002.

15. O'Neill, C.M. and Bancroft, I., Comparative physical mapping of segments of the genome of *Brassica oleracea* var. *alboglabra* that are homologous to sequenced regions of chromosomes 4 and 5 of *Arabidopsis thaliana*, *Plant J.*, 23, 233, 2000.
16. Ryder, C.R., Barker, G.C., and King, G.J., unpublished results.
17. Osborn, T.C. et al., Detection and effects of a homologous reciprocal transposition in *Brassica napus*, *Genetics*, 165, 1569, 2003.
18. Ryder C.D. et al., Contrasting genome organization: two regions of the *Brassica oleracea* genome compared with collinear regions of the *Arabidopsis thaliana* genome, *Genome*, 44, 808–817, 2001.
19. Parkin, I.A.P., Sharpe, A.G., and Lydiate, D.J., Patterns of genome duplication within the *Brassica napus* genome, *Genome*, 46, 291, 2003.
20. Blanc, G., Hokamp, K. and Wolfe, K.H., A recent polyploidy superimposed on older large-scale duplications in the *Arabidopsis* genome, *Genome Res.*, 13, 137–144, 2003.
21. Lan, T. et al., An EST-enriched comparative map of *Brassica oleracea* and *Arabidopsis thaliana*, *Genome Res.*, 10, 776–788, 2000.
22. Snowdon, R.J. and Friedt, W., Molecular markers in *Brassica* oilseed breeding: current status and future possibilities, *Plant Breed.*, 123, 1, 2004.
23. Suwabe, K. et al., Isolation and characterization of microsatellites in *Brassica rapa* L., *Theor. Appl. Genet.*, 104, 1092, 2002.
24. Suwabe, K. et al., Characteristics of microsatellites in *Brassica rapa* genome and their potential utilization for comparative genomics in Cruciferae, *Breed. Sci.*, 85, 2004.
25. Saal, B. et al., Microsatellite markers for genome analysis in *Brassica*, II. Assignment of rapeseed microsatellites to the A and C genomes and genetic mapping in *Brassica oleracea* L., *Theor. Appl. Genet.*, 102, 695, 2001.
26. Lowe, A.J. et al., Efficient large-scale development of microsatellites for marker and mapping applications in *Brassica* crop species, *Theor. Appl. Genet.*, 108, 1103, 2004.
27. Bornet, B. et al., Highly informative nature of inter simple sequence repeat (ISSR) sequences amplified using tri- and tetra-nucleotide primers from DNA of cauliflower (*Brassica oleracea* var. *botrytis* L.), *Genome*, 45, 890, 2002.
28. Tonguc, M. and Griffiths, P.D., Genetic relationships of *Brassica* vegetables determined using database-derived simple sequence repeats, *Euphytica*, 137, 193–201, 2004.
29. Love, C.G. et al., New computational tools for *Brassica* genome research, *Comp. Func. Genom.*, 5, 276–280, 2004.
30. Inoue, H. and Nishio, T., Efficiency of PCR-RF-SSCP marker production in *Brassica oleracea* using *Brassica* EST sequences, *Euphytica*, 137, 233, 2004.
31. Li, G. and Quiros, C.F., Sequence-related amplified polymorphism (SRAP), a new marker system based on a simple PCR reaction: its application to mapping and gene tagging in *Brassica*, *Theor. Appl. Genet.*, 103, 455, 2001.
32. Li, G. et al., Gene for gene alignment between the *Brassica* and *Arabidopsis* genomes by direct transcriptome mapping, *Theor. Appl. Genet.*, 107, 168, 2003.
33. Riaz, A. et al., Genetic diversity of oilseed *Brassica napus* inbred lines based on sequence-related amplified polymorphism and its relation to hybrid performance, *Plant Breed.*, 120, 411, 2001.
34. Brunel, D., Froger, N., and Pelletier, G., Development of amplified consensus genetic markers (ACGM) in *Brassica napus* from *Arabidopsis thaliana* sequences of known biological function, *Genome*, 42, 387–402, 1999.
35. Fourmann, M. et al., From *Arabidopsis* to *Brassica* napus: development of amplified consensus genetic markers (ACGM) for construction of a gene map, *Theor. Appl. Genet.*, 105, 1196, 2002.

36. Kuittinen, H. et al., Primers for 22 candidate genes for ecological adaptations in Brassicaceae, *Molec. Ecol. Not.*, 2, 258, 2002.
37. Jackson, S.A. et al., Comparative fluorescence *in situ* hybridization mapping of a 431-kb *Arabidopsis thaliana* bacterial artificial chromosome contig reveals the role of chromosomal duplications in the expansion of the *Brassica rapa* genome, *Genetics*, 156, 833–838, 2000.
38. Messing, J. and Llaca, V., Importance of anchor genomes for any plant genome project, *Proc. Natl. Acad. Sci. USA*, 95, 2017–2020, 1998.
39. Ziolkowski, P.A. and Sadowski, J., FISH-mapping of rDNAs and *Arabidopsis* BACs on pachytene complements of selected *Brassica*s, *Genome*, 45, 189, 2002.
40. Howell, E.C. et al., Physical organisation of the major duplication on *Brassica oleracea* chromosome O6 revealed through fluorescence *in situ* hybridisation with *Arabidopsis* and *Brassica* BACs, *Genome*, 48, 1093, 2005.
41. Ayele, M. et al., Whole genome shotgun sequencing of *Brassica oleracea* and its application to gene discovery and annotation in *Arabidopsis*, *Genome Res.*, 15, 487–495, 2005.
42. Zhang, X.Y. and Wessler, S.R., Genome-wide comparative analysis of the transposable elements in the related species *Arabidopsis thaliana* and *Brassica oleracea*, *Proc. Natl. Acad. Sci. USA*, 101, 5589, 2004.
43. Ouyang, S. and Buell, C.R., The TIGR plant repeat database: a collective resource for the identification of repetitive sequences in plants, *Nucl. Acids Res.*, 32, D360, 2004.
44. Ermolaeva, M.D. et al., The age of the *Arabidopsis thaliana* genome duplication, *Plant. Mol. Biol.*, 51, 859, 2003.
45. Blanc, G., Hokamp, K., and Wolfe, K.H., A recent polyploidy superimposed on older larger-scale duplications in the *Arabidopsis* genome, *Genome Res.*, 13, 137, 2003.
46. Bolvin, K. et al., The *Arabidopsis* genome sequence as a tool for genome analysis in Brassicaceae. A comparison of the *Arabidopsis* and *Capsella rubella* genomes, *Plant Physiol.*, 135, 735, 2004.
47. Lukens, L.N. et al., Genome redundancy and plasticity within ancient and recent *Brassica* crop species, *Biol. Linnean Soc.*, 82, 665, 2004.
48. Sharpe, A.G. and Lydiate, D.J., Mapping the ancestral genotypes in a cultivar of oilseed rape (*Brassica napus*) selected via pedigree breeding, *Genome*, 46, 461, 2003.
49. Cavell, A.C. et al., Collinearity between a 30-centimorgan segment of *Arabidopsis thaliana* chromosome 4 and duplicated regions within the *Brassica napus* genome, *Genome*, 41, 62, 1998.
50. KowalskiI, S.T. et al., Comparative mapping of *Arabidopsis thaliana* and *Brassica oleracea* chromosomes reveal islands of conserved organization, *Genetics*, 138, 499–510, 1994.
51. Lagercrantz, U., Comparative mapping between *Arabidopsis thaliana* and *Brassica nigra* indicates that *Brassica* genomes have evolved through extensive genome replication accompanied by chromosome fusions and frequent rearrangements, *Genetics*, 150, 1217–1228, 1998.
52. Babula, D. et al., Chromosomal mapping of *Brassica oleracea* based on ESTs from *Arabidopsis thaliana*: complexity of the comparative map, *Molec. Genet. Genom.*, 268, 656, 2003.
53. Lukens, L. et al., Comparison of a *Brassica oleracea* genetic map with the genome of *Arabidopsis thaliana*, *Genetics*, 164, 359, 2003.
54. Yang, Z. and Nielsen, R., Estimating synonymous and nonsynonymous nucleotide substitution rates under realistic evolutionary models, *Mol. Biol. Evol.*, 17, 32–43, 2000.

55. Tiffin, P. and Hahn, M.W., Coding sequence divergence between two closely related plant species: *Arabidopsis thaliana* and *Brassica rapa* ssp. *pekinensis*, *J. Molec. Evol.*, 54, 746, 2002.
56. Gao, M.J. et al., regulation and characterization of four CBF transcription factors from *Brassica napus*, *Plant Molec. Biol.*, 49, 459, 2002.
57. Vision, T.J., Brown, D.G., and Tanksley, S.D., The origins of genomic duplications in *Arabidopsis*, *Science*, 290, 2114–2117, 2000.
58. Liljegren, S.J. et al., *SHATTERPROOF* MADS-box genes control seed dispersal in *Arabidopsis*, *Nature*, 404, 766–770, 2000.
59. Smith, L.B. and King, G.J., The distribution of BoCAL-a alleles in *Brassica oleracea* is consistent with a genetic model for curd development and domestication of the cauliflower, *Mol. Breed.*, 6, 603–613, 2000.
60. Force, A. et al., Preservation of duplicate genes by complementary, degenerative mutations, *Genetics*, 151, 1531–1545, 1999.
61. Fiebig, A., Kimport, R., and Preuss, D., Comparisons of pollen coat genes across Brassicaceae species reveal rapid evolution by repeat expansion and diversification, *Proc. Natl. Acad. Sci. USA*, 101, 3286, 2004.
62. Quiros, C.F. et al., *Arabidopsis* and *Brassica* comparative genomics: sequence, structure and gene content in the ABI1-Rps2-Ck1 chromosomal segment and related regions, *Genetics*, 157, 1321, 2001.
63. Suzuki, T. et al., Structure, sequence and phylogeny of the members of the Ck1 gene family in *Brassica oleracea* and *Arabidopsis thaliana* (Brassicaceae), *Plant Sci.*, 164, 735, 2003.
64. Rana, D. et al., Conservation of the microstructure of genome segments in *Brassica napus* and its diploid relatives, *Plant J.*, 40, 725, 2004.
65. El-Lithy, M.E., Quantitative trait locus analysis of growth-related traits in a new *Arabidopsis* recombinant inbred population, *Plant Phys.*, 135, 444–458, 2004.
66. Kearsey M.J. and Farquhar, A.G.L., QTL analysis in plants. Where are we now? *Heredity*, 80, 137–142, 1998.
67. Koumprogolou, R. et al., STAIRS: a new genetic resource for functional genomic studies of *Arabidopsis*, *Plant J.*, 31, 355–364, 2002.
68. Rae, A.M., Howell, E.C., and Kearsey, M.J., More QTL for flowering time revealed by substitution lines in *Brassica oleracea*, *Heredity*, 83, 586,1999.
69. Greene, E.A. et al., Spectrum of chemically induced mutations from a large-scale reverse-genetic screen in *Arabidopsis*, *Genetics*, 164, 731–40, 2003.
70. Thurling, N. and Depittayanan, V., EMS induction of early flowering mutants in spring rape (*Brassica napus*), *Plant Breed.*, 108, 177–184, 1992.
71. Dunemann, F. and Grunewaldt, J., *In vitro* mutagenesis in *Brassica oleracea* var. *italica* Plenck (Broccoli), *Gartenbauwissenschaft*, 55, 155–8, 1990.
72. King, G.J., Morphological development in *Brassica oleracea* is modulated by *in vivo* treatment with 5-azacytidine, *J. Hort. Sci.*, 70, 333–342, 1995.
73. Bohuon, E.J.R. et al., The association of flowering time quantitative trait loci with duplicated regions and candidate loci in *Brassica oleracea*, *Genetics*, 150, 393, 1998.
74. Sebastian, R.L., Kearsey, M.J., and King, G.J., Identification of quantitative trait loci controlling developmental characteristics of *Brassica oleracea* L., *Theor. Appl. Genet.*, 104, 601–609, 2002.
75. Robert, L.S. et al., Conserved structure and function of the *Arabidopsis* flowering time gene CONSTANS in *Brassica napus*, *Plant Molec. Biol.*, 37, 763, 1998.
76. Osterberg, M.K. et al., Naturally occurring indel variation in the *Brassica nigra* COL1 gene is associated with variation in flowering time, *Genetics*, 161, 299, 2002.

77. Lagercrantz, U., Osterberg, M.K., and Lascoux, M., Sequence variation and haplotype structure at the putative flowering-time locus COL1 of *Brassica nigra*, *Mol. Biol. Evol.*, 19, 1474, 2002.
78. Kole, C. et al., Evidence for homology of flowering-time genes VFR2 from *Brassica rapa* and FLC from *Arabidopsis thaliana*, *Theor. Appl. Genet.*, 102, 425, 2001.
79. Amasino, R., Vernalization, competence, and the epigenetic memory of winter, *Plant Cell,* 16, 2553–2559, 2004.
80. Schranz, M.E. et al., Characterization and effects of the replicated flowering time gene FLC in *Brassica rapa*, *Genetics*, 162, 1457, 2002.
81. Lee H. et al., The AGAMOUS-LIKE 20 MADS domain integrates floral inductive pathways in *Arabidopsis*, *Genes Dev.*, 14, 2366–2376, 2000.
82. Kim, K.W. et al., The function of the flowering time gene AGL20 is conserved in crucifers, *Molec. Cells*, 16, 136, 2003.
83. Byzova, M. et al., Transforming petals into sepaloid organs in *Arabidopsis* and oilseed rape: implementation of the hairpin RNA-mediated gene silencing technology in an organ-specific manner, *Planta*, 218, 379–87, 2004.
84. Fray, M.J. et al., The genetics of stamenoid petal production in oilseed rape (*Brassica napus*) and equivalent variation in *Arabidopsis thaliana*, *Theor. Appl. Genet.*, 94, 731–736, 1997.
85. Jiang, L. and Becker, H.C., Inheritance of apetalous flowers in a mutant of oilseed rape, *Crop Sci.,* 43, 508–510, 2003.
86. Boutilier, K. et al., Ectopic expression of BABY BOOM triggers a conversion from vegetative to embryonic growth, *Plant Cell*, 8, 1737–49, 2002.
87. Shin, J. et al., Transformation of rice OsMADS1 gene causes homeotic mutations in floral organs of Chinese cabbage (*Brassica campestris*), *J. Plant. Biol.*, 46, 46, 2003.
88. Delseny, M. et al., Genomics and plant lipid metabolism, *OCL-Oleagineux Corps Gras Lipides*, 9, 130, 2002.
89. Hobbs, D.H., Flintham, J.E., and Hills, M.J., Genetic control of storage oil synthesis in seeds of *Arabidopsis*, *Plant Physiol.,* 136, 1–9, 2004.
90. Baud, B. et al., Multifunctional acetyl-CoA carboxylase 1 is essential for very long chain fatty acid elongation and embryo development in *Arabidopsis*, *Plant J.*, 33, 75, 2003.
91. Yao, K., Expression of the *Arabidopsis ADS1* gene in *Brassica juncea* results in a decreased level of total saturated fatty acids, *Plant Biotech. J.*, 1, 221, 2003.
92. Burns, M.J. et al., QTl analysis of an intervarietal set of substitution lines in *Brassica napus*: (i) seed oil content and fatty acid composition, *Heredity,* 90, 39–48, 2003.
93. Ecke, W., Uzunova, M., and Weissleder, K., Mapping the genome of rapeseed (*Brassica napus* L.). 2. Localization of genes controlling erucis acid synthesis and seed oil content, *Theor. Appl. Genet.*, 91, 972–977, 1995.
94. Das, S. et al., Cloning and characterization of the fatty acid elongase 1 (FAE 1) gene from high- and low-erucic acid lines of *Brassica campestris* and *Brassica oleracea*, *Plant Sci.*, 162, 245, 2002.
95. O'Hara, P., Slabas, A.R., and Fawcett, T., Fatty acid and lipid biosynthetic genes are expressed at constant molar ratios but different absolute levels during embryogenesis, *Plant Physiol.*, 129, 310, 2002.
96. Ke, J.S. et al., Coordinate regulation of the nuclear and plastidic genes coding for the subunits of the heteromeric acetyl-coenzyme a carboxylase, *Plant Physiol.*, 122, 1057–1071, 2000.
97. Girke, T. et al., Microarray analysis of developing *Arabidopsis* seeds, *Plant Physiol.*, 124, 1570–1581, 2000.

98. Schein, M. et al., Rapid evolution of a pollen-specific oleosin-like gene family from *Arabidopsis thaliana* and closely related species, *Molec. Biol. Evol.*, 21,659, 2004.
99. O'Neill, C.M. et al., Natural variation for seed oil composition in *Arabidopsis thaliana*, *Phytochemistry*, 64, 1077–1090, 2003.
100. Stoewsand G.S., Bioactive organosulfur phytochemicals in *Brassica oleracea* vegetables—a review, *Food Chem. Toxicol.,* 33, 537–543, 1995.
101. Verhoeven, D.T. et al., A review of mechanisms underlying anticarcinogenicity by brassica vegetables, *Chem. Biol. Interact.,* 103, 79–129, 1997.
102. Mithen, R.F., Glucosinolates and their degradation products, *Adv. Botan. Res.*, 35, 213–262, 2001.
103. Wittstock, U. and Hailkier, B.A., Glucosinolate research in the *Arabidopsis* era, *Trends Plant Sci.*, 7, 263, 2002.
104. Li, G. et al., Inheritance of three major genes involved in the synthesis of aliphatic glucosinolates in *Brassica oleracea*, *J. Am. Soc. Hort. Sci.*, 126, 2001.
105. Li, G.Y. and Quiros, C.F., Genetic analysis, expression and molecular characterization of BoGSL-ELONG, a major gene involved in the aliphatic glucosinolate pathway of *Brassica* species, *Genetics*, 162, 1937, 2002.
106. Gao, M.Q. et al., Comparative analysis of a *Brassica* BAC clone containing several major aliphatic glucosinolate genes with its corresponding *Arabidopsis* sequence, *Genome*, 47, 666, 2004.
107. Howell, P.M., Sharpe, A.G., and Lydiate, D.J., Homologous loci control the accumulation of seed glucosinolates in oilseed rape (*Brassica napus*), *Genome*, 46, 454, 2003.
108. Mahmood, T. et al., Molecular mapping of seed aliphatic glucosinolates in *Brassica juncea*, *Genome*, 46, 753, 2003.
109. Li, G. and Quiros, C.F., *In planta* side-chain glucosinolate modification in *Arabidopsis* by introduction of dioxygenase *Brassica* homolog BoGSL-ALK, *Theor. Appl. Genet.*, 106, 1116, 2003.
110. Thangstad, O.P. et al., Cell specific, cross-species expression of myrosinases in *Brassica napus*, *Arabidopsis thaliana* and *Nicotiana tabacum*, *Plant Mol. Biol.*, 54, 597, 2004.
111. Faulkner, K., Mithen, R., and Williamson, G., Selective increase of the potential anticarcinogen 4-methylsulphinylbutyl glucosinolate in broccoli, *Carcinogenesis*, 19, 605–609, 1998.
112. Watanabe, M. et al., Recent progress on self-incompatibility research in *Brassica* species, *Breed. Sci.,* 53, 199, 2003.
113. Charlesworth, D. et al., Diversity and linkage of genes in the self-incompatibility gene family in *Arabidopsis lyrata*, *Genetics*, 164, 1519, 2003.
114. Nasrallah, M.E., Liu, P., and Nasrallah, J.B., Generation of self-incompatible *Arabidopsis thaliana* by transfer of two S locus genes from *A. lyrata*, *Science*, 297, 247, 2002.
115. Ekuere, U.U. et al., Latent S alleles are widespread in cultivated self-incompatible *Brassica napus*, *Genome*, 47, 257, 2004.
116. Amagai, M. et al., Identification of anther-specific genes in a cruciferous model plant, *Arabidopsis thaliana*, by using a combination of *Arabidopsis* macroarray and mRNA derived from *Brassica oleracea*, *Sex. Plant Reprod.*, 15, 213–220, 2003.
117. Lee, H.-S., Sensitivity of 70-mer oligonucleotides and cDNAs for microarray analysis of gene expression in *Arabidopsis* and its related species, *Plant Biotech. J.*, 2, 45–58, 2004.
118. Vicient, C.M., The *Arabidopsis AtEm1* promoter is active in *Brassica napus* L. and is temporally and spatially regulated, *J. Exp. Bot.*, 52, 1587–1591, 2001.

119. Pylatuik, J.D. et al., Elements regulating AGAMOUS expression are conserved in *Arabidopsis thaliana*, *Brassica napus* and *Linum usitatissimum*, *Canad. J. Bot.*, 81, 523, 2003.
120. Pylatuik, J.D. et al., Isolation and characterization of a *Brassica napus* cDNA corresponding to a B-class floral development gene, *J. Exp. Bot.*, 54, 2385, 2003.
121. Hulzink, R.J. et al., *In silico* identification of putative regulatory sequence elements in the 5′-untranslated region of genes that are expressed during male gametogenesis, *Plant Physiol.*, 132, 75–83, 2003.
122. Margulies, E.H. et al., Identification and characterization of multispecies conserved sequences, *Genome Res.,* 13, 2507–2518, 2003.
123. Colinas, J., Birnbaum, K., and Benfey, P.N., Using cauliflower to find conserved noncoding regions in *Arabidopsis*, *Plant Physiol.*, 1229, 451–454, 2002.
124. Sjakstem, T.G., Rashalm I., and Roder, M.S., Inheritance of microsatellite alleles in pedigrees of Latvian barley varieties and related European ancestors, *Theor. Appl. Genet.*, 106, 539–549, 2003.
125. Maggioni, L., Activities and achievements of the ECP/GR *Brassica* working group, *Acta Hort.,* 459, 243–248, 1998.
126. http://www.brassica.info/diversity/diversity_sets.htm.
127. http://www.oregin.info.
128. Yamazaki, Y. and Jaiswal, P., Biological ontologies in rice databases. An introduction to the activities in gramene and oryzabase, *Plant Cell Physiol.*, 46, 63–68, 2005.
129. Parkin, I. et al., Segmental structure of the *Brassica napus* genome based on comparative analysis with *Arabidopsis thaliana*, *Genetics*, 171: 765–781, 2005.
130. Koch, M. et al., Comparative evolutionary analysis of chalcone synthase and alcohol dehydrogenase loci in *Arabidopsis*, *Arabis*, and related genera, *Mol. Biol. Evol.*, 17, 1482–1498, 2000.
131. Francia, E. et al., Marker assisted selection in crop plants, *Plant Cell Tissue Org. Cult.*, 82, 317–342, 2005.
132. Lysak, M.A. et al., Chromosome triplication found across the tribe Brassiceae, *Genome Res.*, 15, 516–525, 2005.
133. Alix, K. et al., The genomic organization of retrotransposons in *Brassica oleracea*, *Plant Mol. Biol.*, 59, 839–851, 2005.
134. Gupta, P.K. et al., Linkage disequilibrium and association studies in higher plants: present status and future prospects, *Plant Mol. Biol.*, 57, 461–485, 2005.
135. Gilchrist, E.J. and Haughn, G.W., TILLING without a plough: a new method for reverse genetics, *Curr. Opin. Plant Biol.*, 8, 1–5, 2005.
136. Henikoff, S. and Comai, L., Single-nucleotide mutations for plant functional genomics, *Annu. Rev. Plant Biol.*, 54, 375–401, 2003.
137. Yang, T.J. et al., The Korea *Brassica* Genome Project: a glimpse of the *Brassica* genome based on comparative genome analysis with *Arabidopsis*, *Comp. Func. Genom.*, 6, 138–146, 2005.
138. Qi, B. et al., Production of very long chain polyunsaturated omega-3 and omega-6 fatty acids in plant, *Nature Biotech.*, 22, 739–745, 2004.
139. Abbadi, A. et al., Biosynthesis of very-long-chain polyunsaturated fatty acids in transgenic oilseeds: constraints on their accumulation, *Plant Cell*, 16, 2734–2748, 2004.
140. Wu, G. et al., Stepwise engineering to produce high yields of very long-chain polyunsaturated fatty acids in plants, *Nature Biotech.*, 23, 1013–1017, 2005.
141. Glazebrook, J., Use of *Arabidopsis* for genetic dissection of plant defense responses, *Annu. Rev. Genet.*, 31, 547–569, 1997.

142. Nimchuk, Z. et al., Recognition and response in the plant immune system, *Annu. Rev. Genet.*, 37, 579–609, 2003.
143. Hammond-Kosack, K.E. and Parker, J.E., Deciphering plant–pathogen communication: fresh perspectives for molecular resistance breeding, *Curr. Opin. Biotech.*, 14, 177–193, 2003.
144. Dangl, J.L. and Jones, J.D.G., Plant pathogens and integrated defense responses to infection, *Nature*, 411, 826–833, 2001.
145. Meyers, B.C., Kaushik, S., and Nandety, R.S., Evolving disease resistance genes, *Curr. Opin. Plant Biol.*, 8, 129–134, 2005.
146. Feys, B.J. and Parker, J.E., Interplay of signaling pathways in plant disease resistance, *Trends Genet.,* 16, 449–455, 2000.
147. Williams, P.H. and Hill, C., Rapid-cycling populations of brassicas, *Science*, 232, 1385–89, 1986.
148. Musgrave, M.E., Realizing the potential of rapid-cycling *Brassica* as a model system for use in plant biology research, *J. Plant Growth Regul.*, 19, 314–25, 2000.
149. Walsh, J.A. et al., Characterization of resistance to turnip mosaic virus in oilseed rape (*Brassica napus*) and genetic mapping of *TuRB01*, *Theor. Appl. Genet.*, 99, 1149–1154, 1999.
150. Hughes, S.L. et al., Genetic mapping of a novel *turnip mosaic virus* resistance gene in *Brassica napus*, *Theor. Appl. Genet.,* 107, 1169–1173. 2003.
151. Vicente, J.G. et al., Inheritance of race specific resistance to *Xanthomonas campestris* pv. *campestris* in *Brassica* genomes, *Phytopathology*, 92, 1134–1141, 2002.
152. Ferreira, M.E. et al., Mapping of a locus controlling resistance to *Albugo candida* in *Brassica napus* using molecular markers, *Phytopathology*, 85, 218–220, 1995.
153. Somers, D.J. et al., *Brassica napus* DNA markers linked to white rust resistance in *Brassica juncea*, *Theor. Appl. Genet.*, 104, 1121–1124, 2002.
154. Kole, C. et al., Linkage mapping of genes controlling resistance to white rust (*Albugo candida*) in *Brassica rapa* (syn. *campestris*) and comparative mapping to *Brassica napus* and *Arabidopsis thaliana*, *Genome*, 45, 22–27, 2002.
155. Sillito, D. et al., *Arabidopsis thaliana*: a source of candidate disease-resistance genes for *Brassica napus*, *Genome*, 43, 452–460, 2000.
156. Vicente, J.G. and King, G.J., Characterization of disease resistance gene-like sequences in *Brassica oleracea* L., *Theor. Appl. Genet.*, 102, 555–563, 2001.
157. Grant, M.R. et al., Independent deletions of a pathogen-resistance gene in *Brassica* and *Arabidopsis*, *PNAS (USA)*, 95, 15843–15848, 1998.
158. Malvas, C.C. et al., A homolog of the RPS2 disease resistance gene is constitutively expressed in *Brassica oleracea*, *Genet. Molec. Biol.*, 26, 511–516, 2003.
159. Beckett, P. et al., Computational tools for *Brassica–Arabidopsis* comparative genomics, *Comp. Func. Genom.*, 6, 147–152, 2005.
160. Lan, T.-H. et al., An EST-enriched comparative map of *Brassica oleraceae* and *Arabidopsis thaliana*, *Genome Res.*, 10, 776–788, 2000.
161. Kole, C. et al., Evidence for homology of flowering-tree genes *VFR2* from *Brassica napus* and *FLC* from *Arabidopsis thaliana, Theor. Appl. Genet.*, 102, 425–430, 2001.
162. Fourmann, M. et al., Expression, mapping, and genetic variability of *Brassica napus* disease resistance gene analogues, *Genome*, 44, 1083, 2001.

4 Characterization of the Completed Rice Genome Sequence and Scope of Its Utilization in Cereal Improvement

Baltazar A. Antonio and Takuji Sasaki

CONTENTS

4.1 INTRODUCTION

Rice is considered a model cereal crop because its genome may serve as the key in understanding the structure of other grass species. Therefore, achievements in rice genome analysis will have a tremendous impact not only on rice production but also on improvement of other cereal crops. In collaboration with the International Rice Genome Sequencing Project (IRGSP), the Rice Genome Research Program (RGP) in Japan has successfully completed sequencing of the rice genome. A high-quality and map-based sequence of the entire genome is now available in the public domain. In addition, outputs of the large-scale genome analysis, including the genetic map, physical map, and transcript map, have become indispensable tools in many areas of cereal genomics.

Decoding the rice genome sequence with high accuracy and deciphering genetic information in the genome will have a great impact on understanding the biology of the number-one food crop consumed by more than half the world's population. As the regions where rice is primarily consumed—namely, Asia, Africa, and Latin America—continue to experience rapid population growth, rice production must also increase proportionally to cope with ever increasing demand. It is projected that global rice production must increase by 30% in the next 20 years to cope with the impending population increase. This will not be possible without new discoveries in breeding and growing methods as well as development of new plant types with high yield potential.

For rice, revealing the genome sequence is an achievement of unparalleled proportions, particularly in what can be accomplished with other cereal crops such as maize, wheat, barley, and sorghum. Extensive rice genome analysis based on the principles of molecular biology has been pursued since the 1980s with advances in tools for analysis of biological phenomena at the molecular level. However, the significance of rice in Japanese agriculture as well as in the context of maintaining a stable world food supply prompted the Japanese government to embark on a large-scale analysis of the rice genome in the early 1990s [1,2]. Although the original target was simply to characterize the rice genome based on genetic and physical structure of the genome, the project expanded 7 years later into full-scale sequencing of the entire rice genome.

The RGP was joined in this initiative by a consortium of publicly funded laboratories from countries in Asia, Europe, and North and South America, with the common aim of accelerating the completion of sequencing and immediate release of the sequence data in public databases [3]. Similar sequencing efforts have been pursued in rice, resulting in publication of two whole-genome shotgun assemblies of draft-quality rice sequences [4,5] at a time when the IRGSP was just halfway through sequencing efforts. However, the scientific community recognized that for an important crop such as rice, nothing more than a completed sequence could substitute for applying the information in crop improvement programs, particularly in determining the function of many agronomically useful genes. Thus, the IRGSP worked to generate a high-quality, finished sequence of the rice genome using a clone-by-clone approach and eventually finished the sequence several years ahead of schedule.

This chapter will focus on the comprehensive analysis of the rice genome—from mapping to sequencing—that facilitated a thorough understanding of its structure and function. The scope of utilization of the map-based rice genome sequence will be described based on extensive studies conducted in applied and functional genomics. These efforts may eventually lead to needed improvement of current varieties of various cereal crops, as well as development of novel crops carrying agronomically important traits essential to meeting the growing demands for food production in the years to come.

4.2 MAPPING THE RICE GENOME

Initial efforts in rice genome analysis focused on genetic and physical mapping of the genome. The fundamental tools derived from these studies have been extremely useful in the ultimate goal of sequencing the rice genome. They have various applications in many areas of rice genomics as well as in understanding the genome structure of other grass species.

4.2.1 Genetic Mapping

The first step in comprehensive analysis of the genome is construction of a map to provide an overall view of the structure of the entire genome (Figure 4.1). Large-scale genetic mapping has been carried out for rice from the latter part of the 1980s because of rapid advances in the development of molecular markers such as restriction fragment length polymorphism (RFLP), amplified fragment length polymorphism (AFLP), and random amplified polymorphic DNA (RAPD). The first rice molecular map based on RFLPs was developed using an F_2 population derived from an *indica* × *javanica* cross [6]. Successive studies generated molecular maps based on different mapping populations [7,8].

As part of the first phase of the RGP, a high-density linkage map was constructed using a single F_2 population derived from the cross between a *japonica* cultivar, Nipponbare, and an *indica* cultivar, Kasalath [9]. The DNA markers, which consisted mostly of partially sequenced cDNA clones derived from various tissues, organs, and cultured cells, were mapped at an average distance of 1.1 cM or approximately 300 kb. With additional markers, a more saturated map with 2275 DNA markers was established [10]. At present, the high-density linkage map for rice consists of 3267 markers (http://rgp.dna.affrc.go.jp/publicdata/geneticmap2000/index.html), with additional PCR-based markers such as sequence tag site (STS) markers and cleaved amplified polymorphic sequence (CAPS). In addition, simple sequence repeat (SSR) markers have also been mapped in the rice genome [11]. The availability of a highly saturated molecular map for rice is one of the major achievements in rice and has become indispensable in many breeding strategies as well as for map-based cloning of agronomically important genes.

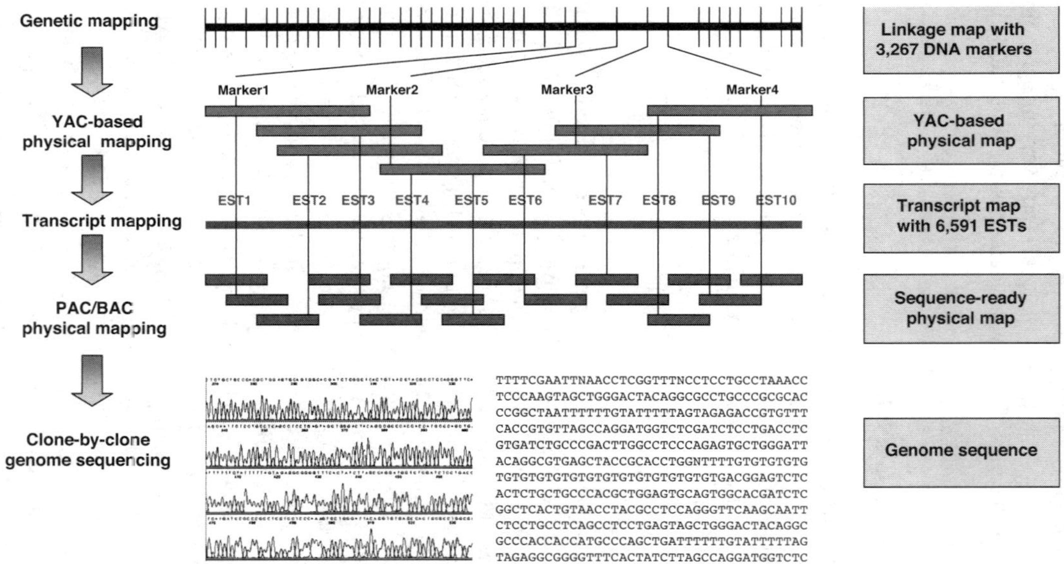

FIGURE 4.1 Color Figure 4.1 follows p. 144.) An overview of the strategy for generating a map-based sequence of the rice genome. A high-density linkage map was initially established and the mapped DNA markers were used to screen a YAC genomic library. The resulting YAC-based physical map was used as a template for transcript mapping. The mapped ESTs were then used to screen genomic libraries in PAC and BAC vectors to construct a sequence-ready physical map. The aligned clones were used for sequencing so that all sequenced clones could be assigned to specific positions in the chromosomes.

4.2.2 Physical Mapping

The overall strategy of the RGP included the construction of a physical map by anchoring cloned DNA fragments on the molecular genetic map (Figure 4.1). A genomic library of cultivar Nipponbare was initially constructed using yeast artificial chromosome (YAC) as vector. The resulting YAC library consisted of 7000 clones with an average insert length of 350 kb [12]. This library was screened with the RFLP markers in the genetic map by colony/southern hybridization and the positive YAC clones were aligned based on the map position of the markers. The genetic map with 1383 markers generated a YAC-based physical map with 52% coverage corresponding to 222 Mb of the genome [13].

Subsequent anchoring of YAC clones using 1439 additional DNA markers generated a YAC-based physical map with 297 contigs and 142 islands with a 63% coverage corresponding to 270 Mb of the genome [14]. Although the YAC clones were not suitable as substrates for genome sequencing, the YAC-based physical map has become an effective tool for positional cloning of many agronomically important traits and construction of a sequence-ready physical map.

4.2.3 Transcript Mapping

The RGP has developed a catalog of expressed rice genes using cDNA libraries derived from various tissues [15]. However, only a limited number of these sequences have been anchored on the genetic map. Using the YAC-based physical map, large-scale mapping of rice ESTs was carried out to increase the genomic anchors necessary for construction of a sequence-ready physical map (Figure 4.1). Paired PCR primers were designed from the 3′-terminal sequences of unique cDNA clones and these 3′-ESTs were mapped by PCR screening using the physically mapped YAC clones as template [16].

The resulting transcript map consisted of 6591 ESTs evenly distributed throughout the 12 chromosomes (except the centromeric and telomeric regions) and provides an overall view of the distribution of active genes in the genome. Such information may be useful in map-based cloning, particularly in clarifying the correlation between genetically identified loci and gene candidates expressed within a specific region.

4.2.4 Sequence-Ready Physical Mapping

Sequence-ready physical maps of the 12 rice chromosomes were constructed as the template for genome sequencing. A Nipponbare PAC (P1-derived artificial chromosome) library with 70,000 clones [17] and a BAC (bacterial artificial chromosome) library with 50,000 clones [18] were used as the main resources for establishing a sequence-ready physical map of the entire genome. The pooled PAC/BAC clones were used as templates for PCR screening using the PCR primers that were generated for mapping ESTs. Additional genome resources such as two BAC libraries (total of 90,000 clones) from the Clemson University Genomics Institute [19] and BAC clones with draft sequences donated by Monsanto Co. were used to increase map coverage.

After all genomic libraries were exhausted, the remaining physical gaps were filled using various strategies. The draft sequences of the seed clones were used to search for minimally overlapping clones from the BAC-end sequence database. The FPC contigs with end sequences that matched the sequences of seed BACs were added to the physical maps. A PCR screening method was also used to search for clones that filled the remaining gaps. Two 10-kb insert genomic libraries and a 40-kb fosmid library were also constructed as additional resources to facilitate extension of the contigs [20].

4.3 SEQUENCING THE GENOME

The IRGSP adopted a hierarchical clone-by-clone genome sequencing strategy, which involved sequencing individual cloned DNA fragments anchored on the physical map using molecular markers. In contrast, Beijing Genomics Institute and Syngenta Company used a whole genome shotgun sequencing strategy, which relied on sequencing many short, unmapped clones simultaneously and assembling the pieces into longer coherent units of sequence using computers, in establishing the draft sequence of *indica* cultivar 93-11 and *japonica* cultivar Nipponbare, respectively [4,5]. This strategy is particularly disadvantageous for rice because it cannot establish the relationships of the genes to each other and to the genetic map.

On the other hand, the clone-by-clone sequencing strategy allowed for efficient management of sequencing involving various groups because specific chromosomes or regions of the chromosome could be assigned independently. Japan was in charge of sequencing 6 of the 12 chromosomes corresponding to almost 55% of the genome. Although sequencing protocols differed slightly in each laboratory, standard procedures were followed to assure 99.99% accuracy of the sequence.

The PAC/BAC clones comprising the minimum tiling path were subjected to shotgun sequencing using universal primers and the dye-terminator or dye-primer methods for approximately 500 bases. Each PAC/BAC clone was randomly sheared and small insert libraries were constructed. The sequences were analyzed using capillary sequencers and assembled by PHRED and PHRAP packages or with the TIGR Assembler (http://www.tigr.org/software/assembler/).

Each clone comprised 3840 sequences from 1920 subclones, half of which were from 2-kb insert libraries and half from 5-kb libraries to produce tenfold shotgun sequences. The sequence ambiguities indicated by low PHRAP scores were resolved by confirming the sequence data using alternative chemistries or different polymerases. The gaps were filled by full sequencing of gap-bridge clones. The finished assemblies were verified by comparing sizes of virtual restriction digests with the experimental data.

The assembled sequence contigs for each PAC/BAC clone were joined by inserting successive Ns in the sequence gap region and submitted to the public databases. The PAC/BAC sequences were then analyzed using an automated annotation system and manual curation. The gene predictions using GENESCAN, FGENESH, and Genemark were integrated with the results of BLAST homology search against protein sequence database (nr), rice ESTs, and full-length cDNAs. These results were manually curated to construct the most plausible gene models based on existing evidence. The annotated sequences with gene models, as well as the suggested protein functions, were resubmitted to public databases (DDBJ, GenBank, and EMBL).

4.4 CHARACTERIZATION OF THE COMPLETED SEQUENCE

Detailed analysis of the complete rice genome sequence was based on the Build 3.0 pseudomolecules (http://rgp.dna.affrc.go.jp/IRGSP/Build3/build3.html). The nucleotide sequences representing entire chromosomes were constructed by joining the sequence of PAC/BAC/fosmid clones comprising the minimum tiling path in the physical map of the 12 chromosomes. The major features of the genome [21] will be summarized here.

4.4.1 Genome Coverage

The complete sequence of the rice genome was obtained from 3401 PAC and BAC clones [21]. Chromosomes 8 and 12 have been reduced to two contigs each and chromosomes 6 and 7 to three contigs each; the rest of the chromosomes range from four to seven contigs. The centromeres of chromosomes 4 and 8 have been completely sequenced, whereas the centromere chromosome of 5 has been partially sequenced. So far, 62 physical gaps, including nine centromeres and 17 telomere gaps, remain on the 12 chromosomes.

To facilitate characterization of the entire genome, reference molecules or pseudomolecules of the 12 chromosomes were constructed. The nucleotide sequence of each PAC/BAC/fosmid clone was joined based on the order of the clones on the physical map. The overlapping sequences were removed and physical gaps were replaced by successive Ns. The total nucleotide sequence of the 12 pseudomolecules is 370,733,456 bp, excluding the ambiguous nucleotides. This corresponds to more than 97% of the entire length of each chromosome. With the remaining gaps estimated to total 19.6 Mb, the total chromosomal length of the rice genome was calculated to be 390 Mb. Therefore, the pseudomolecules so far cover 95.3% of the entire genome and an estimated 98.9% of the euchromatin (Figure 4.2).

4.4.2 Gene Content

The statistics of the rice genome sequence based on the 12 pseudomolecules are summarized in Table 4.1 [21]. The *ab initio* gene finder FGENESH was used to discover 55,296 genes, including those related to transposable elements (TE-related), which were captured as "genes" in the annotation process. The overall density of one gene per 6.7 kb is similar to gene densities reported for chromosomes 1 [22], 3 [23], 4 [24], and 10 [25]. To identify only nontransposable element-related genes, the pseudomolecules were masked for repetitive sequences. Thus, 37,544 nontransposable element protein-coding sequences were predicted, resulting in a density of one gene per 9.9 kb. The average gene length is 2.7 kb with an average exon size of 254 bp and an average intron size of 413 bp.

Rice shows a lower gene density than *Arabidopsis*, which shows a gene density of one gene for every 4.5 kb [26]. Average gene size is also higher for rice compared with *Arabidopsis* with an average gene size of about 2 kb. Although the average exon size is almost the same in rice and *Arabidopsis*, the average intron size is about 2.6 times larger in rice than in *Arabidopsis*. This means that although the longer

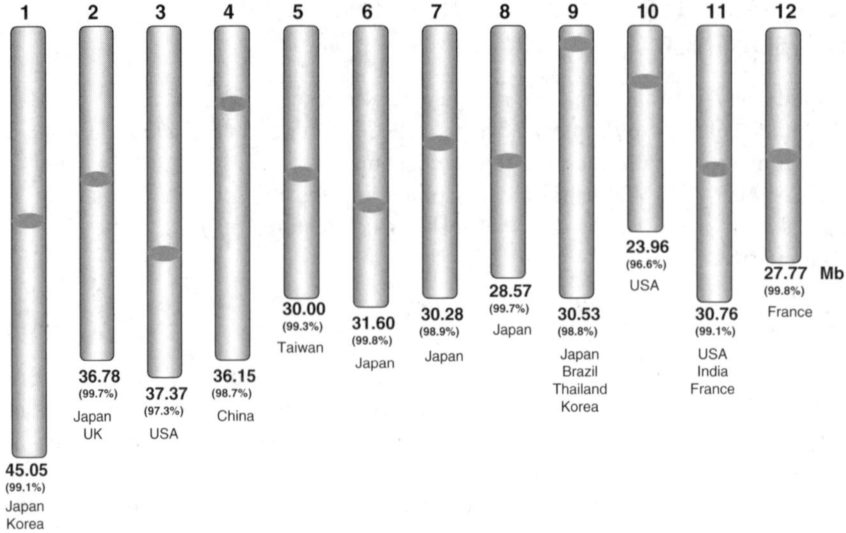

FIGURE 4.2 Extent of completion of the 12 chromosomes sequenced by participating countries of the IRGSP. The numbers below each chromosome correspond to the size of each pseudomolecule; the coverage of the completed sequence is in parentheses.

introns engender larger gene sizes in rice, the average transcriptome size is similar in both species.

In terms of base composition, the average GC content is about 54% for coding region and 38% for noncoding region. These values are relatively higher than the corresponding values for *Arabidopsis*, suggesting that, overall, rice is more GC rich than *Arabidopsis* is. In general, density of expressed genes is greatest on the distal portions of the chromosome arms compared with the regions around the centromeres. Among the 12 chromosomes, chromosomes 3 and 1 had the highest percentage of expressed genes and chromosomes 11 and 12 had a lower percentage. More detailed analysis of chromosomes 11 and 12 showed a higher proportion of disease resistance genes than other rice chromosomes had [27].

4.4.3 Gene Composition

About 61% (22,840) of the predicted gene models showed high identity matches with a combination of rice ESTs and full-length cDNAs [21,28]. A comparison with the predicted genes in *Arabidopsis* indicates that 88% of the rice gene models with some transcription evidence have corresponding homologs in the *Arabidopsis* genome. A total of 2859 gene models with no homologs in *Arabidopsis* include a few genes with homologs to known proteins such as prolamins, proteinase inhibitors, chitinases, pathogenesis-related proteins, and seed allergens. The rest of these genes have no protein hits or most closely matched unknown or hypothetical proteins.

A significant number of genes are duplicated and arrayed in tandem. Of the gene models, 29% are amplified at least once in tandem. Among them, 15% are duplicated

TABLE 4.1
Profile of the Rice Genome Based on the Complete Map-Based Sequence

Total genome size (bp)	370,733,456
Predicted genes	
Number	37,544
Gene density (kb/gene)	9.9
Average gene length (bp)	2,699
Exons	
Number	175,203
Total length (bp)	44,492,676
Average per gene	4.7
Average size (bp)	254
Introns	
Number	137,659
Total length (bp)	56,841,388
Average per gene	3.7
Average size (bp)	413
Base composition (GC%)	
Exon	54.2
Intron	38.3
Intergenic	42.9
Gene	45.3
Transposable elements	
Copy number	249,300
Coverage	129,019,300
Fraction of genome	44.79
Chloroplast inserts	
Number of inserts	453
Total length	703,086
Genome equivalent	5.22
Mitochondrial inserts	
Number of inserts	909
Total length	630,457
Genome equivalent	1.29

and more than 50% are represented by three to ten copies. There are also 31 rice gene arrays, which contained 24 to 134 members.

The rice genome is characterized by a large number of transposons and transposable elements [21]. Representatives from all known transposon superfamilies have been identified in the rice genome. The class I elements, which include the non-LTR retrotransposons (such as LINEs and SINEs) and long terminal-repeat (LTR) retrotransposons (*Ty1/copia*, *Ty3/gypsy*, and *TRIM*) with a total coverage of 72 Mb, correspond to almost 19% of the genome. The class II elements characterized by terminal inverted repeats and including the *h*AT, CACTA, IS*256/Mutator*, IS*5/Tourist*, and IS*630*/Tc1/*mariner* superfamilies have coverage of 48 Mb, corresponding to almost 13% of the genome.

In total, the transposon content of the *Oryza sativa* ssp. *japonica* genome is about 129 Mb and corresponds to at least 35% of the genome. Organellar DNA fragments representing chloroplast and mitochondrial DNA correspond to 0.38 to 0.43% of the nuclear genome. The distribution of ct and mt insertions for the 12 chromosomes indicates that mitochondrial and chloroplast transfers occurred independently [21]. A total of 763 tRNA genes, 158 micro-RNAs (miRNAs), 215 small nuclear RNA (snoRNA), and 93 spliceomal RNA genes were identified on the 12 pseudomolecules.

4.4.4 Centromere Sequence

The rice genome sequence provided a detailed characterization of the centromeres of chromosomes 4 [31] and 8 [32]. These rice centromeres are characterized by the highly repetitive 155- to 165-bp satellite DNA, CentO, and centromere-specific retrotransposons [29,30]. The core region of centromere 8 has a 68.5 kb of 155-bp CentO satellite repeats, which were divided into three clusters [32]. More than 220 transposable element-related sequences (most of which belong to RIRE family) were also found in the same region. On the other hand, the centromeric region of chromosome 4 contains a 59-kb cluster of CentO repeats and a much larger number of inserted retroelements [31]. A large variation of CentO contents in the 12 chromosomes has also been detected, suggesting a unique centromere structure for each rice chromosome.

4.4.5 Complete versus Draft Sequence

Initial comparison of the finished *japonica* sequence with *indica* sequence contigs assembled from the whole genome shotgun sequences of Beijing Genomics Institute [5] was conducted using the *japonica* sequence as a query for BLASTN analysis. The results showed that a corresponding *indica* sequence could be detected in about 78% of the *japonica* sequence, although 65 gaps occurred in the aligned contigs and a total of 110,389 bases (22%) of *japonica* sequence could not be identified in the *indica* assembly [22]. This may partly reflect the sequence difference between the two subspecies, although some artifacts in the whole-genome shotgun assembly cannot be ruled out. Furthermore, some predicted genes in the complete *japonica* sequence were only partially predicted or were not predicted at all in the corresponding *indica* draft sequence, thus preventing an accurate prediction of the number of genes.

The two whole-genome shotgun (WGS) assemblies of draft-quality rice sequences have been recently reassembled with the intact single nucleotide polymorphism information for *japonica* and *indica* sequences [33]. This resulted in an assembly of 6.28× coverage of *O. sativa* ssp. *indica* cv. 93-11 and ~6× coverage of *O. sativa* ssp. *japonica* cv. Nipponbare. These assemblies predicted genome sizes of 433 Mb for *O. sativa* ssp. *japonica* and 466 Mb for *O. sativa* ssp. *indica*; these differ from the IRGSP genome sequence estimation of 389 Mb for the *japonica* genome [21]. When the contigs from the WGS assemblies were aligned with the IRGSP pseudomolecules, the nonredundant coverage of the pseudomolecules by the *indica* assembly varied from 78% for chromosome 3 to 59% for chromosome 12; overall coverage was 69% [21].

Furthermore, when genes supported by full-length cDNA coverage were aligned to the covered regions, about 65.2% were completely covered by the *indica* sequences. Although the whole sequencing strategy provides an overall picture of the rice genome structure, accuracy is very much compromised. Thus, there is no substitute for the clone-by-clone strategy in terms of accuracy of the sequence.

4.4.6 Subspecies *japonica* versus *indica* Sequence

The degree of polymorphism between subspecies *japonica* and subspecies *indica* was determined by *in silico* mapping of BAC-end sequences from the *indica* rice cultivar Kasalath using the complete genome sequence of Nipponbare as a reference standard [34]. Based on analysis of 26,632 paired Kasalath BAC-end sequences mapped to the 12 rice pseudomolecules, 80,127 sites differed in the corresponding regions in Nipponbare and Kasalath [21]. Single nucleotide polymorphism (SNP) and insertion/deletion (INDEL) were observed throughout the genome. SNP frequency ranged from 0.53 to 0.80% among the 12 chromosomes, whereas the frequency of INDELs was 1.23 sites per kilobase. These results may provide insights in understanding the differences and similarities that define the two major subspecies of cultivated rice.

4.5 UTILIZATION IN CEREAL GENOMICS

The ultimate goal for developing various genomics tools in rice, such as the genetic map, physical map, transcript map, and the map-based genome sequence, lies in the area of applied genomics, particularly in the improvement of major cereal crops. This involves breeding strains of the crop with specific characteristics such as increased yield, good-eating quality, stress tolerance, or disease resistance more quickly than through traditional methods. The information on the genome sequence and the accompanying genomics tools are indispensable to achieve this goal. However, thorough understanding of the sequence of all the genes and their functions is just as important to exploit the full potential of the genome resources at hand for cereal improvement. The following sections describe some approaches to characterizing many agronomic characters in rice that could, in turn, lead to crop improvement.

4.5.1 Marker-Assisted Approaches

Even before completion of sequencing of the entire genome, the sequences of individual PAC/BAC clones that correspond to specific regions of the genome could be accessed through public databases. This allowed many researchers to use the genome sequence information in developing molecular markers for genetic analysis. The use of markers tightly linked to a trait of interest can accelerate screening, particularly in cases of recessive traits or traits expressed at specific stages of development.

Several types of markers, such as restriction fragment length polymorphism (RFLP) markers, random amplified polymorphic DNAs (RAPDs), sequence tag sites (STS), cleaved amplified polymorphic sequences (CAPS), amplified fragment length polymorphism (AFLP), and single nucleotide polymorphisms (SNP), have been developed for rice. When all the molecular markers from various genetic maps of rice are combined, more than 6000 markers are currently available; this corresponds to a genome distribution of approximately 1 marker for every 0.25 cM or every 75 to 100 kb [35]. These molecular markers have also been useful for comparative mapping among various cereal species [36–38] and marker-assisted selection [39].

In addition, 2740 experimentally confirmed simple sequence repeat (SSR) markers are available for rice [11]. The complete sequence of the genome has revealed 18,828 class 1 di-, tri-, and tetranucleotide SSRs representing 47 distinctive motif families [21]. Several thousand of these SSRs have already been shown to amplify well and be polymorphic in mapping populations; they will therefore provide a rich source of molecular markers, particularly in genetic analysis of closely related genotypes.

A random analysis of polymorphism showed that corresponding regions in Nipponbare and Kasalath differed in 80,127 sites [21,34]. These SNPs could be very valuable resources for marker-aided selection involving *japonica* and *indica* crosses. A more extensive study aligning the draft sequences of *indica* and *japonica* showed 384,341 SNPs and 24,557 single-base INDELs [40]. In addition to marker-aided selection, the SNP/INDEL information for rice can be used for further enhancing the current genetic maps for rice [41] and understanding the linkage disequilibrium in plants [42]. Using the *japonica* cultivar Nipponbare and the *indica* cultivar 93-11 genome sequence, a database with 1,703,176 SNPs and 479,406 InDels has been developed [43]. This rice DNA polymorphism database will provide additional markers for genetic analysis of other *japonica* and *indica* cultivars.

4.5.2 Map-Based Cloning

The most fundamental application of a high-density genetic map and an accompanying clone-based physical map is in the area of map-based cloning, which relies on finding molecular markers very closely linked to the gene of interest. Map-based cloning starts from narrowing down the genetic interval containing the gene, saturating that region with many molecular markers, constructing a physical map, and isolating the gene from a BAC or PAC clone.

Several agronomically important genes involved in disease resistance, plant architecture, and grain production have been isolated by this strategy. These include resistance genes against bacterial blight, such as *Xa21* [44], *Xa1* [45], and *Xa26* [46], and rice blast resistance genes such as *Pib* [47], *Pi-ta2* [48], and *Pi5(t)* [49]. In terms of plant architecture and development, rice gibberellin-insensitive dwarf mutant gene *Dwarf 1* [50], a timekeeper of leaf initiation in rice *PLASTOCHRON1* [51], a leaf-spotted leaf gene *Spl7* [52], a pentatricopeptide repeat-containing gene that promotes the processing of aberrant *atp6* RNA of cytoplasmic male-sterile rice [53], and a fertility restorer gene, *Rf-1* [54] have been isolated as well.

Similarly, many agricultural traits that directly affect grain production have been characterized and cloned through map-based cloning strategy. An important gene in the control of rice tillering, *MOC1* (monoculm 1), has been isolated and found to encode a putative GRAS family nuclear protein expressed mainly in the axillary buds; it functions in initiating and promoting the outgrowth of axillary buds [55]. Another important gene isolated by map-based cloning is the gene *BC1* (brittle culm 1), which encodes a COBRA-like protein and is expressed mainly in developing schlerenchyma cells and in vascular bundles of rice [56]. The same strategy has also led to characterization and isolation of the major semidwarfing allele, *sd1*, which induces height reduction associated with significant yield increase [57].

Among other cereals, information on synteny has been very useful in map-based cloning of important genes. Successful cloning of barley stem rust resistance gene *Rpg1* has been facilitated by the synteny between the short arm of barley chromosome 1 and short arm of rice chromosome 6, although the gene was not found in the syntenous position in rice [58,59]. The complete rice genome sequence will further accelerate map-based cloning strategies in rice and other cereal crops by providing a major source of a wide range of molecular markers for many agronomic traits.

4.5.3 QTL Approach

Analysis of polygenic characteristics more commonly known as quantitative trait loci (QTL), including yield, heading date, culm length, grain quality, and stress tolerance, has been greatly facilitated by genome sequence information. A large array of molecular markers as well as genetic populations with well characterized chromosomal segments, such as recombinant inbred lines (RILs), backcross inbred lines (BILs), doubled haploid lines (DHLs), nearly isogenic lines (NILs), and chromosome segment substitution lines (CSSLs), has been very useful in detecting specific positions of many agronomic traits and high-resolution mapping of QTLs controlling these target traits [60–62].

The heading date loci in rice that are involved in photoperiod sensitivity have been widely characterized. These include *Hd1* [63], *Hd3a* [64,65], *Hd3b* [65], *Hd4* [66], *Hd5* [66], *Hd6* [67], *Hd8* [68], and *Hd9* [69]. Among them, *Hd1* has been clarified as closely related to the *Arabidopsis* flowering time gene *CONSTANS* [63] and *Hd6* has been found to encode the alpha subunit of protein kinase *CK2* [67]. Other QTLs have been identified and characterized:

- QTLs for seed dormancy such as *Sdr1* [68], *qSD1* [70], *qSD7-1* [70], and *qSD12* [70]
- QTLs for Na+ and K+ uptake of shoots and roots controlling salt tolerance of rice [71]
- QTL SKC1, which maintained K(+) homeostasis in the salt-tolerant variety under salt stress [72]
- *qUVR-10*, which confers resistance to ultraviolet-B radiation in rice [73]
- A QTL controlling low-temperature germinability [74]
- A QTL for plant growth of rice in paddy fields flooded with salt water [75]
- A QTL for cytosolic glutamine synthetase content and panicle number [76]

These studies have been facilitated by the publicly available rice genome sequence information tied to the genetic map. As more QTLs are identified, pinpointing the crucial genes will expedite transfer of beneficial traits into locally adapted elite lines and will permit plant breeders to search for genes necessary to grow higher yielding rice strains in various environmental conditions.

4.5.4 Synteny Approach

Rice has the smallest genome size among the major cereal crops, including corn, wheat, barley, rye, sorghum, oats, and millet, which have similar arrangements of genes on the chromosomes. The syntenic relationships between rice and these cereal grasses have long been supported by comparative mapping using DNA markers [36,77]. Extensive conservation of gene order and gene content along the chromosomes of various grasses has led to establishment of the so-called "circle diagram" with rice as the reference genome [78]. As a consequence, information about the rice genome will be indispensable in understanding genome structure of much larger genomes such as sorghum (750 Mb), corn (2500 Mb), barley (4900 Mb), and wheat (16,000 Mb) with an evolutionary divergence time of 60 million years [79].

It has been assumed that the larger genomes differ from each other as a result of chromosomal inversions, translocations, or duplications involving portions or entire chromosome arms of the ancestral genome. However, although collinearity between cereal genomes is very pronounced at the megabase level, extensive evidence indicates numerous small rearrangements at the submegabase level. These include frequent insertions of transposable elements and duplications or deletion of genes that occur without rearrangement of adjacent sequences [80].

Disruption of microcollinearity has also been reported between rice and barley [81], rice and wheat [82], and rice and maize [83]. Dubcovsky et al. reported one of the most detailed examinations of the level of microcollinearity between barley and rice [84]. Although orientation and number of genes differed, no extensive similarity was found beyond the exon structures, untranslated regions, and promoter sequences. A mosaic organization of orthologous regions in which conserved sequences were interspersed with nonconserved sequences was clarified based on comparison of sequences among maize, barley, and rice [85]. Thus, assessment of the level of microcollinearity between rice and other cereal genomes suggests that synteny of cereal genomes is more complex than expected.

With elucidation of the complete rice genome sequence, variation in sequences among cereal genomes can be scrutinized with more accuracy. The complete map-based rice sequence provides an opportunity to clarify the extent of collinearity between cereal genomes at the nucleotide level and may provide new insights into grass genome evolution.

4.5.5 Functional Approach

The accurate public map-based sequence now serves as a unifying platform for discovering all the genes that comprise the rice plant and establishing their functions. Among the 37,544 predicted genes, a large proportion with unknown function may correspond to rice-specific genes or novel cereal genes. Therefore, the next focus in rice genomics would be to characterize and assign function to all these genes. Several laboratories are already pursuing this goal using forward and reversed genetics strategies. The vast amount of genetic resources for rice, particularly the mutant lines induced through transposable elements, retrotransposons, and T-DNA insertions, is a major advantage for functional characterization of rice genes.

An endogenous copia-like retrotransposon in rice, *Tos17*, has been used effectively as a gene tag generating a collection of mutant lines carrying about 500,000 insertions [86]. Among them, a total of 5000 lines have been analyzed for flanking sequences. Using the *Tos17* tagging strategy, the causative genes for mutant phenotypes such as viviparous, dwarf, semidwarf, brittle culm, pale green, and narrow leaf mutations have been cloned [86]. Similarly, mutation in the homeobox gene *OSH15* [87], rice zeaxanthin epoxidase gene *OSABA1* [88], and rice phytochrome A [89] has been isolated and characterized using *Tos17* mutant lines.

Other mutant populations include insertion lines employing T-DNA [90,91] and *Ac/Ds* [92,93] as well as populations of deletion mutants developed in elite cultivars grown around the world. Location of insertional elements in these populations will depend on comparing flanking sequences with the complete genome sequence as has been done for the *Tos17* insertion lines.

Researchers from around the world are now addressing functional analysis of the rice genome through the International Rice Functional Genomics Consortium [94]. The collective goals of the consortium include sharing genomic materials, integrating databases, bilateral or multilateral partnerships, implementing initiatives for the cooperative elucidation of gene function, and accelerating delivery of research results to benefit rice production (http://www.iris.irri.org:8080/IRFGC/). Specifically, various functional genomics programs should focus on tagging rice genes using a variety of mutant collections, establishing a global Internet network of rice functional genomics databases, developing a high-throughput verification system, and characterizing the function of at least 50% of rice genes by 2010 [95].

The Rice Genome Resource Center (RGRC) in Japan is contributing to this initiative by providing access to biological materials developed from various projects in rice genome analysis that could be useful for functional characterization of rice genes [96]. With rapid accumulation of genomics resources and researchers' growing interest in rice, these could be achievable goals, especially if the cooperation of the rice scientific community can be maximized.

4.6 FUTURE PROSPECTS

The rice genome sequence will be the most important tool in cereal genomics in the years to come. It will provide the basic framework for whole-system approaches to understanding the biology of rice, including gene expression, proteome dynamics, and metabolite interactions. Comparisons across the cereal genomes using rice as the standard should provide the basis for understanding similarities and differences among cereal genomes; this could provide important clues in clarifying the evolution of grass species and a viable platform for designing future crop improvement programs. The genome sequence linked to a genetic map has already proven especially useful for identification of genes underlying diverse agronomic traits such as flowering time, plant architecture and development, fertility restoration, and disease resistance.

With availability of the complete map-based genome sequence, it is expected that more genes useful in agriculture will be elucidated and used to improve major agronomic traits. Furthermore, knowing the sequence and location of the gene will allow breeders to look for more useful variants of a gene in other rice strains or in distant relatives. In particular, the wild rice germplasm, which remains basically unexplored, can be a rich source of many useful genes. From now on, cereal crop improvement will increasingly rely on genomic technology.

Although worldwide efforts for sequencing other cereal crops such as maize (http://www.maizegenome.org/) and wheat (http://www.wheatgenome.org/) are ongoing, the high-quality, map-based sequence of the rice genome will probably remain the gold standard for an efficient and productive cereal crop improvement program.

REFERENCES

1. Sasaki, T., Rice genome analysis: Understanding the genetic secrets of the rice plant, *Breed. Sci.*, 53, 281, 2003.
2. Sasaki, T., Rice genome sequence analysis and the development of rice science, *Farming Jpn.*, 38, 18, 2004.
3. Sasaki, T. and Burr, B., International Rice Genome Sequencing Project: the effort to completely sequence the rice genome, *Curr. Opin. Plant Biol.*, 3, 138, 2000.
4. Goff, S.A. et al., A draft sequence of the rice genome (*Oryza sativa* L. ssp. *japonica*), *Science*, 296, 92, 2002.
5. Yu, J. et al., A draft sequence of the rice genome (*Oryza sativa* L. ssp. *indica*), *Science*, 296, 79, 2002.
6. McCouch, S.R. et al., Molecular mapping of rice chromosomes, *Theor. Appl. Genet.*, 76, 815, 1988.
7. Saito, A. et al., Linkage map of restriction fragment polymorphism loci in rice, *Jpn. J. Breed.*, 41, 665, 1991.
8. Causse, M.A. et al., Saturated molecular map of the rice genome based on an interspecific backcross population, *Genetics*, 138, 1251, 1994.
9. Kurata, N. et al., A 300-kilobase interval genetic map of rice including 883 expressed sequences, *Nat. Genet.*, 8, 365, 1994.
10. Harushima, Y. et al., A high-density rice genetic linkage map with 2275 markers using a single F_2 population, *Genetics*, 148, 479, 1998.
11. McCouch, S.R. et al., Development and mapping of 2240 new SSR markers for rice (*Oryza sativa* L.), *DNA Res.*, 9, 199, 2002.
12. Umehara, Y. et al., Construction and characterization of a rice YAC library for physical mapping, *Mol. Breed.*, 1, 79, 1995.
13. Kurata, N. et al., Physical mapping of the rice genome with YAC clones, *Plant Mol. Biol.*, 35, 101, 1997.
14. Saji, S. et al., A physical map with yeast artificial chromosomes (YAC) clones covering 63% of the 12 rice chromosomes, *Genome*, 44, 32, 2001.
15. Yamamoto, K. and Sasaki, T., Large-scale EST sequencing in rice, *Plant Mol. Biol.*, 35, 135, 1997.

16. Wu, J. et al., A comprehensive rice transcript map containing 6591 expressed sequence tag sites, *Plant Cell*, 14, 525, 2002.
17. Baba, T. et al., Construction and characterization of rice genomic libraries: PAC library of *japonica* variety, Nipponbare and BAC library of *indica* variety, Kasalath, *Bull. Natl. Inst. Agrobiol. Res.*, 14, 41, 2000.
18. Wu, J. et al., Physical maps and recombination frequency of six rice chromosomes, *Plant J.*, 36, 720, 2003.
19. Mao, L. et al., Rice transposable elements: a survey of 73,000 sequence-tagged-connectors, *Genome Res.*, 10, 982, 2000.
20. Yang, T.J. et al., Construction and utility of 10-kb libraries for efficient clone-gap closure for rice genome sequencing, *Theor. Appl. Genet.*, 107, 652, 2003.
21. International Rice Genome Sequencing Project, The map-based sequence of the rice genome, *Nature*, 436, 793, 2005.
22. Sasaki, T. et al., The genome sequence and structure of rice chromosome 1, *Nature*, 420, 312, 2002.
23. Rice Chromosome 3 Sequencing Consortium, Sequence, annotation, and analysis of synteny between rice chromosome 3 and diverged grass species, *Genome Res.*, 15, 1284, 2005.
24. Feng, Q. et al., Sequence and analysis of rice chromosome 4, *Nature*, 420, 316, 2002.
25. Rice Chromosome 10 Sequencing Consortium, In-depth view of structure, activity and evolution of rice chromosome 10, *Science*, 300, 1566, 2003.
26. *Arabidopsis* Genome Initiative, Analysis of the genome sequence of the flowering plant *Arabidopsis thaliana*, *Nature*, 408, 796, 2000.
27. Rice Chromosomes 11 and 12 Sequencing Consortia, The sequence of rice chromosomes 11 and 12, rich in disease resistance genes and recent gene duplications, *BMC Biol.*, 3, 20, 2005.
28. Kikuchi, S. et al., Collection, mapping, and annotation of over 28,000 cDNA clones from *japonica* rice, *Science*, 301, 376, 2003.
29. Dong, F. et al., Rice (*Oryza sativa*) centromeric regions consist of complex DNA, *Proc. Natl. Acad. Sci. USA*, 95, 8135, 1998.
30. Cheng, Z. et al., Functional rice centromeres are marked by a satellite repeat and a centromere-specific retrotransposon, *Plant Cell*, 14, 1691, 2002.
31. Zhang, Y. et al., Structural features of the rice chromosome 4 centromere, *Nucleic Acids Res.*, 32, 2023, 2004.
32. Wu, J. et al., Composition and structure of the centromeric region of rice chromosome 8, *Plant Cell*, 16, 967, 2004.
33. Yu, J. et al., The genomes of *Oryza sativa*: a history of duplications, *PloS Biol.*, 3, e28, 2005.
34. Katagiri, S. et al., End sequencing and chromosomal *in silico* mapping of BAC clones derived from an *indica* rice cultivar, Kasalath, *Breed. Sci.*, 54, 273, 2004.
35. Tyagi, A.K. et al., Structural and functional analysis of rice genome, *J. Genet.*, 83, 79, 2004.
36. Kurata, N. et al., Conservation of genome structure between rice and wheat, *Bio/Technology*, 12, 276, 1994.
37. Dunford, R.P. et al., Conservation of fine-scale DNA marker order in the genomes of rice and the Triticeae, *Nucleic Acids Res.*, 23, 2724, 1995.
38. Gale, M.D. and Devos, K.M., Comparative genetics in the grasses, *Proc. Natl. Acad. Sci. USA*, 95, 1971, 1998.
39. Mohan, M. et al., Genome mapping, molecular markers and marker-assisted selection in plants, *Mol. Breed.*, 3, 87, 1997.

40. Feltus, F.A. et al., An SNP resource for rice genetics and breeding based on subspecies *indica* and *japonica* genome alignments, *Genome Res.*, 14, 1812, 2004.
41. Nasu, S. et al., Search for and analysis of single nucleotide polymorphisms (SNPs) in rice (*Oryza sativa*, *Oryza rupifogon*) and establishment of SNP markers, *DNA Res.*, 9, 163, 2002.
42. Ching, A. et al., SNP frequency, haplotype structure and linkage disequilibrium in elite maize inbred lines, *BMC Genet.*, 3, 19, 2002.
43. Shen, Y.J. et al., Development of genome-wide DNA polymorphism database for map-based cloning of rice genes, *Plant Physiol.*, 135, 1198, 2004.
44. Song, W.Y. et al., A receptor kinase-like protein encoded by the rice disease resistance gene, *Xa21*, *Science*, 270, 1804, 1995.
45. Yoshimura, S. et al., Expression of *Xa1*, a bacterial blight-resistance gene in rice, is induced by bacterial inoculation, *Proc. Natl. Acad. Sci. USA*, 95, 1663, 1998.
46. Sun, X. et al., *Xa26*, a gene conferring resistance to *Xanthomonas oryzae* pv. *oryzae* in rice, encodes an LRR receptor kinase-like protein, *Plant J.*, 37, 517, 2004.
47. Wang, Z.X. et al., The *Pib* gene for rice blast resistance belongs to the nucleotide binding and leucine-rich repeat class of plant disease resistance genes, *Plant J.*, 19, 55, 1999.
48. Bryan, G.T. et al., A single amino acid difference distinguishes resistant and susceptible alleles of the rice blast resistance gene *Pi-ta*, *Plant Cell*, 12, 2033, 2000.
49. Jeon, J.S. et al., Genetic and physical mapping of *Pi5(t),* a locus associated with broad-spectrum resistance to rice blast, *Mol. Genet. Genomics*, 269, 280, 2003.
50. Ashikari, M. et al., A rice gibberellin-insensitive dwarf mutant gene *Dwarf 1* encodes the alpha-subunit of GTP-binding protein, *Proc. Natl. Acad. Sci. USA*, 96, 10284, 1999.
51. Miyoshi, K. et al., PLASTOCHRON1, a timekeeper of leaf initiation in rice, encodes cytochrome P450, *Proc. Natl. Acad. Sci. USA*, 101, 875, 2004.
52. Yamanouchi, U. et al., A rice spotted leaf gene, *Spl7*, encodes a heat stress transcription factor protein, *Proc. Natl. Acad. Sci. USA*, 99, 7530, 2002.
53. Kasama, T. and Toriyama, K., A pentatricopeptide repeat-containing gene that promotes the processing of aberrant atp6 RNA of cytoplasmic male-sterile rice, *FEBS Lett.*, 544, 99, 2003.
54. Komori, T. et al., Map-based cloning of a fertility restorer gene, *Rf-1*, in rice (*Oryza sativa* L.), *Plant J.*, 37, 315, 2004.
55. Li, X. et al., Control of tillering in rice, *Nature*, 422, 618, 2003.
56. Li, Y. et al., *BRITTLE CULM1*, which encodes a COBBA-like protein, affects the mechanical properties of rice plants, *Plant Cell*, 15, 2020, 2003.
57. Spielmeyer, W., Ellis, M., and Chandler, P., Semidwarf (*sd-1*), green revolution rice, contains a defective gibberellin 20-oxidase gene, *Proc. Natl. Acad. Sci. USA*, 99, 9043, 2002.
58. Han, F. et al., Sequence analysis of a rice BAC covering the syntenous barley *Rpg1* region, *Genome*, 42, 1071, 1999.
59. Brueggeman, R. et al., The barley stem rust-resistance gene *Rpg1* is a novel disease-resistance gene with homology to receptor kinases, *Proc. Natl. Acad. Sci. USA*, 99, 9328, 2002.
60. Yano, M. and Sasaki, T., Genetic and molecular dissection of quantitative traits in rice, *Plant Mol. Biol.*, 35, 145, 1997.
61. Yano, M. et al., Identification of quantitative trait loci controlling heading date in rice using a high-density linkage map, *Theor. Appl. Genet.*, 95, 1025, 1997.

62. Yano, M., Genetic and molecular dissection of naturally occurring variation, *Curr. Opin. Plant Biol.*, 4, 130, 2001.
63. Yano, M. et al., *Hd1*, a major photoperiod sensitivity quantitative trait locus in rice, is closely related to the *Arabidopsis* flowering time gene *CONSTANS*, *Plant Cell*, 12, 2473, 2000.
64. Kojima, S. et al., *Hd3a*, a rice ortholog of the *Arabidopsis FT* gene, promotes transition to flowering downstream of *Hd1* under short-day conditions, *Plant Cell Physiol.*, 43, 1096, 2002.
65. Monna, L. et al., Genetic dissection of a genomic region for a quantitative trait locus, *Hd3*, into two loci, *Hd3a* and *Hd3b*, controlling heading date in rice, *Theor. Appl. Genet.*, 104, 772, 2002.
66. Lin, H. et al., Fine mapping and characterization of quantitative trait loci *Hd4* and *Hd5* controlling heading date in rice, *Breed. Sci.*, 53, 51, 2003.
67. Takahashi, Y. et al., *Hd6*, a rice quantitative trait locus involved in photoperiod sensitivity, encodes the alpha subunit of protein kinase *CK2*, *Proc. Natl. Acad. Sci. USA*, 98, 7922, 2001.
68. Takeuchi, Y. et al., Fine linkage mapping enables dissection of closely linked quantitative trait loci for seed dormancy and heading in rice, *Theor. Appl. Genet.*, 107, 1174, 2003.
69. Lin, H. et al., Identification and characterization of a quantitative trait locus, *Hd9*, controlling heading date in rice, *Breed. Sci.*, 52, 35, 2002.
70. Gu, X.Y., Kianian, S.F., and Foley, M.E., Isolation of three dormancy QTLs as Mendelian factors in rice, *Heredity* advance online publication, September 28, 2005, doi:10.1038/sj.hdy.6800757. 2005.
71. Lin, H.X. et al., QTLs for Na+ and K+ uptake of the shoots and roots controlling rice salt tolerance, *Theor. Appl. Genet.*, 108, 253, 2004.
72. Ren, Z.H. et al., A rice quantitative trait locus for salt tolerance encodes a sodium transporter, *Nat. Genet.*, 37, 1141, 2005.
73. Ueda, T. et al., Delimitation of the chromosomal region for a quantitative trait locus, *qUVR-10*, conferring resistance to ultraviolet-B radiation in rice (*Oryza sativa* L.), *Theor. Appl. Genet.*, 108, 385, 2004.
74. Fujino, K. et al., Mapping of quantitative trait loci controlling low-temperature germinability in rice (*Oryza sativa* L.), *Theor. Appl. Genet.*, 108, 794, 2004.
75. Takehisa, H. et al., Identification of quantitative trait loci for plant growth of rice in paddy field flooded with salt water, *Field Crops Res.*, 89, 85, 2004.
76. Obara, M. et al., Identification and characterization of a QTL on chromosome 2 for cytosolic glutamine synthetase content and panicle number in rice, *Theor. Appl. Genet.*, 110, 1, 2004.
77. Ahn, S. and Tanksley, S.D., Comparative linkage maps of the rice and maize genomes, *Proc. Natl. Acad. Sci. USA*, 90, 7980, 1993.
78. Moore, G. et al., Cereal genome evolution. Grasses line up and form a circle, *Curr. Biol.*, 5, 737, 1995.
79. Keller, B. and Feuillet, C., Collinearity and gene density in grass genomes, *Trends Plant Sci.*, 5, 246, 2000.
80. Bennetzen, J.L., Comparative sequence analysis of plant nuclear genomes: microcollinearity and its many exceptions, *Plant Cell*, 12, 1021, 2000.
81. Kilian, A. et al., Towards map-based cloning of the barley stem rust resistance genes *Rpg1* and *rpg4* using rice as an intergenomic cloning vehicle, *Plant Mol. Biol.*, 35, 187, 1997.

82. Li, W.L. and Gill, B.S., The collinearity of the *Sh2/A1* orthologous region in rice, sorghum and maize is interrupted and accompanied by genome expansion in the Triticeae, *Genetics*, 160, 1153, 2002.
83. Tarchini, R. et al., The complete sequence of 340 kb of DNA around the rice *Adh1–Adh2* region reveals interrupted collinearity with maize chromosome 4, *Plant Cell*, 12, 381, 2000.
84. Dubcovsky, J. et al., Comparative sequence analysis of collinear barley and rice bacterial artificial chromosomes, *Plant Physiol.*, 125, 1342, 2001.
85. Song, R., Llaca, V., and Messing, J., Mosaic organization of orthologous sequences in grass genomes, *Genome Res.*, 12, 1549, 2002.
86. Hirochika, H. et al., Retrotransposons of rice as a tool for the functional analysis of genes, in *Rice Genet. IV. Proc. 4th Int. Rice Genet. Symp.*, 279, 2001.
87. Sato, Y. et al., Loss-of-function mutations in the rice homeobox gene *OSH15* affect the architecture of internodes resulting in dwarf plants, *EMBO J.*, 18, 992, 1999.
88. Agrawal, G.K. et al., Screening of the rice viviparous mutants generated by endogenous retrotransposon *Tos17* insertion—tagging of a zeaxanthin epoxidase gene and a novel *OsTATC* gene, *Plant Physiol.*, 125, 1248, 2001.
89. Takano, M. et al., Isolation and characterization of rice phytochrome A mutants, *Plant Cell*, 13, 512, 2001.
90. Sallaud, C. et al., Highly efficient production and characterization of T-DNA plants for rice (*Oryza sativa* L.) functional genomics, *Theor. Appl. Genet.*, 106, 1396, 2003.
91. Lee, S. et al., Systematic reverse genetic screening of T-DNA tagged genes in rice for functional genomic analyses: MADS-box genes as a test case, *Plant Cell Physiol.*, 44, 1403, 2003.
92. Chin, H.G. et al., Molecular analysis of rice plants harboring an Ac/Ds transposable element-mediated gene trapping system, *Plant J.*, 19, 615, 1999.
93. Greco, R. et al., Transposon insertional mutagenesis in rice, *Plant Physiol.*, 125, 1175, 2001.
94. Hirochika, H. et al., Rice mutant resources for gene discovery, *Plant Mol. Biol.*, 54, 325, 2004.
95. Phillips R.L., Leung, H., and Cantrell, R.P., An international genetic platform for the assessment of gene function in rice, in *New Directions for a Diverse Planet: Proc. 4th Int. Crop Sci. Congr.*, Brisbane, Australia. CDROM publication, 2004.
96. Antonio, B.A. et al., The Rice Genome Resource Center as an outlet for distribution of biological materials from the rice genome project, *Rice Genet. Newslett.*, 20, 10, 2003.

5 Model Legume *Medicago truncatula*

Henk Franssen and René Geurts

CONTENTS

5.1 UNIQUE FEATURES OF LEGUMES

Legumes belong to the taxonomic family Fabaceae, containing over 18,000 species divided into the three subfamilies Mimosoideae, Caesalpinoideae, and Papilionoideae. Legume species have been cultivated for millennia all over the world because of the nutritional value of their seeds. Nowadays, legumes contribute about 27% of the world's primary crop production [1]; the major single contributing species is soybean (*Glycine max*), which is used for multiple applications in the food and feed industries. Others, such as cowpea (*Vigna unguiculata*), common bean (*Phaseolus vulgaris*), and chickpea (*Cicer arietinum)* contribute significantly to the diets of large numbers of people in Asia, Africa, and South America.

The high-quality nutrition of legumes is achieved by the presence of a wealth of secondary metabolites and in the capacity of legumes to live in symbiosis with the nitrogen-fixing bacterium *Rhizobium* [2,3]. This symbiosis only occurs under nitrogen limiting conditions and results in formation of complete new organs: the root nodules. Nodules host the *Rhizobium* bacteria, which differentiate in the nodules into symbiotic bacteroids and are the site of catalysis of dinitrogen into ammonia by nitrogenase. As an energy source to achieve N fixation, the bacteria obtain dicarboxylic acids from the host plant. By a complex amino-acid cycle, the reduced nitrogen is provided to the plant [4], where it is accumulated into proteins.

The importance of legumes as a protein source for feed and food and their independence of an external nitrogen supply thanks to the symbiosis with *Rhizobium* have encouraged a genomics-led molecular characterization to facilitate applied crop research. For this purpose, the development of model species has been imperative; *Medicago truncatula* (*Medicago*) and *Lotus japonicus* (*Lotus*) have been selected. In this chapter, we review ongoing *Medicago* research and refer to *Lotus* when it is relevant. We discuss how current progress can already help the characterization of loci in crop species and contribute to the identification of the genes required for critical steps in the establishment of the legume–*Rhizobium* symbiosis.

5.2 *MEDICAGO TRUNCATULA*

Medicago originates from the Mediterranean basin and many accessions have been collected from this region [5]. Phylogenetically, it belongs to the galegoid clade and is closely related to alfalfa (the major world forage legume), lentil, pea, faba bean, and clover (Figure 5.1). Unlike these species, *Medicago* has all characteristics of a plant model species: a simple diploid genome (Table 5.1), self-fertility, a short generation time (3 to 4 months from seed to seed), and good genetic transformability.

The size of the *Medicago* genome has been estimated to be about 500 Mbp, divided across eight chromosomes and equivalent to four to five times the size of the *Arabidopsis thaliana* (*Arabidopsis*) genome and similar to that of rice (*Oryza sativa*). The *Medicago* genome is more simply organized than that of rice, as visualized during the pachytene stage of meiosis (Figure 5.2). At this stage, chromosomes are fully

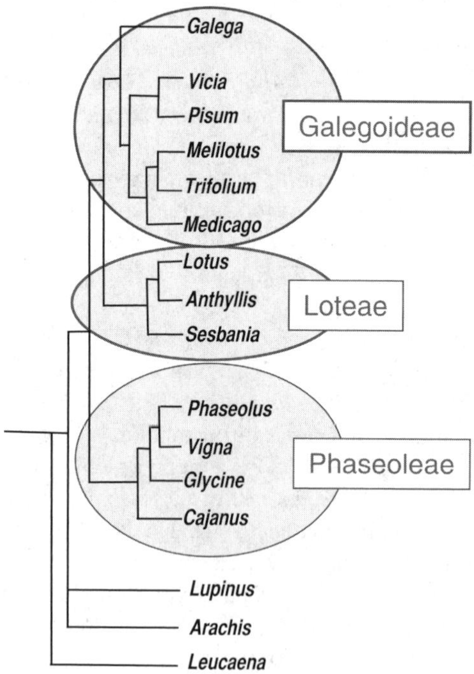

FIGURE 5.1 Phylogentic relationship of the legume subfamily Papilionoideae. Crop species belong mainly to the Galegoideae or Phaseoleae clade.

TABLE 5.1
Genome Size of Crop Legumes

Common name	Scientific name	Phylogenetic clade	Genome size (Mbp/1 C)[a]
Peanut	*Arachis hypogaea*	Aeschynomeneae	2,813
Cowpea	*Vigna unguiculata*	Phaseoloid	588
Mung bean	*Vigna radiata*	Phaseoloid	515
Common bean	*Phaseolus vulgaris*	Phaseoloid	588
Soybean	*Glycine max*	Phaseoloid	1,103
Pigeon pea	*Cajanus cajan*	Phaseoloid	858
Chickpea	*Cicer arietinum*	Galegoid	931
Lentil	*Lens culinaris*	Galegoid	4,116
Pea	*Pisum sativum*	Galegoid	4,778
Faba bean	*Vicia faba*	Galegoid	26,852
Alfalfa	*Medicago sativa*	Galegoid	1,715
Medicago	*Medicago truncatula*	Galegoid	466
Lotus	*Lotus japonicus*	Loteae	466

[a] Information concerning genome size is obtained from the plant DNA C-values database (release 3.0) www.rbgkew.org.uk/cval/homepage.html.

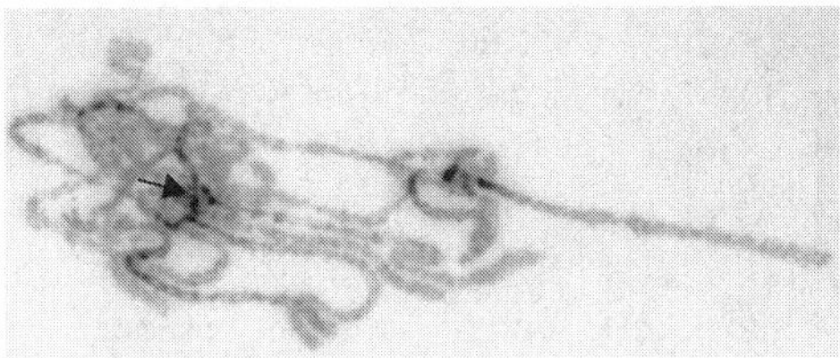

FIGURE 5.2 *Medicago* chromosomes in the pachytene stage of meiosis I. Chromosomes are stained with DAPI, thereby visualizing differences in DNA condensation. The chromosome arms are euchromatic, whereas the pericentromeric regions contain large heterochromatic blocks. Note that chromosome 6 (marked by arrow) has more heterochromatic knobs along its chromosome arms.

paired and are about 20 times longer than at mitotic metaphase. Seven of the eight *Medicago* pachytene chromosomes have large blocks of heterochromatin around the centromere (the pericentromeric region), whereas the distal parts of the arms are euchromatic (Figure 5.2) [6].

Only *Medicago* chromosome 6 displays a more complex organization, with several heterochromatic chromomeres along both chromosome arms. (In general, pachytene chromosomes display a clear differentiation between euchromatin and heterochromatin, which correspond, respectively, to gene-rich and gene-poor regions in the genome.) The overall pattern of heterochromatic euchromatic blocks is reminiscent of the conspicuous heterochromatic blocks in *Arabidopsis* pachytene chromosomes [7]. In contrast, rice exhibits numerous smaller heterochromatic knobs distributed along all its chromosome arms, which is the result of interspersion of gene-rich and gene-poor regions along the chromosome arms [8].

Medicago has a relatively low gene density, based on the number of gene-based molecular markers that can be generated within a particular linkage group [9,10]. To estimate the size of the gene-rich portion of the *Medicago* genome, two methods have been used:

1. By determining the condensation degree at different positions of euchromatin, the amount of DNA in a euchromatic conformation can be extrapolated [6,11,12]. This has resulted in a euchromatin size estimation of 100 to 200 Mb.
2. In the second approach, the difference in genome size between the two accessions Jemalong A17 and R108-1 has been used. To this end the length of pachytene chromosomes and distribution of the hetero/euchromatin were determined. The two genotypes do not differ in the length of euchromatin present, but do in the size of the heterochromatic pericentromeric blocks. Thus, the difference in genome size between both accessions is contributed by the heterochromatic fraction. The size of the heterochromatic blocks of Jemalong A17 has been estimated to be 320 Mb, leaving the remaining 180 to 230 Mb as euchromatin [13].

Although both calculations are rather indirect, they support the hypothesis that the major part of the *Medicago* genome is composed of repeats within heterochromatic regions. The current *Medicago* sequencing projects are focusing specifically on the euchromatic part of the genome. The sequence of this region will be determined in a BAC-by-BAC approach, enabling a full integration of the physical, genetic, and cytogenetic maps [14].

Functional genomic initiatives have resulted in the generation of >226,000 expressed sequence tags (ESTs) originating from 35 different libraries. These ESTs have been assembled into 18,600 tentative consensus sequences (TCs) of more than 1 EST and 18,200 singleton sequences (Release 8.0 of the TIGR *Medicago* Gene Index, January 2005) [15–17]. Assuming that *Medicago* has a similar number of genes as *Arabidopsis* does, the vast majority of these are represented by at least one EST (see also the following links: http://www.medicago.org/MtDB2, http://www.tigr.org/tigr-scripts/tgi/T_reports.cgi?species=medicago, and http://medicago.toulouse.inra.fr/Mt/EST). This data set has been used to construct cDNA-based microarrays [18], as well as oligonucleotide-based chips [19]. In addition, the EST collection has been mined to identify legume-specific genes.

To this end, the *Medicago* ESTs plus similar sets from soybean and *Lotus* were compared to sequence data generated for non-legume plant species, resulting in the identification of 2525 legume-specific EST contigs [20]. Among these are genes specifically induced during *Rhizobium* symbiosis (so-called nodulin genes) [21] and genes encoding for legume-specific seed storage proteins. However, for the vast majority of genes, the function of the encoded proteins remains largely unknown and probably will need to be elucidated by other methods.

To unravel gene function, forward and reverse genetic tools have been applied to *Medicago*. Reverse genetic approaches became possible with the development of stable transgenic *Medicago* lines, although transformation is significantly less efficient and more time consuming than, for example, for *Arabidopsis*. Of several protocols that have been developed, protocols based on regeneration of transgenic callus have been shown to be the most effective [22–26]. Lines with increased regeneration efficiency have been selected. However, crosses between the best line in this respect—*Medicago* R108-1—and lines most widely used for genetic studies (selections from the cultivar Jemalong, e.g., A17 and J5) result in severe segregation distortion, most likely due to genomic incompatibility [13]. These disadvantages of *Medicago* R108-1 mean that it can be used only in applications that do not require forward genetics.

Genome wide T-DNA and transposon tagging approaches have been initiated for *Medicago*. To effect the latter, the tobacco (*Nicotiana tobaccum*) retrotransposon element Tnt1 was introduced into *Medicago*. The Tnt1 transposition is only activated during tissue culture, during which its copy number remains relatively low. Insertions seem to occur preferentially in the gene-rich portion of the genome, generating gene disruptions. Currently, a collection of 8000 independent lines representing over 150,000 Tnt1 insertions is being created by an international consortium.

In addition to stable transformation, efficient protocols have been developed based on *Agrobacterium rhizogenes*-mediated root transformation [27,28]. In contrast to stable lines, compound plants are generated in which a nontransformed shoot carries transgenic roots. The advantage of this system is mainly through a reduction in the time required for acquiring transgenic material; therefore, it is an attractive method to study gene function in roots. Compound plants can be obtained within 4 to 6 weeks and can be nodulated by *Rhizobium* as well as be infected by other symbionts (e.g., mycorrhizal fungi) or root pathogens. Roots generated with this system can be propagated in culture independently of the shoot due to the presence of the *root inducing locus* (*rol*) genes of *A. rhizogenes*. The drawbacks of an *A. rhizogenes*-mediated root system are:

1. The introduced *rol* genes interfere with plant hormone balance (particularly in overproduction of cytokinin), making this system unsuitable for studies of plant growth regulators.
2. The transformation remains transient because no transgenic offspring are generated.
3. Because the roots are primary transformed tissue, significant variation is observed in expression levels of the introduced transgenes.

In addition, roots can be chimerical because root formation occurs from a group rather than from a single cell. Careful selection is therefore required, and methods based on antibiotic resistance and/or fluorescent markers have been developed [27,28].

RNA interference (RNAi) has proven to be a powerful tool to unravel gene function. RNAi can be triggered by generating transgenic lines that express RNAs capable of forming a double-stranded hairpin [29]. For *Medicago* (and legumes in general), the generation of transgenic lines is time consuming, so RNAi has been applied in *A. rhizogenes*-mediated root transformation and shown to be functional [28,30,31]. However, systemic spreading of the silencing signal is limited in compound plants. It is transmitted very inefficiently from the transgenic root system to the nontransgenic shoot, and transport from transgenic roots to nontransgenic roots does not occur. Furthermore, because the roots are primary transformed tissue, variation in silencing efficiency tends to occur.

In addition to reverse genetics methods that require transformation, TILLING ("*t*argeted *i*nduced *l*ocal *l*esions *in* *g*enomes") has been applied in *Medicago* and in *Lotus* [32; D. Cook, personal communication]. This technique combines ethyl methanesulfonate (EMS)-induced mutagenesis with the ability to detect base pair changes by heteroduplex analysis by PAGE of Cel1 digests [33,34] and generates a range of mutant alleles. The identification of mutant alleles obtained by TILLING is expected in the near future.

5.3 THE *MEDICAGO* GENOME AS REFERENCE FOR LEGUME CROP SPECIES

To implement knowledge generated from model species for legume crop improvement, comparative genetic maps between model species and economical important crop species need to be constructed. To develop cross-species genetic markers, an intron-targeted marker strategy has been shown to represent a powerful approach [35]. For this purpose, PCR primer pairs are designed to anneal in exon regions conserved between *Medicago* and *Lotus*, soybean, or *Arabidopsis*, designed to amplify across introns. Because introns are significantly more frequently polymorphic than coding regions, these markers are often informative and have been used to integrate the genetic maps of various legume species [10,35].

Comparative genetic maps have been created between *Medicago* and the galegoid crop species alfalfa (*Medicago sativa*) and pea, between model species *Medicago* and *Lotus*, and between *Medicago* and the phaseoloid crop species soybean, cowpea, and common bean [10,35]. A comparison of *Medicago* and alfalfa based on 68 sequence-characterized genetic markers indicates that the two *Medicago* genomes are highly similar [10]. Pea is more differentiated from *Medicago* than is alfalfa. It has a significantly larger genome and contains one chromosome less than both *Medicago* species (seven vs. eight). Despite these differences, a high degree of synteny exists between pea and the *Medicago* species, and only two major chromosomal translocations have been identified.

Medicago linkage group two is distributed over linkage groups III and VI of pea (Figure 5.3) [35,36]. Similar comparisons have been conducted between *Medicago* and *Lotus*. The *Lotus* genome is about the same size as that of *Medicago* and is divided over six chromosomes [14]. Both species show a significant level of macrosynteny, but several chromosome arm translocations have occurred during evolution. This synteny is also reflected on the microsynteny scale because the order and orientation of genes have been shown to be conserved significantly [35]. Loteae is a sister group of the Galegoideae, and therefore *Lotus* is more closely related to *Medicago* than to the phaseoloid species.

Alignment of the genetic maps of phaseoloid species and *Medicago* shows significant distortions due to translocations, duplications, and loss of synteny. However, the gene repertoire in orthologous regions in *Medicago* and soybean still displays a degree of conserved gene order [35,37]. Based on these studies, it has

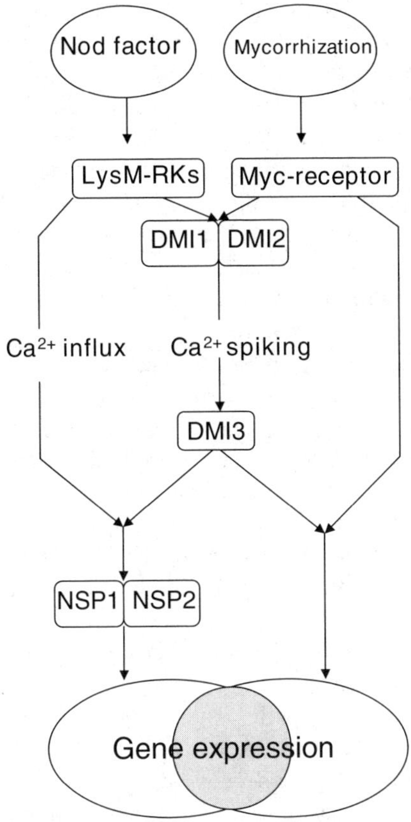

FIGURE 5.3 A consensus map of *Medicago*, alfalfa (*Medicago sativa*), and pea (*Pisum sativum*). The three species are highly syntenic. Note that *Medicago* (and alfalfa) linkage group two is represented by pea linkage groups III and VI. (The figure is based on Choi, H.K. et al., *Genetics,* 166, 1463, 2004; Choi, H.K. et al., *Proc. Natl. Acad. Sci. USA,* 101, 15289, 2004; and Kalo, P. et al., *Mol. Genet. Genomics,* 272, 235, 2004.)

been concluded that the *Medicago* genome can be effectively used as a reference for the galegoid species. For the phaseoloid species, *Medicago* and *Lotus* can be used, but both show a reduced level of macrosynteny because of translocations.

5.3.1 *Rhizobium*–Legume Symbiosis

The interaction between *Rhizobium* and legumes has been the subject of many studies over the last century. These have culminated in a detailed description of the steps involved, including infection and primordium formation; a good understanding of the physiology of nitrogen fixation, including the requirement for low oxygen tension; the mechanism of nitrogen assimilation; and the identification of plant genes that are regulated during the interaction [38,39].

When rhizobia have colonized the root surface of their legume host, they induce morphological changes in root hairs, a phenomenon referred to as root hair deformation.

In some root hairs, rhizobia induce curling as a result of which the bacteria become entrapped in the pocket of the curl. At this point, the plant cell wall is locally modified, the plasma membrane invaginates, and new plant material is deposited. In this way, a tube-like structure, the infection thread, is formed containing the bacteria. The infection thread grows toward the base of the root hair cell and subsequently to the nodule primordium that has simultaneously been formed in the cortex of the root. There, the infection thread ramifies and bacteria are released into primordial cells. The released bacteria—now called bacteroids—remain surrounded by a membrane of plant origin in a similar fashion as in mitochondria and chloroplasts. Subsequently, bacteroids differentiate and begin to fix nitrogen. Likewise, the nodule primordium differentiates into a mature nodule.

Strikingly, genetic analyses have shown that part of the *Rhizobium* symbiosis has evolved from the much older mycorrhizal symbiosis [40–42]. In contrast to *Rhizobium* symbiosis, which is restricted mainly to legume species, the majority of higher plants have the ability to interact with arbuscular endomycorrhiza, producing a symbiotic association between the plant root and fungi belonging to the order of Glomales. These fungi grow toward the inner cortical cells of the root, where they differentiate into highly branched structures, the arbuscules. Because the fungus retains hyphae outside the plant it provides the host better access to nutrients such as phosphate.

Genetic approaches in *Rhizobium* have been very successful in identifying the bacterial genes crucial to establishing a proper symbiosis. Many of these genes encode proteins required for production and secretion of a bacterial signaling molecule, the Nod factors. Nod factors for which the structure has been elucidated all share a β-1,4-linked *N*-acyl-D-glucosamine backbone of three to six subunits. The nonreducing end of this glucosamine backbone is substituted with a fatty acid of variable structure. Furthermore, at both ends of the backbone, substitutions may be present that include acetyl, sulfuryl, fucosyl, mannosyl, or arabinosyl groups (for review, see, for example, Spaink [43]).

Purified Nod factors applied in the nano- to picomolar range are able to induce developmental processes needed for root nodule formation. These responses are provoked at spatially separated sites—specifically, the epidermis, cortical cells, and pericycle. In some species (e.g., alfalfa), Nod factors can even trigger the formation of a complete nodule (lacking bacteria). Just as genetic approaches have led to identification of genes crucial for signaling organogenesis in *Arabidopsis*, mutants impaired in nodulation have proven to be instrumental in identification of plant genes essential for *Rhizobium*–legume symbiosis and to a better understanding of the underlying mechanisms.

Although several mutants that fail to establish proper symbiosis have been identified for some years in pea and soybean, the identity of the mutated genes has only become possible following establishment of mutagenesis programs in *Medicago* and *Lotus*. Emerging mutants can be grouped roughly into three classes: class I, non-nodulators (nod-); class II, hypernodulators (nod+++); and class III, nodulators with impaired fixation (nod+, fix-) (Table 5.2). Here, we will focus on the current knowledge obtained after identification of the genes corresponding to class I and II mutants in *Medicago* and *Lotus*.

TABLE 5.2
***Medicago* Symbiotic Loci Identified by Forward Genetics**

Symbol	Gene name	Protein	Ref.
	Class I		
BIT	Branching infection threads	Not cloned	72
DMI1	Does not make infections 1	Putative cation channel	54
DMI2	Does not make infections 2	LRR-RK	53
DMI3	Does not make infections 3	Ca^{2+} calmodulin kinase	55
HCL	Hair curling	Not cloned	73
LIN	Lumpy infections	Not cloned	74
NIP	Numerous infections with polyphenolics	Not cloned	75
NFP	Nod factor perception	Likely ortholog of *Lotus* NFR5 encoding a LysM-RK	44
NSP1	Nodulation signaling pathway 1	Transcription factor	82
NSP2	Nodulation signaling pathway 2	Transcription factor	83
PDL	Poodle	Not cloned	76, 77
RIT	Root hairs infection threads trichomes	Not cloned	72
SYM1	Symbiosis 1	Not cloned	78
SYM16	Symbiosis 16	Not cloned	79
	Class II		
SKL	Sickle	Not cloned	80
SUNN	Supernumeric nodules	Likely homolog of *Lotus* HAR encoding an LRR-RK	12
	Class III		
DNF1	Defective in nitrogen fixation 1	Not cloned	72
DNF2	Defective in nitrogen fixation 2	Not cloned	72
DNF3	Defective in nitrogen fixation 3	Not cloned	72
DNF4	Defective in nitrogen fixation 4	Not cloned	72
DNF5	Defective in nitrogen fixation 5	Not cloned	72
DNF6	Defective in nitrogen fixation 6	Not cloned	72
DNF7	Defective in nitrogen fixation 7	Not cloned	72
SYM6	Symbiosis 6	Not cloned	81
SYM17	Symbiosis 17	Not cloned	79
SYM18	Symbiosis 18	Not cloned	79
SYM19	Symbiosis 19	Not cloned	79
SYM20	Symbiosis 20	Not cloned	79
SYM21	Symbiosis 21	Not cloned	79

Notes: Mutants are classified into three groups: non-nodulators (class I), supernodulators (class II), and nodulators with impaired fixation (class III). Detailed information concerning phenotypes and references can be obtained from the Nodulation Mutant Database (NodMutDB) http://nodmutdb.vbi.vt.edu. Allelism studies between *dnf* and *sym* class III mutants (fix-) have not been reported.

5.3.1.1 Non-Nodulators Enclose Nod Factor Signaling Genes

Various genetic approaches have been used to unravel the Nod factor-signaling cascade. Selection has been made for *Lotus* and *Medicago* mutants that are impaired

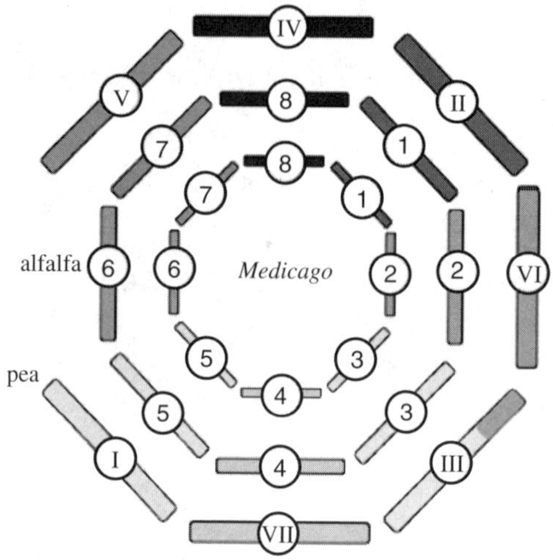

FIGURE 5.4 **(Color Figure 5.4 follows p. 144.)** Nod factor signaling cascade as identified by forward genetics. Nod factors are perceived by LysM receptor kinases. These activate at least two downstream signaling pathways: one depending on the DMI proteins and a DMI independent pathway for which no specific genes have yet been identified. Both pathways can be discriminated based on distinct Ca^{2+} signals, external Ca^{2+} influx, and perinuclear Ca^{2+} spiking. The DMI pathway is shared mycorrhizal-secreted signal perceived by a hypothetical plant Myc receptor. For mycorrhizal-based signaling, a DMI independent pathway is predicted.

in the first visible Nod factor-provoked responses (Figure 5.4). In a second approach, naturally occurring variation within legumes has been exploited and characterized at the molecular level using *Medicago* in a synteny-based approach. Both methods will be discussed.

Among the non-nodulating mutants involving six loci in *Medicago*, only a few are disturbed in most of the Nod factor-induced responses. Cloning these genes in *Medicago* and *Lotus* has shown that a similar set of genes was mutated, suggesting that mutation screens for impairment in Nod factor signaling are close to saturation.

A knock-out mutation in a Nod factor receptor is expected to be impaired in all Nod factor-induced responses. In *Medicago*, only one mutant displaying such a phenotype has been identified—namely, *nfp*; in *Lotus*, two such loci have been found: *NFR1* and *NFR5* [44–46]. The other mutants disturbed in Nod factor signaling identified in either species show at least some Nod factor-induced responses in root hair morphology. In *Medicago* there are three *dmi* and two *nsp* mutants (Table 5.2). Strikingly, all these genes (including *NFP*) are essential for Nod factor-induced changes in gene expression, as demonstrated by microarray analysis [47].

NFR1 and *NFR5* of *Lotus* have been cloned and encode distinct LysM domain-containing receptor kinases (LysM-RK) that, based on their sequence, are localized in the plasma membrane [45,46]. The putative extracellular regions of both proteins contain LysM domains, which previously have been found in proteins binding peptidoglycans [48]. Thus, these LysM-RKs are good candidates as Nod factor binders because they contain an *N*-acetyl glucosamine backbone. Because the extracellular domains of NFR1 and NFR5 are markedly different from one another, it seems unlikely that the two receptors function independently and recognize the same Nod factor structure.

More probably, a heterodimer involving both receptors is needed for Nod factor perception, a model consistent with the loss of Nod factor responses in both mutants [45,46]. This is further supported by the atypical serine/threonine kinase in NFR5, which lacks an activation loop that generally regulates kinase activity. Therefore, activation of NFR5-type kinases probably occurs upon phosphorylation by an interacting kinase. *Medicago* NFP is the most likely ortholog of *Lotus* LjNFR5.

The *DMI* and *NSP* genes are positioned downstream of LysM-RK(s) because growth responses in root hairs can be triggered upon Nod factor application (Figure 5.4) [49,50]. In this respect, the *dmi* mutant root hairs mainly show root hair swelling and only very limited tip growth upon Nod factor perception; *nsp1* and *nsp2* mutants show root hair responses more similar to that of the wild-type. The functioning of the three DMI proteins can be dissected based on Nod factor-induced oscillation of Ca^{2+} concentration. This Ca^{2+} spiking occurs in the perinuclear region of epidermal cells and is induced within a few minutes [51]. Of the *dmi* mutants, only *dmi3* shows this response (Figure 5.4) [52].

Although the function of this intracellular Ca^{2+} signaling is not yet well understood, pharmacological studies have shown that that it is essential for Nod factor-induced gene expression. DMI1 has a low global similarity to ligand-gated cation channels, whereas DMI2 is a receptor kinase in which the putative extracellular region contains three LRR domains [53,54]. *DMI3* encodes a Ca^{2+} calmodulin-dependent protein kinase (CCaMK) and is assumed to respond to this Ca^{2+} signal [55,56]. Genes orthologous to *Medicago DMI1* and *DMI2* have been identified in *Lotus* [57,58]. Downstream of the DMI-module, NSP1 and NSP2 are functional (Figure 5.4) [49,50].

Because all these genes are essential for Nod factor-induced gene expression, it is probable that either of the *NSP* genes encodes a transcription factor that is activated upon Nod factor signaling. Indeed, cloning of NSP1 and NSP2 shows that these genes encode transcription factors belonging to the GRAS family of plant-specific transcription factors [82,83].

The three DMI genes are essential not only for *Rhizobium*-induced nodulation, but also for mycorrhizal symbiosis, whereas the putative Nod factor receptors are not [49,50]. Because the mycorrhizal and *Rhizobium* symbioses only in part trigger expression of a common set of genes [59,60], signaling cascades in addition to the DMI module must exist; these (together with the DMI genes) will be essential to trigger mycorrhizal or *Rhizobium* Nod factor specific transcriptional changes. In the case of Nod factor signaling, the existence of a second such pathway is supported by the different Ca^{2+} response. Apart from the intracellular Ca^{2+} spiking that occurs upon Nod factor perception in a DMI1- and DMI2-dependent fashion, an influx of extracellular Ca^{2+} occurs in a DMI-independent manner [61].

Ca^{2+} influx is one of the first responses in the root epidermis upon Nod factor signaling and is essential for at least some induced transcriptional changes [62]. Similarly to the Ca^{2+} influx following Nod factor signaling, it is possible that mycorrhizal fungi also trigger an alternative DMI-independent signaling cascade that, together with the DMI module, is required for mycorrhizal specific transcriptional changes (Figure 5.4).

As mentioned earlier, a second strategy based on naturally occurring variation was used to clone a putative Nod factor receptor specifically involved in bacterial infection. In pea accessions originating from the Middle East, the *SYM2* locus was identified as specifically involved in controlling infection thread formation in relation to Nod factor structure. In the pea accession Afghanistan, this locus inhibits infection by *Rhizobium leguminosarum* bv. *viciae* strains that are unable to add an additional acetate at the reducing end of the sugar backbone of the Nod factor [63,64]. Thus, the activity of SYM2 depends on the structure of Nod factors secreted by the infecting rhizobia and is part of the mechanism that controls bacterial entry.

Likewise, structure–function relationship studies derived from bacterial genetics have shown that, in *Medicago*, bacterial infection is more dependent on Nod factor structure than are other responses (e.g., nodule primordium formation) [65]. To clone a Nod factor receptor essential for *Rhizobium* infection, a synteny-based approach was used to characterize the pea *SYM2* orthologous region in *Medicago* [31,66]. This region contains several genes encoding LysM-RKs that have been named *LYK*. Knockdown of *LYK3* and *LYK4* by means of *A. rhizogenes*-mediated RNAi has shown that both genes are essential for *Rhizobium* infection in a Nod factor structure-dependent manner [31]. *Medicago LYK3* and *LYK4* are homologous to *Lotus* NFR1, although their loss-of-function phenotypes are strikingly different.

In addition to *LYK3* and *LYK4* in *Medicago*, the gene *NODULE INCEPTION* (*NIN*) is also essential for infection initiation. *NIN* encodes a protein with homology to transcription factors and was originally cloned in *Lotus* by insertion of an AC transposable element. The insertion mutant showed excessive root hair deformation and curling, but no infection [67]. In contrast to the genes described previously, *NIN* is induced upon Nod factor perception and therefore cannot be primarily involved in the Nod factor signaling pathway.

5.3.1.2 Supernodulators

In the 1980s, soybean and pea mutants were identified that formed nodules independent of the nitrogen status of the soil. Apart from this characteristic, these mutant plants formed more nodules than did wild-type plants grown in the absence of nitrogen. As a result, these mutants have been termed supernodulators. The discovery of supernodulators supports the notions that nodule formation is suppressed by the presence of nitrogen and that legumes have an autoregulatory mechanism that controls the number of nodules formed. Strikingly, supernodulation does not lead to any increase in plant biomass, indicating that nodule formation and nitrogen fixation of the hosted *Rhizobium* are established at the expense of the plant.

The recent cloning of the orthologous genes *HAR1* in *Lotus*, *SYM29* in pea, and *NARK* in soybean has allowed for the characterization of an important key regulator of the autoregulatory mechanism [68–70]. The signature of the protein encoded predicts that it functions as a receptor kinase because of the presence of extracellular leucine-rich repeats (LRRs) and an intracellular serine/threonine kinase domain. Based on its homology to *Arabidopsis* CLAVATA1 and the observation that the number of lateral roots in the *Lotus har1* mutant is affected after inoculation with *Rhizobium*, it has been suggested that this protein has a role in mediating control over root organ formation, including lateral roots as well as nodules.

5.4 PERSPECTIVE

The capacity to form root nodules in which bacteria convert nitrogen into ammonia allows the seed of a legume crop to accumulate high protein content. This is a unique feature among plants. To keep pace with the growing world demand for protein-rich food, *Rhizobium* symbiosis must be exploited to its limits. This need has led to development of the legume model species *Lotus* and *Medicago*, of which the latter is presented in this review in more detail. In addition, nonsymbiotic traits, like tolerance to biotic and abiotic stress, seed quality, plant architecture, and flowering behavior, are important for legume crop performance. Therefore, these traits are important targets for legume breeders.

Series of gene-based genetic markers that can be used across legume species have been developed and used to integrate genetic linkage maps. The possibility to exploit the synteny between model and crop legumes will certainly be instrumental in future legume breeding. This has been recognized; several well-funded programs covering many aspects of legume biology have been implemented worldwide. This will ultimately lead to full integration of the *Medicago* and *Lotus* genome sequence with high-density genetic maps of crop legumes.

Development of model systems has speeded up identification of genes encoding key players in *Rhizobium* symbiosis. Initially, these studies have been focused on genes involved in Nod factor signaling because perception of this bacterial signal molecule by the plant forms the main trigger for root nodule development. Strikingly, the number of genes essential for Nod factor signaling that can be identified genetically is low and is conserved among all legume species studied so far. Furthermore, mutations in these genes mainly affect symbiosis, suggesting that they do not play important roles in other plant processes. It can be expected that in the near future the link between the Nod factor signaling network and common cellular processes will be elucidated and thereby will make available knowledge of how plants rewire processes for organ formation.

Homologs of genes affected in non-nodulating mutants such as Nod factor receptor LysM-RKs or *DMI* genes are present also in non-legumes. This indicates that the processes needed for nodule formation could be, in part, already present in non-legume species and suggests that *Rhizobium* has recruited genes involved in general plant development for nodule formation. The observation that *dmi* mutants are also impaired in the interaction with arbuscular mycorrhiza has led to the hypothesis that nodule formation evolved from this more widespread symbiosis [40–42]. This suggests that non-legumes may lack a spectrum of the genes that enable the establishment of a symbiotic relationship with rhizobia. The longstanding dream of nodulated rice—or other important non-legume crop species—might be feasible.

However, transferring the capacity of nitrogen fixation to non-legume species will be a difficult task that will depend largely on how much processes needed for nodule development are present in non-legumes and can be geared to each other in the way in which the process occurs in legumes. This implies that we need to know how many additional components must be transferred to non-legumes [71]. An in-depth analysis of the 2500+ legume-specific genes identified could provide clues in this direction [20]. A major challenge is to uncover the extent to which non-legume homologs of genes such as LysM-RKs and *DMI* are able to complement corresponding legume mutants. Such studies would give insight into the extent to which the functioning of Nod factor signaling genes are unique to legumes. The availability of model species will definitely prove their value in finding answers to these exciting questions.

REFERENCES

1. Graham, P.H. and Vance, C.P., Legumes: importance and constraints to greater use, *Plant Physiol.*, 131, 872, 2003.
2. Dixon, R.A. and Sumner, L.W., Legume natural products: understanding and manipulating complex pathways for human and animal health, *Plant Physiol.*, 131, 878, 2003.
3. Desbrosses, G.G., Kopka, J., and Udvardi, M.K., *Lotus japonicus* metabolic profiling. Development of gas chromatography-mass spectrometry resources for the study of plant–microbe interactions, *Plant Physiol.*, 137, 1302, 2005.
4. Lodwig, E.M. et al., Amino-acid cycling drives nitrogen fixation in the legume–*Rhizobium* symbiosis, *Nature*, 422, 722, 2003.
5. Bataillon, T., Ronfort, J., and Prosperi, J.M., Natural genetic variation in *Medicago trunctula*: understanding the structure of genetic variation for the marker assisted building of a core collection of inbred lines, in *1st International Conference on Legume Genetics and Genomics: Translation to Crop Improvement* (St. Paul, MN), 2002.
6. Kulikova, O. et al., Integration of the FISH pachytene and genetic maps of *Medicago truncatula*, *Plant J.*, 27, 49, 2001.
7. Koornneef, M., Fransz, P., and de Jong, H., Cytogenetic tools for *Arabidopsis thaliana*, *Chromosome Res.*, 11, 183, 2003.
8. Cheng, Z. et al., Resolution of fluorescence *in situ* hybridization mapping on rice mitotic prometaphase chromosomes, meiotic pachytene chromosomes and extended DNA fibers, *Chromosome Res.*, 10, 379, 2002.
9. Zhu, H. et al., Phylogeny and genomic organization of the TIR and non-TIR NBS-LRR resistance gene family in *Medicago truncatula*, *Mol. Plant Microbe Interact.*, 15, 529, 2002.
10. Choi, H.K. et al., A sequence-based genetic map of *Medicago truncatula* and comparison of marker collinearity with *M. sativa*, *Genetics*, 166, 1463, 2004.
11. Ane, J.M. et al., Genetic and cytogenetic mapping of *DMI1*, *DMI2*, and *DMI3* genes of *Medicago truncatula* involved in Nod factor transduction, nodulation, and mycorrhization, *Mol. Plant Microbe Interact.*, 15, 1108, 2002.
12. Schnabel, E. et al., An integrated physical, genetic and cytogenetic map around the *sunn* locus of *Medicago truncatula*, *Genome*, 46, 665, 2003.
13. Kulikova, O. et al., Satellite repeats in the functional centromere and pericentromeric heterochromatin of *Medicago truncatula*, *Chromosoma*, 113, 276, 2004.
14. Young, N.D. et al., Sequencing the genespaces of *Medicago truncatula* and *Lotus japonicus*, *Plant Physiol.*, 137, 1174, 2005.

15. Quackenbush, J. et al., The TIGR gene indices: analysis of gene transcript sequences in highly sampled eukaryotic species, *Nucleic Acids Res.*, 29, 159, 2001.
16. Journet, E.P. et al., Exploring root symbiotic programs in the model legume *Medicago truncatula* using EST analysis, *Nucleic Acids Res.*, 30, 5579, 2002.
17. Lamblin, A.F. et al., MtDB: a database for personalized data mining of the model legume *Medicago truncatula* transcriptome, *Nucleic Acids Res.*, 31, 196, 2003.
18. Kuster, H. et al., Construction and validation of cDNA-based Mt6k-RIT macro- and microarrays to explore root endosymbioses in the model legume *Medicago truncatula*, *J. Biotechnol.*, 108, 95, 2004.
19. Barnett, M.J. et al., A dual-genome symbiosis chip for coordinate study of signal exchange and development in a prokaryote–host interaction, *Proc. Natl. Acad. Sci. USA*, 101, 16636, 2004.
20. Graham, M.A. et al., Computational identification and characterization of novel genes from legumes, *Plant Physiol.*, 35, 1179, 2004.
21. Van Kammen, A., Suggested nomenclature for plant genes involved in nodulation and symbiosis, *Plant Mol. Biol. Rep.*, 2, 43, 1984.
22. Trieu, A.T. et al., Transformation of *Medicago truncatula* via infiltration of seedlings or flowering plants with *Agrobacterium*, *Plant J.*, 22, 531, 2000.
23. Hoffmann, B. et al., A new *Medicago truncatula* line with superior *in vitro* regeneration, transformation, and symbiotic properties isolated through cell culture selection, *Mol. Plant Microbe Interact.*, 10, 307, 1997.
24. Kamaté, K. et al., Transformation of floral organs with GFP in *Medicago truncatula*, *Plant Cell Rep.*, 19, 647, 2000.
25. Chabaud, M., De Carvalho-Niebel, F., and Barker, D.G., Efficient transformation of *Medicago truncatula* cv. Jemalong using the hypervirulent *Agrobacterium tumefaciens* strain AGL1, *Plant Cell Rep.*, 22, 46, 2003.
26. Zhou, X., Chandrasekharan, M.B., and Hall, T.C., High rooting frequency and functional analysis of GUS and GFP expression in transgenic *Medicago truncatula* A17, *New Phytol.*, 162, 813, 2004.
27. Boisson-Dernier, A. et al., *Agrobacterium rhizogenes*-transformed roots of *Medicago truncatula* for the study of nitrogen-fixing and endomycorrhizal symbiotic associations, *Mol. Plant Microbe Interact.*, 6, 695, 2001.
28. Limpens, E. et al., RNA interference in *Agrobacterium rhizogenes*-transformed roots of *Arabidopsis* and *Medicago truncatula*, *J. Exp. Bot.*, 55, 983, 2004.
29. Chuang, C. and Meyerowitz, M., Specific and heritable genetic interference by double-stranded RNA in *Arabidopsis thaliana*, *Proc. Natl. Acad. Sci. USA*, 97, 4985, 2000.
30. Kumagai, H. and Kouchi, H., Gene silencing by expression of hairpin RNA in *Lotus japonicus* roots and root nodules, *Mol. Plant Microbe Interact.*, 16, 663, 2003.
31. Limpens, E. et al., LysM domain receptor kinases regulating rhizobial Nod factor-induced infection, *Science*, 302, 630, 2003.
32. Perry, J.A. et al., A TILLING reverse genetics tool and a Web-accessible collection of mutants of the legume *Lotus japonicus*, *Plant Physiol.*, 131, 866, 2003.
33. McCallum, C.M. et al., Targeted screening for induced mutations, *Nat. Biotechnol.*, 18, 455, 2000.
34. Till, B.J. et al., Mismatch cleavage by single-strand specific nucleases, *Nucleic Acids Res.*, 11, 2632, 2004.
35. Choi, H.K. et al., Estimating genome conservation between crop and model legume species, *Proc. Natl. Acad. Sci. USA*, 101, 15289, 2004.

36. Kalo, P. et al., Comparative mapping between *Medicago sativa* and *Pisum sativum*, *Mol. Genet. Genomics*, 272, 235, 2004.
37. Yan, H.H. et al., Comparative physical mapping reveals features of microsynteny between *Glycine max*, *Medicago truncatula*, and *Arabidopsis thaliana*, *Genome*, 47, 141, 2004.
38. Stougaard, J., Regulators and regulation of legume root nodule development, *Plant Physiol.*, 124, 532, 2000.
39. Hirsch, A.M., Lum, M.R., and Downie, J.A., What makes the rhizobia–legume symbiosis so special? *Plant Physiol.*, 127, 1484, 2001.
40. Gianinazzi-Pearson, V., Plant cell responses to arbuscular mycorrhizal fungi: getting to the roots of the symbiosis, *Plant Cell,* 8, 1871, 1996.
41. Albrecht, C., Geurts, R., and Bisseling, T., Legume nodulation and mycorrhizae formation; two extremes in host specificity meet, *EMBO J.*, 18, 281, 1999.
42. Gualtieri, G. and Bisseling, T., The evolution of nodulation, *Plant Mol. Biol.*, 42, 181, 2000.
43. Spaink, H.P., Regulation of plant morphogenesis by lipo-chitin oligosaccharides, *Crit. Rev. Plant Sci.*, 15, 559, 1996.
44. Ben Amor, B. et al., The *NFP* locus of *Medicago truncatula* controls an early step of Nod factor signal transduction upstream of a rapid calcium flux and root hair deformation, *Plant J.*, 34, 495, 2003.
45. Madsen, E.B. et al., A receptor kinase gene of the LysM type is involved in legume perception of rhizobial signals, *Nature*, 425, 637, 2003.
46. Radutoiu, S. et al., Plant recognition of symbiotic bacteria requires two LysM receptor-like kinases, *Nature*, 425, 585, 2003.
47. Mitra, R.M., Shaw, S.L., and Long, S.R., Six non-nodulating plant mutants defective for Nod factor-induced transcriptional changes associated with the legume–rhizobia symbiosis, *Proc. Natl. Acad. Sci. USA*, 101, 10217, 2004.
48. Bateman, A. and Bycroft, M., The structure of a LysM domain from *E. coli* membrane-bound lytic murein transglycosylase D (MltD), *J. Mol. Biol.*, 299, 1113, 2000.
49. Catoira, R. et al., Four genes of *Medicago truncatula* controlling components of a Nod factor transduction pathway, *Plant Cell*, 12, 1647, 2000.
50. Oldroyd, G.E. and Long, S.R., Identification and characterization of nodulation-signaling pathway 2, a gene of *Medicago truncatula* involved in Nod actor signaling, *Plant Physiol.*, 131, 1027, 2003.
51. Ehrhardt, D.W., Wais, R., and Long, S.R., Calcium spiking in plant root hairs responding to *Rhizobium* nodulation signals, *Cell*, 85, 673, 1996.
52. Wais, R.J. et al., Genetic analysis of calcium spiking responses in nodulation mutants of *Medicago truncatula*, *Proc. Natl. Acad. Sci. USA*, 97, 13407, 2000.
53. Endre, G. et al., A receptor kinase gene regulating symbiotic nodule development, *Nature*, 417, 962, 2002.
54. Ane, J.M. et al., *Medicago truncatula* DMI1 required for bacterial and fungal symbioses in legumes, *Science*, 303, 1364, 2004.
55. Levy, J. et al., A putative Ca^{2+} and calmodulin-dependent protein kinase required for bacterial and fungal symbioses, *Science*, 303, 1361, 2004.
56. Mitra, R.M. et al., A Ca^{2+}/calmodulin-dependent protein kinase required for symbiotic nodule development: gene identification by transcript-based cloning, *Proc. Natl. Acad. Sci. USA*, 101, 4701, 2004.
57. Stracke, S. et al., A plant receptor-like kinase required for both bacterial and fungal symbiosis, *Nature*, 417, 959, 2002.

58. Imaizumi-Anraku, H. et al., Plastid proteins crucial for symbiotic fungal and bacterial entry into plant roots, *Nature*, 433, 527, 2005.
59. Manthey, K. et al., Transcriptome profiling in root nodules and arbuscular mycorrhiza identifies a collection of novel genes induced during *Medicago truncatula* root endosymbioses, *Mol. Plant Microbe Interact.*, 17, 1063, 2004.
60. Weidmann, S. et al., Fungal elicitation of signal transduction-related plant genes precedes mycorrhiza establishment and requires the dmi3 gene in *Medicago truncatula*, *Mol. Plant Microbe Interact.*, 17, 1385, 2004.
61. Shaw, S.L. and Long, S.R., Nod factor elicits two separable calcium responses in *Medicago truncatula* root hair cells, *Plant Physiol.*, 131, 976, 2003.
62. Pingret, J.L., Journet, E.P., and Barker, D.G., Rhizobium nod factor signaling. Evidence for a G protein-mediated transduction mechanism, *Plant Cell*, 10, 659, 2003.
63. Firmin, J.L. et al., Resistance to nodulation of cv. Afghanistan peas is overcome by *nodX*, which mediates an O-acetylation of the *Rhizobium leguminosarum* lipo-oligosaccharide nodulation factor, *Mol. Microbiol.*, 10, 351, 1993.
64. Geurts, R. et al., Sym2 of pea is involved in a nodulation factor-perception mechanism that controls the infection process in the epidermis, *Plant Physiol.*, 115, 351, 1997.
65. Ardourel, M. et al., *Rhizobium meliloti* lipo-oligosaccharide nodulation factors: different structural requirements for bacterial entry into target root hair cells and induction of plant symbiotic developmental responses, *Plant Cell*, 6, 1357, 1994.
66. Gualtieri, G. et al., Microsynteny between pea and *Medicago truncatula* in the *SYM2* region, *Plant Mol. Biol.*, 50, 225, 2002.
67. Schauser, L. et al., A plant regulator controlling development of symbiotic root nodules, *Nature*, 402, 191, 1999.
68. Krusell, L. et al., Shoot control of root development and nodulation is mediated by a receptor-like kinase, *Nature*, 420, 422, 2002.
69. Nishimura, R. et al., HAR1 mediates systemic regulation of symbiotic organ development, *Nature*, 420, 426, 2002.
70. Searle, I.R. et al., Long-distance signaling in nodulation directed by a CLAVATA1-like receptor kinase, *Science*, 299, 109, 2003.
71. Parniske, M., Intracellular accommodation of microbes by plants: a common developmental program for symbiosis and disease? *Curr. Opin. Plant Biol.*, 3, 320, 2000.
72. Mitra, R.M. and Long, S.R., Plant and bacterial symbiotic mutants define three transcriptionally distinct stages in the development of the *Medicago truncatula/Sinorhizobium meliloti* symbiosis, *Plant Physiol.*, 134, 595, 2004.
73. Catoira, R. et al., The *HCL* gene of *Medicago truncatula* controls *Rhizobium*-induced root hair curling, *Development*, 128, 1507, 2001.
74. Kuppusamy, K.T. et al., *LIN*, a *Medicago truncatula* gene required for nodule differentiation and persistence of rhizobial infections, *Plant Physiol.*, 136, 3682, 2004.
75. Veereshlingam, H. et al., *Nip*, a symbiotic *Medicago truncatula* mutant that forms root nodules with aberrant infection threads and plant defense-like response, *Plant Physiol.*, 136, 3692, 2004.
76. Cohn, J.R. et al., Differential regulation of a family of apyrase genes from *Medicago truncatula*, *Plant Physiol.*, 125, 2104, 2001.
77. Navarro-Gochicoa, M.T. et al., Characterization of four lectin-like receptor kinases expressed in roots of *Medicago truncatula*. Structure, location, regulation of expression, and potential role in the symbiosis with *Sinorhizobium meliloti*, *Plant Physiol.*, 133, 1893, 2003.
78. Benaben, V. et al., TE7, An inefficient symbiotic mutant of *Medicago truncatula* Gaertn. cv. Jemalong, *Plant Physiol.*, 107, 53, 1995.

79. Morandi, D. et al., Characterisation of new symbiotic *Medicago truncatula* (Gaertn.) mutants, and phenotypic or genotypic complementary information on previously described mutants, *Mycorrhiza*, in press.
80. Penmetsa, R.V. and Cook, D.R., A legume ethylene-insensitive mutant hyperinfected by its rhizobial symbiont, *Science*, 275, 527, 1997.
81. Tirichine, L., de Billy, F., and Huguet, T., *Mtsym6*, a gene conditioning *Sinorhizobium* strain-specific nitrogen fixation in *Medicago truncatula*, *Plant Physiol.*, 123, 845, 2000.
82. Kalo, P. et al., Nodulation signaling in legumes requires NSP2, a member of the GRAS family of transcriptional regulators, *Science*, 308, 1786, 2005.
83. Smit, P. et al., NSP1 of the GRAS protein family is essential for rhizobial Nod factor-induced transcription, *Science*, 308, 1789, 2005.

6 *Brachypodium distachyon*: A New Model System for Structural and Functional Analysis of Grass Genomes

David F. Garvin

CONTENTS

6.1 INTRODUCTION

The releases of the genome sequences of the dicot model plant *Arabidopsis thaliana* (*Arabidopsis*) and the monocot crop rice represent a revolutionary advance in plant biology. These sequences have provided a potent source of information for genetics, evolution, development, and all other fields of plant biology at a resolution that was previously unavailable. The choice of these species as templates for genomic sequencing was based on their small genomes. Furthermore, *Arabidopsis* has long served as a tool for exploration of basic plant processes, and rice is a member of the most important plant family (Poaceae) from the standpoint of human subsistence.

The success of the *Arabidopsis* and rice genome sequencing projects, coupled with advances in strategies and technologies, has led to efforts to sequence all or parts of the genomes of many other plant species representing diverse plant families.

Given the rapid progress in genome sequencing of crops, one question that emerges is whether future investment in new model species to serve as surrogates for crops is necessary. The answer to this question is largely dependent upon the crop or crops that would be represented by such a new model species. In this regard, a cogent argument in favor of a model crop species can be made for the cool-season grass crops. Although it is clear that the rice genome sequence has been profitably exploited by researchers studying wheat, barley, and rye, it is equally clear that the rice genome sequence has limitations as a template for use in isolating genes from the cool-season cereals. This is largely due to the nearly 50 million years of evolution that separate rice from the cool-season grasses [1], which is reflected in profound differences in phenology, morphology, physiology, and biotic and abiotic stress susceptibility and tolerance. These differences impose limitations on the use of rice as a model for exploring the gene structure–function interface in the cool-season grasses.

An alternative model species that, like rice, is diploid, possesses a small genome, and is transformable, but more closely related to the cool-season grasses, can be expected to find a place in the laboratories of scientists interested in cool-season grass crop improvement. The species *Brachypodium distachyon* possesses all of these characteristics and is emerging as a new model species. The intent of this review is to provide an introduction to evolutionary, genetic, genomic, and morphological attributes of *B. distachyon* that make it such an attractive new model plant system.

6.2 EVOLUTIONARY RELATIONSHIPS BETWEEN *BRACHYPODIUM DISTACHYON* AND OTHER GRASS CROPS

The grass family Poaceae (Gramineae) comprises approximately 10,000 species, including most of the world's most important crops such as wheat (*Triticum* spp.), rice (*Oryza sativa*), maize (*Zea mays*), barley (*Hordeum vulgare*), sorghum (*Sorghum bicolor*), and many forage and turf species [2]. The members of this large plant family are distributed around the world in highly diverse environments. Results developed from the joint efforts of a consortium of grass systematicists support the presence of eight subfamilies within the Poaceae [3]. The largest of these subfamilies is the Pooideae which, depending upon the authority, includes about 10 tribes and over 3000 species. This one subfamily includes most of the important cool-season grain crops, forage grasses, and turfgrasses [4]. The genus *Brachypodium* belongs to the subfamily Pooideae as well and is considered distinct enough from other members of the Pooideae that it has been placed within its own tribe, the Brachypodieae [2].

Presumably because of the economic importance of the subfamily, many studies have been undertaken to resolve evolutionary relationships between the tribes in the

Pooideae, with some including *Brachypodium* in analyses [5]. A recent analysis based on chloroplast gene (*ndhF*) sequence data indicated that the tribe Brachypoideae is sister to the evolutionary lineage that subsequently gave rise to the tribes Triticeae, Aveneae, Bromeae, and Poeae [6]. As such, of the thousands of species comprising the entire family Poaceae (Gramineae), those of the genus *Brachypodium* are considered to be the closest ancestors to these four "core pooid" tribes [3,7] that encompass most of the major cool-season grasses. Thus, simply based on evolutionary considerations, we can predict that members of the genus *Brachypodium* will possess a genome structure and composition more similar to the "core pooids" than other genera in the entire grass family. This is one feature that makes *Brachypodium* attractive as a genus from which to select a possible model species for the cool-season grasses.

The systematics of *Brachypodium* has been examined a number of times in an attempt to resolve the evolutionary relationships among the species in this genus. The genus has not been unambiguously delineated, but it is likely to include 12 or more species [8]. Several members of the genus are endemic to Europe, and one or more different species are found in other parts of the world including regions of Africa, Central and South America, and Asia [2]. Two unusual characteristics of the genus are that the different species of *Brachypodium* exhibit significant variation in base chromosome number, including 5, 7, 8, or 9, and that polyploid series are encountered within species in the genus, so ecotypes in a given *Brachypodium* species may range from diploid ($2n = 2x$) to octaploid ($2n = 8x$) in some cases [9].

The species within the genus *Brachypodium* are perennial in nature with one exception: the annual species *B. distachyon*, often referred to as purple false brome. This species appears to have diverged from the other perennial *Brachypodium* species early in the evolution of the genus [10]. The evolutionary divergence of *B. distachyon* from other *Brachypodium* species is bolstered by results of experimental hybridizations between *B. distachyon* and other *Brachypodium* species. Although these crosses can produce F_1 progeny, they are sterile and exhibit abnormal meiosis. This contrasts with results from crosses among perennial European *Brachypodium* species, which generally produce fertile F_1 hybrids [11].

With one exception, *B. distachyon* is the only species in the genus with a base chromosome number of 5. The observed ploidy levels reported in *B. distachyon* ecotypes include diploidy ($2n = 2x = 10$), tetraploidy ($2n = 4x = 20$), and hexaploidy ($2n = 6x = 30$) [9]. This species is considered to be endemic to regions surrounding the Mediterranean Sea; however, it has spread to other regions of the world, where it can be a weed. In terms of its breeding system, *B. distachyon* is characterized as highly self-compatible with a floral morphology that encourages inbreeding [11].

An annual habit, self-compatibility, and availability of diploid ecotypes make *B. distachyon* the most desirable of the *Brachypodium* species from the standpoint of broad potential as a model species for functional and structural genomics research. The first publication that touted *B. distachyon* as a possible model species was that of Bablak et al. [12], in which it was noted that various parallels between *B. distachyon* and *Arabidopsis* might support research into the former species as a new model system.

6.3 STRUCTURAL FEATURES OF *BRACHYPODIUM DISTACHYON* GENOME

In diploid ecotypes of *B. distachyon*, the 5 chromosomes, while small, can be distinguished morphologically, and the major ribosomal gene tandem repeat loci (5S and 45S) have been localized to their respective chromosomes by *in situ* hybridization [13]. Several publications have reported size estimates of the *B. distachyon* genome. Using flow cytometry, Draper et al. [13] measured the genome size of diploid ecotypes of the species and reported a c-value of 0.21 pg that they considered equal to 172 megabase pairs—a size approximately the same as what they obtained for *Arabidopsis*. However, this value is in conflict with other reports.

Shi et al. [14] report that diploid *B. distachyon* has a c-value of 0.3 pg, which is similar to recent results of Bennett and Leitch [15], who reported a c-value of 0.36 pg. These latter values are similar to the average c-value of 0.39 pg obtained from three separate analyses by flow cytometry for a reference inbred diploid line of *B. distachyon* that we have developed from the accession PI 185134 and is deposited in the United States Department of Agriculture's National Plant Germplasm System (NPGS) [16]. It is unclear why the estimates of Draper et al. [13] are not congruent with other reports because they used various plant species as calibration standards in that report. At this time, it is recommended that the c-value estimate of 0.36 provided by Bennett and Leitch [15] be accepted as the best estimate for the genome size of diploid *B. distachyon*. This indicates that the genome of diploid ecotypes of this species is approximately 6.5% the size of barley and approximately 2% the size of the hexaploid bread wheat genome.

Thus, the entire genome of diploid *B. distachyon* is slightly smaller than an "average" chromosome arm in barley or wheat, is smaller than the genome of rice, and is only approximately twice as large as the genome of *Arabidopsis*. The extremely small genome of diploid *B. distachyon* is another important feature desired in a model plant species.

The derivation of polyploidy in *B. distachyon* is an interesting topic worth discussing. Given that the base chromosome number of the diploid ecotypes is 5 and that this doubled or tripled in the tetraploid and hexaploid ecotypes, respectively, one might assume that this is the signature of autopolyploidy rather than allopolyploidy. This assumption would be supported by the fact that no other European species of *Brachypodium* with a base chromosome number of 5 are known to exist. Thus, this rules out the likelihood that the polyploids are in fact allopolyploids derived from interspecific hybridization between diploids with a base number of 5, followed by chromosome doubling in a manner analogous to the evolution of tetraploid durum wheat (*T. turgidum*) and hexaploid bread wheat [17].

However, the observation that 15 bivalents form during meiosis in hexaploid ecotypes of *B. distachyon* [18] is contrary to expectations in an autopolyploid, in which multivalent pairing for some of the chromosomes is expected between some proportion of the chromosomes. Furthermore, our recent observations that a large set of putative hexaploid ecotypes of *B. distachyon* from the NPGS appear to exhibit c-values approximately two and not three times the size of diploid ecotypes [16] are not concordant with an autopolyploid origin of the polyploid ecotypes.

A recent publication [18] reported on similarities and differences in chromosome karyotypes, location, and abundance of ribosomal DNA loci, and results of genomic *in situ* hybridization between ecotypes representing the three ploidy levels. From these results, the authors postulated that the tetraploid ecotypes may actually represent a distinct diploid subtype (perhaps even a different species) with a base chromosome number of 10 and with chromosomes more similar to those of the related perennial species *B. sylvaticum* (slender false brome). They also proposed that the hexaploid ecotypes may represent the product of hybridization and chromosome doubling between diploid *B. distachyon* and a plant similar to the putative tetraploid that they proposed as a distinct subtype. As such, the polyploid ecotypes of *B. distachyon* may well have value as a system for exploring the impact of polyploidy on genome organization—a topic of great interest to wheat researchers because of the polyploid derivation of that crop.

6.4 GENOME COMPOSITION

The emergence of *B. distachyon* as a new model grass species is quite recent and thus information on the genome composition of the species is scarce. Nonetheless, data gleaned from available studies help to shed some light on this topic. In an early molecular phylogenetic analysis of the genus *Brachypodium*, Catalan et al. [8] produced a *Hin*dIII library of genomic DNA from a diploid *B. distachyon* ecotype to obtain probes for RFLP analysis. Hybridization of labeled genomic DNA against a series of randomly selected clones from this library revealed that less than 12% of them gave a strong hybridization signal indicative of the presence of high copy genomic sequences. Hybridization of several of these high copy clones to *Brachypodium* species DNA and DNA from other diverse grass species revealed that they hybridized exclusively to *Brachypodium* DNA. Furthermore, the hybridization patterns obtained from the different high copy probes suggested the presence of tandemly repeated DNA as well as interspersed repetitive DNA [8]. In contrast, a set of "low copy" *B. distachyon* clones hybridized to DNA of most other grass species included in the hybridizations, suggesting conservation of such sequences across the Poaceae.

These results are consistent with those obtained in *B. sylvaticum* (slender false brome). Because of its small genome size, Moore et al. [19] proposed that this member of the genus *Brachypodium* may be a useful model genome for studying the organization of grass genomes in general. Moore et al. [20] found that 100% of the short (<2 kb) clones from *B. sylvaticum Hpa*II and *Hin*dIII genomic libraries that were tested cross-hybridized to wheat genomic DNA, and 60% of these clones gave simple hybridization patterns on wheat genomic DNA. In contrast, a set of comparably sized single copy rice *Pst*I genomic clones hybridized to wheat just 64% of the time, with less than 60% of the hybridizing sequences revealing simple hybridization patterns on wheat genomic DNA.

Furthermore, their results indicated that the genome of *B. sylvaticum* is less methylated than those of wheat and barley. Also, a presumed centromere repeat sequence (CCS1) isolated from *B. sylvaticum* was found to hybridize to different grass species with different intensities; hybridization was strongest in wheat and rye, while maize and rice showed hybridization intensities less than 10% of that seen in the former species [21].

Taken as a whole, the results obtained from these analyses of the genomes of *B. distachyon* and its close relative *B. sylvaticum* support the notion that the small genome of *B. distachyon* harbors significantly less repetitive DNA than large genome grasses such as wheat and that the genomes of rice and *B. distachyon* are similar in complexity. However, the *B. distachyon* genome exhibits higher sequence similarity to the Triticeae than does rice for low-copy clones and at least one repetitive sequence.

6.5 GENE CONTENT AND ORGANIZATION VERSUS THOSE OF OTHER GRASSES

One of the principal interests in *B. distachyon* is its potential as a compact grass genome model that can be exploited to isolate genes from larger grass crop genomes because of conservation of genome organization and gene order. Various research groups are exploring this topic, though results have yet to be published.

In one study, the structure and abundance of C-repeat binding factor genes (*CBF*s), which encode transcription factors implicated in conditioning cold tolerance, were examined in *B. distachyon* with the purpose of comparison to the organization of the *CBF* gene family on chromosome 5H of barley. In barley, approximately ten *CBF* genes are present in a tightly linked gene cluster on chromosome 5H coincident with QTLs for cold tolerance, and they are estimated to be present at a density of less than one gene per 15 kb [22]. In contrast, seven *CBF* genes in a diploid *B. distachyon* line derived from NPGS accession PI 185134, which include presumed orthologs and paralogs of many sequenced barley 5H *CBF* genes based on phylogenetic analysis, have been localized to just two lambda phage clones spanning 33 kb.

Thus, the *CBF* gene density in the region of the *B. distachyon* genome presumed to be syntenic to the *CBF* cluster on barley chromosome 5H is at least three times higher [16]. These results also are in agreement with the expectation that significant compression of coding sequences takes place in the small genome of *B. distachyon* because of a reduced amount of repetitive DNA, but not a large scale loss of coding sequences relative to its large genome cool-season relatives.

The phylogenetic analysis of barley and *B. distachyon CBF* genes mentioned earlier not only provides interesting insights into the evolution of the *CBF* gene family in the grasses, but also offers evidence that orthologs of genes of agronomic interest in small grain cereal crops such as barley are present in the *B. distachyon* genome. This contention is further bolstered by the observation that a presumed ortholog of the *Q* gene, which played an important role in wheat domestication because it confers the free-threshing character, was identified in *B. sylvaticum* [23]. Finally, the genes *A1*, *X1*, *X2*, and *Sh2* have been used to examine the evolutionary dynamics of genome size and organization in many grass crops spanning different subfamilies [24–26]. To date, the presence of orthologs of the first three of these has been identified in a diploid inbred *B. distachyon* line derived from the NPGS accession PI 185134 (unpublished data). Efforts are currently under way to identify the

B. distachyon ortholog of *Sh2* so that the organization of these genes can be compared to that in other grasses.

The limited information available on comparative gene content and genome organization between *B. distachyon* and other grasses is supplemented with results obtained from its sister species *B. sylvaticum*. Foote et al. [27] reported construction and characterization of a bacterial artificial chromosome (BAC) library of *B. sylvaticum* and then used this library to explore synteny relationships among rice, wheat, and *B. sylvaticum*. They screened the BAC library with 48 probes derived from different grass species, and 36 were found to hybridize to one or more BAC clones. Of these, 33 were postulated to exhibit the same gene order found in rice chromosome 9 and/or the Triticeae group 5 chromosomes. The largest *B. sylvaticum* BAC contig spanned 367 kb, and 9 of 11 genes/probes examined on this contig were present as expected in the syntenic region of Triticeae group 5 chromosomes. Furthermore, they sequenced a 163-kb *B. sylvaticum* BAC clone from this library that revealed a minimum of 17 hypothetical genes. This corresponds to a gene density exceeding one gene per 10 kb of genome sequence.

Thus, results from these studies suggest that we should have confidence that orthologs of genes of broad interest to crop productivity will be present in *B. distachyon*. Furthermore, if we accept that the genome of *B. distachyon* will be highly similar in gene content and order to that of *B. sylvaticum*—aside from changes associated with major structural rearrangements, we can also be confident that *B. distachyon* has a genome amenable to service as a surrogate genome to accelerate gene discovery efforts in wheat, barley, and other large genome grasses.

6.6 FUNCTIONAL GENOMICS

The utility of *B. distachyon* to serve as a model grass genome is not limited to serving simply as a static "nucleotide roadmap" to help researchers navigate to genes of interest in grass genomes less tractable to physical analysis. Because *B. distachyon* is expected to have essentially the same ensemble of genes found in crops within the subfamily Pooideae, it may also be a potent resource for accelerating investigations of the role of particular genes in biological processes—for instance, diverse developmental or biochemical pathways and responses to biotic or abiotic stresses—through functional genomics methods.

Initial steps have been taken in this direction. Comparative functional analysis of molecular responses of *B. distachyon* to *Magnaporthe grisea*, the fungal pathogen causing the disease rice blast, has been reported [28]. Microscopic analyses of host responses in the resistant ecotype strongly resembled those of rice. The responses observed included localized cell death likely due to oxidative stress and the rapid induction of pathogenesis-related (PR) gene expression. Interestingly, ecotypes of diploid *B. distachyon* were found to exhibit differential resistance to the pathogen; this will permit the use of these ecotypes for comparative functional analysis of resistance and susceptibility to this pathogen.

Although this example illustrates that *B. distachyon* can serve as a model for functional genomic analysis of this host–pathogen system in rice, it would be desirable to extend similar opportunities to pathogens of relevance in the cool-season grass crops. For instance, the pathogens of greatest impact worldwide in wheat are the rusts belonging to the genus *Puccinia*, including leaf rust (*P. triticina*), stem rust (*P. graminis*), and stripe or yellow rust (*P. striiformis*). Draper et al. [13] inoculated *B. distachyon* with the leaf and stripe rust pathogens to determine its response to these pathogens. In both instances, host responses ranged from flecking (presumably indicative of a resistance response) to no visible response. In no case was an ecotype found to exhibit a typical susceptible response, though stripe rust uredinia were observed within heavily necrotic regions in two ecotypes, indicating successful colonization by the pathogen. In the same publication, reaction of *B. distachyon* ecotypes to *Blumeria graminis*, which causes powdery mildew, was investigated and particular host resistance responses, including the formation of papillae, were found.

These initial studies suggest that *B. distachyon* holds promise for functional genomic analysis of disease resistance in cool-season crops. However, beyond these studies, there is a dearth of information reporting *B. distachyon* as a functional genomics tool. Nonetheless, as this species becomes adopted as a model system, this is expected to change rapidly.

6.7 TRANSFORMATION AND MUTAGENESIS

The ability to transform a plant is an important attribute for a model plant species. One important aspect of transformation competency is the regeneration of plants from tissue culture. Bablak et al. [12] reported the first successful regeneration of *B. distachyon* from tissue culture. Three diploid ecotypes were used in this study. A high frequency of callus induction was observed from cultured seeds, and significant differences in callus induction were present between the ecotypes. The percentage of embryogenic callus obtained with the appropriate culture medium was as high as 60% in some experiments, and the number of regenerants per gram of callus was as high as 26. Subsequently, Draper et al. [13] confirmed the high level of embryogenic callus that could be obtained and the high level of plant regeneration for one of the same diploid ecotypes studied by Bablak et al. [12].

More recently, Christiansen et al. [29] found that it was possible to obtain embryogenic callus from immature embryos of diploid and tetraploid *B. distachyon* ecotypes, but the quality of this callus varied considerably. Nonetheless, they found that the percentage of cultured embryos that produced embryogenic callus for the two diploid ecotypes they examined was approximately 50%, similar to the results of Draper et al. [13]. The frequency of plant regeneration from the callus of their diploids approached 100%, but was dependent on the age of the callus. Thus, it appears that *B. distachyon* is very amenable to regeneration from tissue culture.

Draper et al. [13] and Christiansen et al. [29] also undertook biolistic transformation experiments on *B. distachyon*. Results of transformation for a single hexaploid ecotype were reported by Draper et al. [13]. Callus of this hexaploid ecotype was subjected to microprojectile bombardment with a plasmid carrying the hygromycin resistance gene and a glucuronidase (GUS) marker, resulting in recovery of an average of seven independent hygromycin-resistant calli per gram of bombarded callus. The authors reported a success rate in regenerating plants from the resistant calli of approximately 70%. These plants were subsequently found to exhibit GUS activity. However, transformation of diploid ecotypes was not reported.

Similarly, Christiansen et al. [29] undertook biolistic transformation of diploid and tetraploid ecotypes. In this study, dramatic differences in transformation efficiency between the two diploid ecotypes included were observed. Staining with GUS to examine transient transformation revealed that one diploid ecotype exhibited no apparent GUS-stained sectors in calli after microprojectile bombardment, and the other diploid exhibited an average of nearly 2000 GUS-positive sectors per bombardment experiment. This latter result was roughly similar to those obtained for the two tetraploids included in the experiments. Plant regeneration efficiencies (percent of bombarded calli that produced a transgenic plant) from these experiments were estimated at 5% for the diploid ecotype that exhibited GUS staining of calli and approximately 4% for the two tetraploids. However, individual experiments yielded efficiencies of between 9 and 14% depending upon the ecotype.

An alternative method for plant transformation is through the use of *Agrobacterium tumafaciens*, which has the advantage of producing lower copy and more stable genome integration events than particle bombardment [30]. Vogel et al. [31] studied *Agrobacterium*-mediated transformation efficiency of *B. distachyon* lines derived from accessions deposited in the NPGS and found a wide range of regeneration and transformation efficiencies among lines, which included diploid and polyploid ecotypes. For the three diploids evaluated (derived from the NPGS accessions PI 185133, PI 185134, and PI 254867), the regeneration efficiency ranged from 4 to nearly 11%. However, as reported for microjectile-mediated transformation by Christiansen et al. [29], significant differences in transformation efficiency between diploid lines using *Agrobacterium* were found.

For two of the diploid *B. distachyon lines*, no transformed plants were generated, while the other diploid line derived from PI 254867 exhibited an average plant regeneration frequency of over 3% across different experiments. In contrast, the regeneration frequency for polyploid lines ranged from 0 to 15%, though it should be noted that significantly more polyploids were examined in the experiments. Thus, it is clear that *B. distachyon* is transformable by two different and common methods and that certain diploid ecotypes can be transformed by one or the other method. Presumably, in the future *B. distachyon* transformation can be optimized by building upon existing information provided by these earlier studies.

Mutagenesis provides a strategy to introduce new molecular variation into a genome. In doing so, it generates novel variants of genes that can be used for dissecting the molecular basis of traits of interest. As such, mutagenesis is an important technique in the repertoire of a model species. In *Brachypodium*, little mutagenesis research has been conducted to date. Draper et al. [13] have subjected *B. distachyon* to gamma-irradiation, but no mutation frequencies were reported and no descriptions of the mutants obtained were provided.

The only other report of mutagenesis in *B. distachyon* is that of Engvild [32]. In this report, three diploid accessions of *B. distachyon* were used to assess the relative efficacy of different common mutagens and mutagen concentrations to induce mutations in *B. distachyon*. The mutagens included sodium azide and ethyl methanesulfonate (EMS). The sodium azide treatment appeared to be far more effective in inducing mutant phenotypes (based on the frequency of chlorophyll mutations) in the M1 generation than the EMS treatments. Furthermore, the relative efficacy of the mutagen treatments in inducing mutations appeared to vary depending upon the diploid accession used. Additional attempts to establish conditions that permit EMS or alternative mutagens to be used as effective mutagens in *B. distachyon* are strongly warranted so that mutant pools can be developed to identify mutations of interest based on phenotype or to identify lesions in genes of interest using methods such as TILLING [33].

6.8 GENETIC STOCKS AND GENETIC VARIATION

For any model plant species it is desirable to have a large and diverse collection of ecotypes exhibiting variation in phenotype and variation at the molecular level. Broad variation is particularly useful for successful development of segregating populations to serve as the basis for genetic map construction and for positional cloning endeavors. Furthermore, a diverse collection of ecotypes may be useful for studying the genetic and molecular basis of important crop traits by studying the same traits within the model system.

The NPGS has in its collection approximately 30 accessions of *B. distachyon* collected from various regions of the world over the course of the last several decades. The collection list can be viewed at the NPGS Website (http://www.ars-grin.gov/npgs/ as of June 2005). We used flow cytometry to discriminate between different ploidy levels in lines derived from 27 NPGS accessions. Five of these lines (from PI 170218, PI 185133, PI 185134, PI 245730, and PI 254867) were identified as diploids [16,31]. These were collected in Turkey and Iraq. In addition, a collection of over 50 accessions of *B. distachyon* is maintained by a company (Brachyomics Ltd.) established by the University of Wales [18]. The Brachyomics collection includes many lines collected in Europe and Asia, and elsewhere in the world.

It will be important to cross-reference collections of *B. distachyon* maintained by different entities to provide the scientific community clear information on geographic origin and other passport information associated with each ecotype. To date, the number of unique diploid ecotypes of *B. distachyon* is limited to perhaps ten. Therefore, it would be highly desirable to expand this number further by examining *B. distachyon* germplasm collections in other countries for additional novel diploid ecotypes, as well making collecting trips to regions to which this species is indigenous.

Although the number of diploid *B. distachyon* accessions/ecotypes is limited, it is promising to note that even among the five putative diploid accessions of *B. distachyon* deposited in the NPGS, significant phenotypic variation is evident. For instance, a cursory evaluation revealed clear differences in vernalization requirement, flowering date, plant height, pubescence, shattering, and seed size among these accessions (PI 170218, PI 185133, PI 185134, PI 245730, and PI 254867) [16,31]. These morphological differences are likely a reflection of molecular variation through the genome of *B. distachyon*. Diploid *B. distachyon* held by Brachyomics can be expected to contribute even more genetic diversity than that available in these few NPGS diploids.

Additionally, although little emphasis has been placed on the issue, it may be expected that molecular variation within each of these accessions exists because they very likely are composed of seeds collected from multiple plants in a wild population. In fact, support for this prediction is found in the publication of Christiansen et al. [29], which alludes to evidence for molecular heterogeneity between plants from the same accession as revealed by AFLP analysis.

Because diploid *B. distachyon* is highly self-compatible and rarely exerts anthers, it is likely that individual plants are inbred and thus homozygous for most loci. However, this cannot be assumed for certain; genetic homogeneity with original accessions also cannot be assumed. Thus, reference inbred genetic stocks of *B. distachyon* will be a valuable resource to the *Brachypodium* research community. During the past 3 years, we have developed single seed descent-derived inbred lines from 27 different *B. distachyon* accessions obtained from the NPGS, including five putative diploids (PI 170218, PI 185133, PI 185134, PI 245730, and PI 254867) [16,31]. In most cases, we have developed two independent inbred sister lines from each accession to capture additional molecular variation that may reside in each accession, and in some of these inbred sister lines we have observed phenotypic differences, suggesting some success toward this end.

We are making these inbred lines freely available to interested parties. These inbred genetic stocks will be of benefit for various reasons, particularly at this early stage of research with *B. distachyon*. For instance, the use of common inbred genetic stocks for research will reduce incongruous results that may be due to genetic variation in source materials studied. Also, the use of common inbred lines will improve the process of streamlining and integrating information emerging from research undertaken by different laboratories.

The use of inbred lines of *B. distachyon* to develop segregating populations will result in yet another highly useful resource. Despite the small size of the florets, populations appropriate for genetic mapping purposes will be developed unless unanticipated incompatibilities exist between the inbred lines. The only known report of a segregating population developed in *B. distachyon* is that of Routledge et al. [28], who developed F_2 populations from crosses between two diploid ecotypes and used them to assess the inheritance of resistance to *M. grisea*.

Similarly, we have been working to develop segregating populations derived from crosses among multiple diverse inbred lines. Crosses among all pairwise combinations of four of our inbred diploid *B. distachyon* lines have been completed. Interestingly, we found that diploids appear to possess two anthers per floret, in contrast to reports in authoritative references that indicate that the species has three anthers [2]. This character, therefore, could differentiate all diploids from polyploids (unpublished data). The putative hybrids we now have will serve as the basis of recombinant inbred line development, resulting in a diverse set of segregating populations that can be used for a range of purposes.

6.9 GROWTH CHARACTERISTICS

During our program to develop inbred *B. distachyon* lines, we noted that diploid *B. distachyon* plants are more petite than the polyploids during early vegetative growth.

We initially relied upon vernalization ranging between 4 and 10 weeks to induce flowering in the diploid lines because the lines would not flower without such a treatment or took so long to flower that it was impractical to maintain the plants. These observations are similar to observations of Draper et al. [13], who found that the flowering of all but two of their diploid ecotypes could be synchronized with 6 weeks of vernalization.

Because a vernalization requirement is not desirable if one wants to advance generations rapidly, we were interested in identifying conditions that would obviate the need to vernalize diploids. We subsequently discovered that three of our inbred diploid lines (derived from NPGS accessions PI 185133, PI 185134, and PI 254867) can be induced to flower rapidly without any vernalization simply by growing them under long (>20 h) day-lengths [16,31]. Under such conditions, inflorescence emergence of the most rapidly maturing line (from PI 254867) begins approximately 3 weeks after planting (Figure 6.1), and seeds will be mature 2 months after planting. The two other responsive lines flower approximately 1 and 2 weeks later than this line, respectively. Our last two inbred diploids (from NPGS accessions PI 170218

FIGURE 6.1 **(Color Figure 6.1 follows p. 144.)** A diploid *Brachypodium distachyon* inbred line grown under long day-lengths can, depending upon the genotype, begin the transition to flowering within 3 weeks of planting. In this figure, a diploid genotype was grown under 20-h day-length restrictive conditions, as shown in the left panel, or under nutrient replete conditions, as shown in the right panel. The plant in the left panel is 4 weeks old, and the plant on the right is 8 weeks old, with seeds maturing. Alternatively, by growing this inbred diploid line under short-day conditions, it can be maintained in vegetative phase for long periods of time. (Source: David Garvin, USDA-Agricultural Research Service.)

and PI 245730) appear to be unresponsive to long days and require significant vernalization periods to flower in a synchronized fashion.

The capacity to cycle *B. distachyon* rapidly is highly desired in a model species. At the same time, the capacity to retain these same genotypes in a vegetative state simply by growing them under short days is a useful attribute if one is interested in using the species as a model for forage or turf species. Furthermore, growing plants under conditions that encourage vegetative proliferation, followed by conditions that induce flowering, permits the recovery of large numbers of seeds from single plants. In our experience, we have been able to recover over 500 seeds from plants in this manner. Seed size varies between lines, but some of the inbred diploids we have developed produce seeds that have a mass approximately 15% that of wheat seeds, thus indicating a rather large endosperm that may also be beneficial for studies focusing on seed development.

6.10 CONCLUSIONS

The rice genome sequence is a resource that has great value for structural genomics research in the grasses. However, *B. distachyon* is poised to become a potent model system that complements rice because it also possesses attributes desired in a model system for functional genomics, including a small physical size, a growth habit that can be regulated to control generation times, and, perhaps most importantly, a close evolutionary affinity to the cool-season grain, turf, and forage species in the grass subfamily Pooideae.

There is much to be learned about *B. distachyon* before it can be fully exploited for research purposes. If we use *Arabidopsis* as a template to help determine where energies need to be directed to establish *B. distachyon* firmly as a new model species, this would include assembling genetic resources (genetic stocks, segregating populations), molecular resources (traditional genomic and cDNA libraries, as well as large insert libraries), obtaining sequence information (ESTs, genomic sequence data), optimizing transformation systems, and developing mutagenesis protocols and mutant pools.

Although this review reveals deficiencies in this list that need to be addressed, more importantly it also highlights the significant strides forward that have been made in these areas in the short period of time that has passed since *B. distachyon* was proposed as a potential model grass species. This review also highlights the fact that a significant amount of research now is under way in *B. distachyon*. Such research efforts are a tacit acknowledgment by the grass research community that a new model species for the cool-season grasses is indeed desirable and that *B. distachyon* possesses the attributes to serve in this capacity.

REFERENCES

1. Gaut, B., Evolutionary dynamics of grass genomes, *New Phytol.*, 154, 15, 2002.
2. Watson, L. and Dallwitz, M.J., *The Grass Genera of the World*, CAB International, Wallingford, CT, 1992, chap. 1.

3. Kellogg, E.A., Evolutionary history of the grasses, *Plant Physiol.*, 125, 1198, 2001.
4. Moser, L.E. and Hoveland, C.S., Cool-season grass overview, in *Cool-Season Forage Grasses*, Moser, L.E., Buxton, D.R., and Casler, M.D., Eds., Agronomy Society of America, Madison, WI, 1996, chap. 1.
5. Kellogg, E.A., Relationships of cereal crops and other grasses, *Proc. Natl. Acad. Sci. USA*, 95, 2005, 1998.
6. Catalán, P., Kellogg, E.A., and Olmstead, R.G., Phylogeny of Poaceae subfamily Pooideae based on chloroplast *ndh*F gene sequences, *Molec. Phylogenet. Evol.*, 8, 150, 1997.
7. Davis, J.I. and Soreng, R.J., Phylogenetic structure in the grass family (Poaceae) as inferred from chloroplast DNA restriction site variation, *Am. J. Bot.*, 80, 1444, 1993.
8. Catalán, P. et al., Molecular phylogeny of the grass genus *Brachypodium* P. Beauv. based on RFLP and RAPD analysis, *Bot. J. Linnean Soc.*, 117, 263, 1995.
9. Robertson, I.H., Chromosome numbers in *Brachypodium* Beauv. (Gramineae), *Genetica*, 56, 55, 1981.
10. Catalán, P. and Olmstead, R.G., Phylogenetic reconstruction of the genus *Brachypodium* P. Beauv. (Poaceae) from combined sequences of chloroplast *ndh*F gene and nuclear ITS, *Plant Syst. Evol.*, 220, 1, 2000.
11. Khan, M.A. and Stace, C.A., Breeding relationships in the genus *Brachypodium* (Poaceae: Pooideae), *Nord. J. Bot.*, 19, 257, 1999.
12. Bablak, P. et al., Plant regeneration and micropropagation of *Brachypodium distachyon*, *Plant Cell, Tissue Organ Cult.*, 42, 97, 1995.
13. Draper, J. et al., *Brachypodium distachyon*. A new model system for functional genomics in grasses, *Plant Physiol.*, 127, 1539, 2001.
14. Shi, Y., Draper, J., and Stace, C., Ribosomal DNA variation and its phylogenetic implication in the genus *Brachypodium* (Poaceae), *Plant Syst. Evol.*, 188, 125, 1993.
15. Bennett, M.D. and Leitch, I.J., Nuclear DNA amounts in angiosperms: progress, problems, and prospects, *Ann. Bot.*, 95, 45, 2005.
16. Garvin, D.F. and Stockinger, E.J., Development of a U.S. *Brachypodium distachyon* collection and the evaluation of microsynteny with cool-season grass crops, presented at the Plant and Animal Genome XIII Conference, San Diego, http://www.intl-pag.org/13/abstracts/, Jan. 15–19, 2005, P373.
17. Feldman, M., Origin of cultivated wheat, in *The World Wheat Book*, Bonjean, A.P. and Angus, W.J., Eds., Lavoisier, Paris, 2001, chap. 1.
18. Hasterok, R., Draper, J., and Jenkins, G., Laying the cytotaxonomic foundations of a new model grass, *Brachypodium distachyon* (L.) Beauv., *Chromosome Res.*, 12, 397, 2004.
19. Moore, G. et al., Molecular analysis of small grain cereal genomes: current status and prospects, *Bio/Technology*, 11, 584, 1993.
20. Moore, G. et al., Key features of cereal genome organization as revealed by the use of cytosine methylation-sensitive restriction endonucleases, *Genomics*, 15, 472, 1993.
21. Aragon-Alcaide, L. et al., A cereal centromeric sequence, *Chromosoma*, 105, 261, 1996.
22. Stockinger, E.J., Cheng, H., and Skinner, J.S., Structural organization of barley *CBF* genes conicident with a QTL for cold hardiness, in *Cold Hardiness in Plants*, Chen, T.H.H, Uemura, M., and Fujikawa, S., Eds., CABI Publishing, Wallingford, CT, 2006, 53.
23. Faris, J., personal communication, 2005.
24. Chen, M. et al., Microcolinearity in *sh2*-homologous regions of the maize, rice, and sorghum genomes, *Proc. Natl. Acad. Sci. USA*, 94, 3431, 1997.

25. Chen, M., SanMiguel, P., and Bennetzen, J.L., Sequence organization and conservation in *sh2/a1*-homologous regions of sorghum and rice, *Genetics*, 148, 435, 1998.
26. Li, W. and Gill, B.S., The collinearity of the *Sh2/A1* orthologous region in rice, sorghum and maize is interrupted and accompanied by genome expansion in the Triticeae, *Genetics*, 160, 1153, 2002.
27. Foote, T.N. et al., Construction and analysis of a BAC library in the grass *Brachypodium sylvaticum*: its use as a tool to bridge the gap between rice and wheat in elucidating gene content, *Funct. Integr. Genomics*, 4, 26, 2004.
28. Routledge, A.P.M. et al., *Magnaporthe grisea* interactions with the model grass *Brachypodium distachyon* closely resemble those with rice (*Oryza sativa*), *Molec. Plant Pathol.*, 5, 253, 2004.
29. Christiansen, P. et al., A rapid and efficient transformation protocol for the grass *Brachypodium distachyon*, *Plant Cell Rep.*, 23, 751, 2005.
30. Dai, S. et al., Comparative analysis of transgenic rice plants obtained by *Agrobacterium*-mediated transformation and particle bombardment, *Mol. Breeding*, 7, 25, 2001.
31. Vogel, J. et al., Agrobacterium-mediated transformation and inbred line development in the model grass *Brachypodium distachyon*, *Plant Cell Tissue Organ Cult.*, 84, 199, 2006.
32. Engvild, K.C., Mutagenesis of the model grass *Brachypodium distachyon* with sodium azide, Risø-R Report 1510(EN), Risø National Laboratory, Roskilde, Denmark, 2005, 8 pp.
33. McCallum, C.M. et al. Targeted screening for induced mutations, *Nat. Biotech.*, 18, 455, 2000.

7 The Green Alga *Chlamydomonas* as a Tool to Study the Nitrate Assimilation Pathway in Plants

Aurora Galván, Vicente Mariscal, David González-Ballester, and Emilio Fernández

CONTENTS

7.1 *CHLAMYDOMONAS* AS A MODEL ORGANISM

Model species are generally first selected because they allow novel insights into processes that are poorly understood; they are then maintained and accepted because of their experimental amenability and usefulness in understanding further processes. Plants represent an extraordinarily diverse group of living beings at the molecular, genetic, biochemical, and physiological levels, and their great biodiversity reflects the evolution of complex genetic and biochemical networks [1].

Unicellular organisms lack the complexity and sophistication of higher plant systems. However they remain useful as a tool to understand aspects of fundamental plant biology at the cell level that have not yet been elucidated. In particular, conditions for efficient uptake, metabolism, and regulation of nutrient acquisition, as well as for accumulation of assimilate-derived products are relevant for determining the efficiency of growth at whole plant and thus at crop levels. For this reason, many research groups have focused their attention on understanding key processes in model plant systems.

It is becoming clear that present model systems are not as representative as had been hoped, so, for a proper understanding, each system needs to be studied in itself. At the same time, it seems that, to dissect the major processes at the molecular level, it is necessary to work in a system simple enough to control the influence of extraneous factors. Interestingly, the molecular dissection of metabolic steps (e.g., nitrate assimilation) has shown that the differences between unicellular and land plants can be smaller than feared at the level of complexity of gene and protein families and with respect to cellular strategies adopted to achieve particular end results. An ideal model system needs a set of basic physiological, biochemical, and molecular techniques, along with developed resources in genetics, genomics, and transgenesis. These will allow for the ready genetic dissection of mutant phenotype, together with a straightforward correlation of phenotype to genotype. The *Chlamydomonas* system fulfils most of these requirements, as summarized in Table 7.1.

The use of *Chlamydomonas* as an amenable biological system was first described in Harris's book [3]. The advantages of *Chlamydomonas* as a model unicellular system were further presented by Rochaix et al. [9] and were recently compared to those of *Arabidopsis* [10]. Decoding the nuclear and chloroplast genomes of *Chlamydomonas* and developing molecular tools (see Table 7.1) have strengthened the position of this organism for study of important plant cell processes such as photosynthesis, chloroplast inheritance and biology, mitochondrial genetics, nutrient deficiency, carbon metabolism, and nitrate assimilation.

7.2 THE NITRATE ASSIMILATION PATHWAY AND ITS KEY POINTS

Ammonium and nitrate are the primary inorganic nitrogen sources for plant growth. Though many species use ammonium in preference to nitrate, the majority of plants, algae, and microorganisms is able to use nitrate efficiently because ammonium is about 10 to 1000 times less abundant than nitrate in natural soils, except in a few ecosystems such as coniferous forests [11].

TABLE 7.1
Advantages of the *Chlamydomonas* System in Molecular Plant Biology

Property	Description	Ref.
Genome	Haploid organism with a 10^8-bp genome similar to *Arabidopsis*	2
Genome sequencing project	Mostly sequenced; third assembly released	http://www.chlamy.org/
EST and microarrays	Above 200,000	http://www.chlamy.org/
Genetics	Excellent classical genetics, with standard tetrad and complementation analysis	3
Transformation	Nuclear, chloroplast, and mitochondrial genomes transformed	4
Markers	Array of selectable markers including antibiotic resistance available	5
Interfering gene expression	Antisense and RNAi methodologies	6, 7
Mutant library	A 22,000 mutant library with mostly single insertions	8

In addition, nitrate provides an efficient signal for modulating metabolic processes and plant architecture [12]. The assimilation of ammonium ion has a lower energy cost than that of nitrate [13] and many genes involved in nitrate assimilation are strongly repressed in the presence of ammonium [14,15]. However, many plants, especially herbaceous crop plants, utilize nitrate or a combination of both ions as the preferred nitrogen form for growth [16,17]. Thus, about 75% of the nitrogen in proteins consumed by man is ultimately derived from nitrate assimilated by plants.

The fundamental role of the nitrate assimilation pathway in plant nutrition has been the object of intensive studies for many years and has been reviewed by many authors [12,18–31]. This chapter represents an update of information about nitrate assimilation in *Chlamydomonas* as a source of information for the equivalent process in higher plants.

The basic steps for nitrate utilization in a single photosynthetic eukaryotic cell are as follow (Figure 7.1):

1. The entry of nitrate into the cell by means of specific transport systems
2. A first reduction step from nitrate to nitrite, which occurs in the cytosol and is catalyzed by the nitrate reductase (NR) enzyme
3. Nitrite transport to the chloroplast
4. A second reduction step that occurs in the chloroplast, where nitrite reductase (NiR) catalyses nitrite reduction to ammonium
5. Finally, ammonium incorporation into carbon skeletons by the glutamine synthetase/glutamate synthase cycle (GS/GOGAT)

Chlamydomonas does not use intracellular compartments to store nitrate or ammonium. Thus, under conditions where the cell assimilation capability is exceeded, extrusion systems come into play (Figure 7.1), thereby avoiding any toxic effects of excessive intracellular ammonium or nitrite ions [32–34].

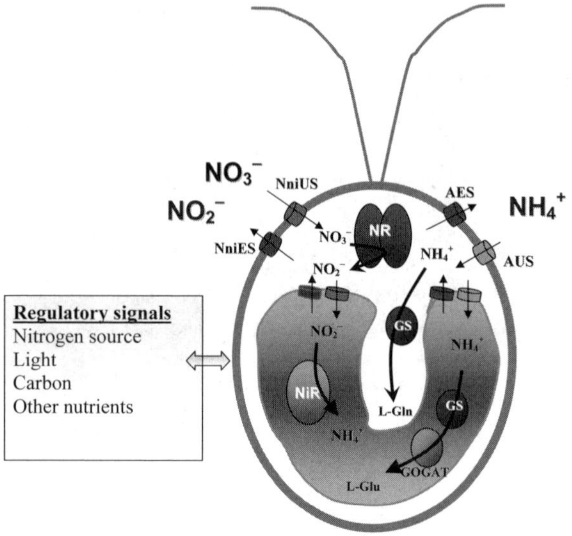

FIGURE 7.1 (**Color Figure 7.1 follows p. 144.**) The nitrate assimilation pathway in *Chlamydomonas*. NniUS = nitrate/nitrite uptake influx systems; NniES = nitrate/nitrite efflux systems; AUS = ammonium uptake influx systems; AES = ammonium efflux systems. Other details are indicated in the text.

The nitrate assimilation pathway is governed by specific regulatory genes mediating positive and negative responses against nitrate and ammonium, respectively. In addition, control of this pathway is connected to a network of stimuli needed to coordinate nitrate assimilation with that of carbon, sulfur, and other nutrients (phosphorous, potassium, etc.). Light is also essential for inorganic nutrient assimilation in photosynthetic organisms [24,26]. A complex network underlying the process of nitrate assimilation has been proposed on the basis of transcriptome analyses in response to nitrate, potassium, etc. [35,36].

The *Chlamydomonas* genes involved in nitrate assimilation are shown in Table 7.2. Some of the direct gene products have been studied and characterized; others have been taken from the recently released genome sequence database (http://www.chlamy.org/) and will require further study. As shown, structural genes responsible for nitrate and nitrite reduction are single copy, while those for ammonium incorporation correspond to plastidial (GS2) and cytosolic (GS1) forms. The large number of genes involved in the transport of nitrate, nitrite, and ammonium is particularly notable.

On *Chlamydomonas* chromosome IX, two gene clusters contain most of the nitrate assimilation genes (Figure 7.2), the majority of which are under control of the regulatory gene *Nit2* (Table 7.2). One occurs in a region of about 36 kb, where the structural genes *Nia1* (encoding NR), *Nii1*(NiR), *Nar1.1* (a plastidial nitrite transporter), *Nar2* (a component of some nitrate transporters), and *Nrt2.1* and *Nrt2.2* (nitrate transporter components) are found. In this cluster, a plastidial malate dehydrogenase gene, *NMdh*, not regulated by nitrate is also present and has been proposed to play an important role in the supply of reducing power for nitrate reduction [38,41,42,50,51].

TABLE 7.2
Major Elements for Nitrate and Ammonium Assimilation in *Chlamydomonas*

Gene product	Gene/name	Scaffold/ linkage group	Protein (aa residues)	Possible and subcellular localization predictions	Ref.
Nitrate reductase	*Nia1* (also Nit1) (C_520041)	52/IX	?	Cyt	37
Nitrite reductase	*Nii1* (C-520008)	52/IX	589	Chl	38
Glutamine synthetase	*GS1* (C-20337)	2/II	382	Cyt	39
	GS2 (also GLN2) (C_380043)	38/XII/XIII	380	Chl	39
	GS3 (C_380117)	38/XII/XIII		Chl	
Glutamate synthase	*NADH-GOGAT* (C_1440026)	144			
	Fd-GOGAT (C-160008)	16/XII–XIII	847	Chl	40
Nitrate transporter NRT1 family	*Nrt1.1* (C_40176)	4/IV			
Nitrate transporter NRT2 family	*Nrt2.1* (also Nar3) (C_520006)	52/IX	547	Pm	41, 42
	Nrt2.2 (also Nar4) (C_520007)	52/IX		Pm	41, 42
	Nrt2.3 (C_330081)	33/IX	572	Pm	
	Nrt2.4-5 (C_1590030-1)	159/III		Pm	
	Nrt2.6 (C-20370)	2/II		Pm	
Nitrate transporter component NAR2	*Nar2* (C_520042)	52/IX		Pm	41, 42
Nitrite transporter NAR1 family	*Nar1.1* (C_520040)	52/IX		Chl	34
	Nar1.2 (also LciA) (C_90197)	9/VI	336	Chl	43, 44
	Nar1.3 (C_8440001)	844,100			43
	Nar1.4 (C_720018)	72/VII			43
	Nar1.5 (C_70011)	7/XII-XIII		Chl	43
	Nar1.6 (C280009)	28/I			
Ammonium transporter AMT1 family	*Amt1.1* (C_110147)	11 / III	539	Pm	45
	Amt1.2 (C_4560001)	456	542	Chl	45
	Amt1.3 (*Amt3*) (C_2680003)	268	579	Pm	45, 46

continued

TABLE 7.2 (continued)
Major Elements for Nitrate and Ammonium Assimilation in *Chlamydomonas*

Gene product	Gene/name	Scaffold/ linkage group	Protein (aa residues)	Possible and subcellular localization predictions	Ref.
	Amt1.4 (*Amt4*) (C_930017)	93	498	Chl	45
	Amt1.5 (*Amt5*) (C_220054)	22/IX	610	Pm	45
	CrAmt1.6 (*Amt6*) (C_980024)	98		Er,Prx	45
	Amt1.7 (*Amt7*) (C_20186)	2/ II	a: 411 b: 487	a: Mit b: Pm	45
	Amt1.8 (*Amt8*) (C_380121)	38/ XII–XIII	481	Pm, Chl	45
Positive regulator NIT2	*Nit2* (C_860001)	86/III		Nuc	47
MoCo carrier protein	*Mcp1* (C_700030)	70	165		48
Alternative oxydase	*Aox1* (C_330029)	33/IX	360	Mit	49
NADP+-malate dehydrogenase	NADP-*Mdh* (C_520009)	52/IX	415	Chl	50

Notes: Gene/name represents the usual denomination of a gene and the specific name annotation in the *Chlamydomonas* Genome Project v. 2. Protein amino acid residues, where indicated, are based on known full-length cDNA information. For *Amt1.7*, two different mature mRNA are known (a and b). The most probable predicted subcellular localizations are indicated as follows: chloroplast (Chl), mitochondria (Mit), plasma membrane (Pm), endoplasmic reticulum (Er), peroxisome (Prx), and nuclear (Nuc). References are indicated where appropriate; otherwise, it refers to information in the *Chlamydomonas* Genome or GenBank database.

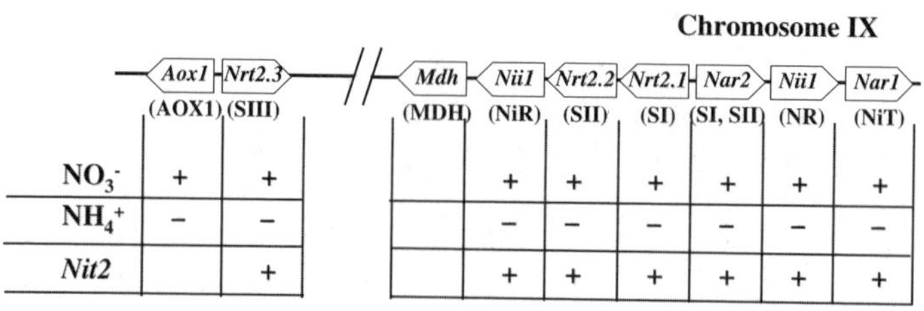

FIGURE 7.2 Nitrate assimilation genes clustered in chromosome IX. Plus and minus symbols refer to positive and negative acting effectors, respectively, on the gene shown above.

The second cluster (about 10 kb in size) contains two nitrate-regulated genes, *Nrt2.3* (a nitrite transporter) and *Aox1* (a mitochondrial alternative oxidase) [51,52]. Clustering of nitrate assimilation genes has also been reported in *Aspergillus nidulans* [53] and in *Hansenula polymorpha* [54], so this may well represent a cellular strategy to optimize the regulation of this pathway. In *Aspergillus*, the intergenic region between the divergently transcribed *niiA* (NiR) and *niaD* (NR) genes contains multiple *NirA* (pathway-specific positive regulator)-binding sites, which act bidirectionally [55].

Interestingly, genes encoding GS2, GS3, fd-GOGAT, and AMT1.8 are located in chromosome XII and chromosome XIII. None of these genes are regulated by nitrate, but they are related to ammonium assimilation. *GS2* and *GS3* are bidirectional clustered genes (Table 7.2).

7.3 THE USEFULNESS OF CHLORATE-RESISTANT MUTANTS TO STUDY OF NITRATE ASSIMILATION

The isolation and characterization of mutants deficient in the nitrate assimilation pathway, mostly through chlorate resistance, have been powerful tools in defining structural and regulatory elements of nitrate metabolism, as well as in understanding the function of each component [12,20,41,47,56–58]. As an analog of nitrate, chlorate can be reduced by NR to produce chlorite, which is cell toxic [59]. Nevertheless, chlorate becomes toxic and causes mutagenesis in the cells by a process dependent on its transport [58]. Thus, mutants incapable of taking up chlorate and those deficient in NR activity could be selected in chlorate media. This strategy has been widely used in fungi, algae, and plants; the mutants generated in this way have been used to study the nitrate assimilation pathway [20,29,60].

The characterization of *Chlamydomonas* chlorate-resistant mutants has led to identification of several loci involved in nitrate assimilation. The function of genes encoding nitrate transporters, nitrate reductase, nitrite reductase, the plastidial nitrite transporter NAR1, and the regulatory gene *Nit2*, was demonstrated by characterization of chlorate-resistant mutants [33,34,38,41,42,47,52,61]. A chlorate-resistant mutant related to light regulation of nitrate assimilation has also been characterized, though the locus affected has yet to be identified [62]. This mutant shares several characteristics with the *Arabidopsis* CR88 mutant that carries a lesion in a gene encoding a chloroplast targeted Hsp90 protein and shows a pleiotropic phenotype [63].

7.4 NITRATE AND NITRITE REDUCTION AND ITS REGULATION

Nitrate reductases from photosynthetic eukaryotes are homodimeric proteins that use pyridine nucleotides as electron donors. Each monomer is a 100- to 120-kDa polypeptide containing three prosthetic groups—flavin adenin dinucleotide (FAD), heme b_{557}, and molybdenum cofactor (Moco)—that are present in three functional domains spaced by two short hinge regions [64,65]. NR sequences from different eukaryotic organisms show a high level of conservation; however, they differ from the cyanobacterial NR, which uses ferredoxin as an electron donor, as do a number of other prokaryotic NRs [25,66].

One or two genes can be identified for NR in different organisms. Barley and *Arabidopsis* contain two structural loci (*Nia1* and *Nia2*), one of which (*Nia1*) encodes the most abundant isoform (NADH-dependent). In other organisms such as *Nicotiana plumbaginifolia*, *Lotus japonicus*, or *Chlamydomonas reinhardtii*, only one gene is responsible for NR activity [25,37].

Molybdenum is a micronutrient essential for nitrate reduction. Its presence in a molybdopterin cofactor (Moco) was first genetically identified in *Aspergillus nidulans* as a cofactor common to NR and xanthine dehydrogenase. Afterwards, Moco was also found to be associated with aldehyde oxidase and sulfite oxidase. Thus, Moco is essential in key metabolic processes: nitrate assimilation, purine catabolism, biosynthesis of phytohormone abscisic acid, and detoxification of sulfite [67,68]. Moco consists of the molybdopterin (MPT), an incompletely alkylated aromatic pterin complexing one Mo atom via a dithiolene group to its four-carbon side chain. The pathway of Moco biosynthesis has been dissected in plants and its steps determined from GTP to Moco [69]. For this purpose, the isolation and characterization of Moco mutants using chlorate resistance was particularly important. The Moco mutants have been classified into six different complementation groups (*CnxA*-*CnxF*) [20,69].

Biochemical and genetic characterization of *Chlamydomonas* NR-deficient mutants has allowed for identification of seven loci (*Nit3* to *Nit7*, *Nit10*, and *Nit11*) related to Moco biosynthesis [62,70,71]. Single mutants defective at the *Nit5* or *Nit6* genes show a wild phenotype, whereas the double mutant *Nit5–Nit6* lacks Moco and molybdate uptake activity [70,72]. Molybdate uptake by *Chlamydomonas* cells is thought to be mediated by specific transporters: a high-capacity system related to the *Nit5* gene function and another system with less capacity [72].

Moco-carrier protein (CP) activity was first identified in *Chlamydomonas* [71] and its presence could also be demonstrated in *Vicia faba* seeds [73]. The corresponding gene, *Mcp1*, was isolated and functionally characterized [48,74]. Both pure Moco-CP from *Chlamydomonas* and the recombinant protein were able to protect Moco from inactivation by oxygen very efficiently. Moco-CP is proposed to participate directly in transfer of the prosthetic group Moco to the apoNR (Figure 7.3). Orthologs of *Mcp1* have been found in prokaryotic organisms, but identification in eukaryotes is not easily deduced from sequence homology data [48].

NR activity is tightly regulated by environmental factors such as light, nitrogen, and carbon availability. In plants, NR [75–77] and GS [78] have been reported to bind 14-3-3 proteins. The mechanism of plant NR reversible inactivation depends on phosphorylation and binding of 14-3-3 [79,80]. The interaction of 14-3-3 proteins with a number of metabolic enzymes is sequence specific and suggests involvement in regulation of complicated daily rhythms in sugar metabolism in coordination with photosynthesis, ATP production, and nitrate reduction [80,81]. *Chlamydomonas* NR is not affected by regulatory interactions with 14-3-3. Nevertheless, GS1 is phosphorylated and binds 14-3-3 [82]. The function of the 14-3-3 regulatory mechanism is not clear and has been related to protein turnover in the cells [83].

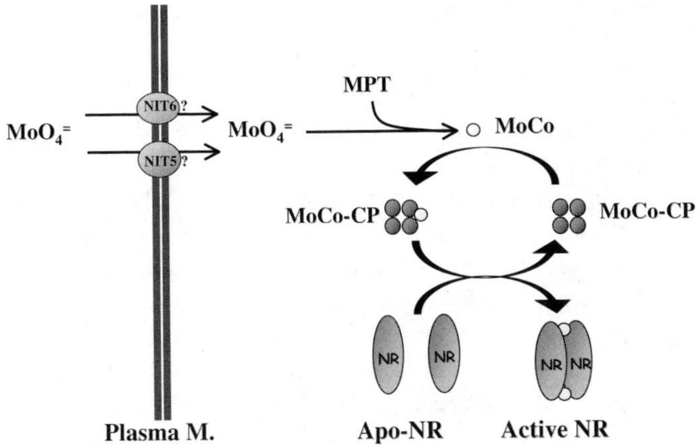

FIGURE 7.3 Model for molybdate transport and Moco-carrier protein function in *Chlamydomonas*. MPT = molybdopterin.

The *Chlamydomonas* NR is subject to a redox interconversion regulatory mechanism. Thus, in the absence of nitrate, NR becomes over-reduced and inactivated [60,84]. This mechanism has been shown to be operative *in vivo* and *in vitro*; it can be reversed *in vivo* by resupply of nitrate and *in vitro* by ferricyanide oxidation. Inactivation of NR results in a decrease in the enzyme's half-life [85].

Nitrite reduction is a six-electron step catalyzed by NiR, which uses reduced ferredoxin (fd) as electron donor. In photosynthetic eukaryotic organisms, NiR is located at the chloroplast stroma and is also present in the plastids of nonphotosynthetic tissues. The holoenzyme is encoded by a nuclear gene. The 63-kDa protein contains two redox centers: a siroheme and a [4Fe-4S] cluster. The N- and C-terminal parts of the protein are proposed to bind ferredoxin and the [4Fe-4S] redox center, respectively [38,64,86]. NiR is encoded by one gene in *Chlamydomonas* [38] as in *Arabidopsis*, but other plants such as tobacco may contain as many as four genes [87].

The *Chlamydomonas* NiR shows a similar regulation pattern as NR and requires absence of ammonium and presence of light and nitrate for maximum expression [38,88]. However, no regulation at the activity level has been shown, so any changes in enzyme amounts appear to be due to transcriptional regulation. Post-transcriptional regulation has been demonstrated in *Nicotiana* and *Arabidopsis* [89].

7.5 NITRATE AND NITRITE TRANSPORTERS

Much attention has been paid over a long period to the reduction of nitrate because this is considered to be the key step in control of the pathway [18–20,29,84]. However, the first transport step through the plasma membrane, together with transport at the chloroplast envelope membrane, has become the focus of present interest because it appears that these steps play an important role in regulation of the overall efficiency of the pathway [27]. As expected for this key role, the transporters are subject to fine regulatory control [12,28,30,31].

7.5.1 Transport Systems

As indicated in Figure 7.1, nitrate transport systems should work in influx as well as efflux processes to provide sufficient nitrate under different nutritional and environmental conditions to satisfy the total demand of nitrogen. These transporters should be operative at the level of the plasma and the plastidial membranes.

Nitrate/nitrite transporters have been classified on the basis of their substrate affinity, specificity and requirements for induction into: *c*onstitutive *h*igh-*a*ffinity *n*itrate *t*ransport *s*ystems (cHANTS), *i*nducible *h*igh-*a*ffinity *n*itrate *t*ransport *s*ystems (iHANTS), *c*onstitutive *l*ow-*a*ffinity *n*itrate *t*ransport *s*ystems (cLANTS), and *i*nducible *l*ow-*a*ffinity *n*itrate *t*ransport *s*ystems (iLANTS) [26,27].

Studies on nitrate transport suggest that in plants, algae, and fungi, nitrate uptake is electrogenic and driven by proton cotransport [90–94]. The proton gradient is maintained by a H^+-ATPase. Thus, the use of H^+-ATPase inhibitors and alkalization of external medium inhibit nitrate uptake [24,94–97].

Other nitrate transport mechanisms have also been proposed. A Na^+/NO_3^- symport system was suggested in the cyanobacteria *Anacystis nidulans* R2 [98] and an active ATP-dependent transport system (ABC transporter) in *Synechoccocus* sp. PCC7942 [99]. In *Escherichia coli*, the nitrate transporters NarK and NarU might be able to catalyze nitrate uptake or nitrate–nitrite antiport [100,101]. Voltage-dependent chloride channels might also participate in nitrate transport. In *Arabidopsis*, AtCLC-a is induced by nitrate and its involvement in the control of the intracellular nitrate status has been suggested [102].

cHANTS have been described in higher plants such as *Nicotiana* [103], barley [104], and *Arabidopsis* [105] and are characterized by low values of Km and Vmax (typically 6 to 20 μM and 0.3 to 0.8 μmol $g^{-1}.h^{-1}$, respectively). The cHATS provides high-affinity, low-capacity activity for NO_3^- entry in uninduced plants. Nevertheless, cHATS activity is upregulated (approximately threefold) by exposure to NO_3^- [24]. The iHANTS have been identified and well characterized in higher plants (*Arabidopsis*, *Nicotiana*, barley, etc.), algae (*Chlamydomonas*, *Chlorella*), yeasts (*Hansenula polymorpha*), and fungi (*Aspergillus*, *Neurospora*) [27,106–108]. The iHANTSs provide high affinity and capacity (Km 20 to 100 μM and Vmax 3–8 μmol $g^{-1}.h^{-1}$) [24]. These transporters require NO_3^- or NO_2^- to be induced and are subject to repression by nitrogen metabolites such as ammonium and glutamine [27,109]. The cLANTS and iLANTSs have been identified in higher plants and can significantly contribute to nitrate uptake at millimolar nitrate concentrations.

The specificity for the nitrate ion has also been used to name the transporter. In *Chlamydomonas,* physiological studies with mutant strains carrying particular transporters have demonstrated their ability to distinguish between NO_3^- and NO_2^-. Thus, the nitrate and nitrite transporters in this alga can be classified into nitrate specific, nitrite specific, and nitrate/nitrite bispecific [27]. In addition, some plant nitrate transporters have been shown to transport amino acids [26,110].

The biochemical characteristics for nitrate transport activities seem to be as complex as the picture for nitrate transporter genes. This gene complexity is observed from comparison of *Arabidopsis* and *Chlamydomonas* genomes. Three families of nitrate/nitrite transporters, *Nrt1*, *Nrt2*, and *Nar1* [12,27,43,109], operate in photosynthetic eukaryotic organisms. For these families, there exist 51 *Nrt1*, 7 *Nrt2*, and no *Nar1*-like genes in *Arabidopsis*, whereas in *Chlamydomonas* there are 1 *Nrt1*, 5 *Nrt2*, and 6 *Nar1* genes. The precise role of each transporter in terms of substrate specificity, capacity, localization, and participation in nitrate assimilation and its efficiency is a challenge that is now starting to be addressed.

7.5.2 Functionality of NRT1 Transporters

The recently released *Chlamydomonas* genome sequence (http://genome.jgi-psf.org/chlre3/chlre3.home.html) suggests the existence of a putative NRT1 transporter (Table 7.2). However, a functional characterization of this system is needed. The following section refers only to data from plants.

The NRT1 transporters (also named PTR transporters) belong to the POT family, which includes numbers of H^+-dependent oligopeptide transporters from mammals, plants, fungi, and bacteria [111,112]. The *Arabidopsis AtNrt1.1* (*CHL1*) gene was the first member of the NRT1 family identified and was cloned on the basis of its function. Mutants in the *AtNrt1.1* gene (*chl1* mutants) were originally isolated in the early 1970s from screens based on ClO_3 resistance [56] and later shown to be defective in ClO_3^- and NO_3^- uptake [113]. The *AtNrt1.1* gene was cloned by T-DNA tagging and found to encode a hydrophobic 65-kDa protein with the characteristic features of a typical membrane transporter [91].

NRT1 proteins are predicted to have 12 transmembrane domains, with a long loop containing many charged residues separating the first six transmembrane domains from the second six, and short N- and C-terminal ends [91,114]. The N- and C-terminal domains are quite short (18 and 28 residues, respectively).

The *Arabidopsis chl1* mutant was the primary source of information about the function of the AtNRT1.1 transporter. CHL1 was initially described as a NO_3^--inducible low-affinity transporter [91,115,116].

The present picture concerning NRT1.1 is complex and shows how a single transporter is involved in regulating multiple functions:

- NRT1.1 is now considered as a dual affinity transporter, both HATS and LATS [117,118]. The phosphorylation of NRT1.1, triggered by limited external NO_3^- availability, is responsible for the shift from low to high affinity, thus adapting the functional properties of the transporter to the resource level in the root environment [119].
- NRT1.1 is strongly expressed in nascent organs of root and shoot (root tips, emerging lateral roots, and nascent leaves) and plays a crucial role in early phases of development of these young organs [120]. In particular, NRT1.1 mutants display altered root architecture in some conditions, with reduced growth of primary and secondary roots, even in the absence of added NO_3^- in the external medium. This suggests an alternative function for NRT1.1, independent of NO_3^- transport [120].
- It has been reported recently that the mutation of NRT1.1 also leads to lower sensitivity to drought, related to a reduced stomatal opening because of impaired NO_3^- transport in stomata guard cells [121b].

Clearly, the view that NRT1.1 behaves only as a transporter in charge of the NO_3^- uptake from the external medium is an oversimplification. This protein appears to fulfill multiple physiological functions, which are becoming evident more than 30 years after identification of the first NRT1.1 mutant [56].

One candidate suggested as LATS is the *AtNRT1.2* gene product. This gene is constitutively expressed in the absence of nitrate, and its functional analysis in *Xenopus* oocytes shows specificity for nitrate as a substrate (*K*m 5.9 m*M*), but it is not able to transport dipeptides or histidine [114]. Transgenic plants containing an antisense *AtNrt1.2* also confirm its role as a LANT [114]. Recently, the *Arabidopsis* NRT1.4 was shown to be a LANT related to nitrate homeostasis in the leaf petiole, so defects in this gene alter leaf development [121a].

Functional studies have been performed over other plant species to demonstrate the functionality of NRT1 transporters. For example, *Brassica napus* NRT1.2 was confirmed to be a LANT [110]. However, in this case the *K*m for nitrate was voltage dependent, increasing from 4 m*M* at a membrane potential of 40 mV to 14 m*M* at 180mV. In addition to its NO_3^- transport activity, BnNRT1.2 was also found to be able to transport L-histidine, generating even larger currents than with NO_3^- [110].

7.5.3 Functionality of NRT2 Transporters

The NRT2 transporters, also named NNP (for nitrate–nitrite porter), belong to the major facilitator superfamily (MFS), which includes sugar transporters from mammals, plants, yeast, and bacteria. MFS is a divergent group of proteins that are typically 500 to 600 amino acids in length and have a characteristic membrane topology of 12 transmembrane domains arranged as two sets of six, connected by a cytosolic loop [112].

The first eukaryotic member from this family was cloned from *Aspergillus* (*Emericella nidulans*, *crnA*). A mutation in *crnA* conferred resistance to ClO_3^- and a partial defect in NO_3^- uptake [122]. Subsequently, two homologous genes were discovered in *Chlamydomonas*. These transporters were located in the nitrate assimilation gene cluster [41,42]. Later, a second cluster was shown to contain a third *Nrt2* gene in the alga [51].

Nrt2 genes have been cloned from a wide range of plant species, fungi, algae, yeast, and bacteria [26,27,108]. On the basis of their structural features, members of the NNP family have been classified into several groups (Figure 7.4). Transporters (from bacteria) are the smallest members of this protein family and have a minimal

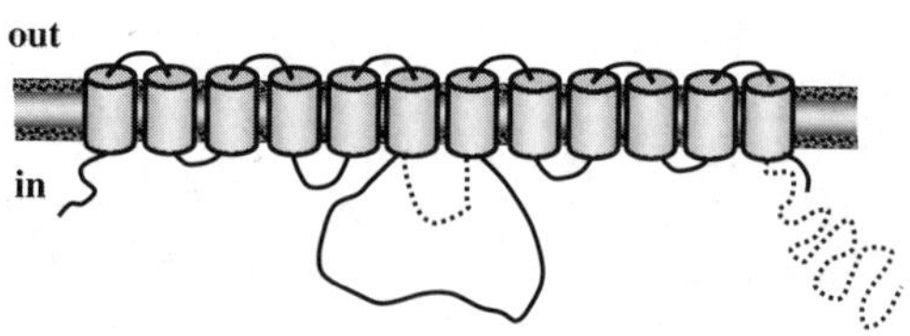

FIGURE 7.4 Models for NRT2 proteins. Central loop and C terminal region are indicated in continuous line for the fungal NRT2 transporter and in discontinuous line for algal/plant transporters.

amount of sequence outside the 12 transmembrane domains. The fungal members of the family have a large hydrophilic central loop of 90 amino acids located between transmembrane domains 6 and 7. The algal and higher plant members of the family have an extended C-terminal domain of ~70 amino acids that can include an N-terminal sequence extension of ~20 amino acids. This N-terminal domain is highly conserved among the NRT2 family but is absent in the algal and barley sequences [26].

In silico structural analysis has been made to identify signature motifs for the NRT2 family and possible substrate recognition [23,112]. Recently, it has been shown experimentally in *Aspergillus crnA* that many conserved glycine residues throughout the protein sequence have a structural role, while certain conserved charged or polar residues within transmembrane domains (for example, arginine residues conserved within transmembrane domains 2 and 8) are involved in a nitrate-binding function [123].

NRT2 transporters seem to involve one or two components (Figure 7.5). In *Chlamydomonas*, a combination of mutant analysis and oocyte expression experiments indicates that some NRT2 members (NRT2.1 and NRT2.2) require an additional protein (NAR2) for their functionality [42,92], and are thus two-component systems. Other NRT2 proteins, such as NRT2.3 from *Chlamydomonas* and CrnA from *Aspergillus*, are single-component systems [52,97]. Until now, NRT2 proteins from plants have not been shown to be functional as single proteins, and they might require a protein homolog to *Chlamydomonas* NAR2, as recently discovered in plants [124]. Nine *Nar2*-type genes have been identified in *Hordeum vulgare* and two in *Arabidopsis* [124]. NAR2 proteins appear to have a single transmembrane domain and may interact with NRT2.1 to modify its function. The amino acid sequence $KX_2KX_2LCYX_2SX_3RXWRX_3DX_4DK$ between amino acids 140 and 180 seems to be characteristic of the higher plant family of NAR2 proteins.

The use of the *Xenopus* oocytes expression system has shown that an NAR2 mRNA (*HvNAR2.3*) was able to reconstitute high-affinity NO_3^- transport activity when co-injected with mRNA for the otherwise inactive *HvNrt2.1* [124]. This result provides strong evidence for the utility of the *Chlamydomonas* model for higher plants.

Chlamydomonas mutant strains defective in several of the nitrate gene clusters have allowed identification of four high-affinity nitrate/nitrite transporters (Figure

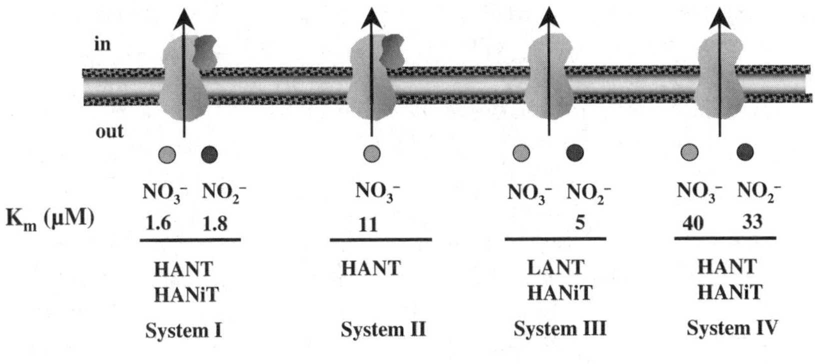

FIGURE 7.5 Single- and two-component nitrate transport systems in *Chlamydomonas*.

7.5). System I corresponds to a bispecific HANT/HANiT encoded by *Nrt2.1/Nar2*; system II to a monospecific HANT encoded by *Nrt2.2/Nar2*; system III to a bispecific HANiT and LANT, probably encoded by *Nrt2.3*; and system IV to a bispecific HANT/HANiT for which the gene responsible has yet to be identified [42,52,61,62]. The pH dependence of the nitrate-elicited currents by NRT2.1–NAR2 expressed in oocytes is consistent with an H^+-cotransport mechanism [92].

These transport systems are differentially regulated by the carbon and nitrogen source. Systems I, II, and III are optimally expressed at high CO_2, and their activity is blocked by ammonium; system IV is expressed optimally under limiting CO_2 and its activity is not inhibited by ammonium. In contrast to systems I, II, and III, system IV is inhibited by CO_2 [52,61]. Concerning the function for each of these systems, mutants deleted in systems I and II and carrying functional systems III and IV are unable to grow efficiently in nitrate media [42]. Thus, under sufficient CO_2, systems I and II have a primary function in the provision of nitrate for growth and systems I and III in nitrite entry.

7.5.4 NAR1 Transporters

The nitrite transport step into the chloroplast is not well documented in plants, probably because of the lack of any molecular evidence and of the long-standing assumption that nitrite can diffuse freely as nitrous acid into the chloroplast [125]. However, nitrite uptake into intact pea chloroplasts shows saturation kinetics, preference for alkaline pH, and sensitivity to protein modifiers; this favors the existence of a nitrite-mediated channel or transporter vs. the permeation of nitrous acid [126,127].

Nitrite transport into chloroplast inner envelope vesicles from pea has been evaluated by Shingles et al. [128]. These authors propose that nitrite rapidly diffuses across the plastid membrane depending on a proton gradient, so the proton-linked NO_2^- transport should be bidirectional. Nitrite concentrations change significantly in roots of barley seedlings, depending on the nitrate availability in the environment [104], and in spinach leaves during the light–dark transitions [129]. Thus, the need for a plastidial nitrite transporter is important for two reasons: (1) to avoid cellular toxicity of nitrite; and (2) to increase efficiency of the nitrite reduction step.

Studies with *Chlamydomonas Nar1.1* have provided the first molecular evidence that nitrite transport to the chloroplast is a regulated process mediated by specific transporters rather than the result of diffusion [34]. In spite of identification of six members of the NAR1 family in *Chlamydomonas* (Table 7.2), further studies will be required to know the role of each NAR1 protein. Available data show that the *Nar1* gene family may be closely associated with carbon and nitrogen metabolism because *Nar1*.1 is nitrate upregulated and under control of the nitrate-pathway-specific regulatory gene *Nit2*, whereas *Nar1*.2 (*LciA*) is upregulated by low CO_2 and under control of the carbon-pathway-specific regulatory gene *ccm1* [44]. In plants, NAR1 proteins cannot be identified on the basis of sequence homology, but the role of NAR1.1 is so fundamental in *Chlamydomonas* that its function needs to be carried out by another protein family in plants. The precise function for *Nar1.1*, as a chloroplast nitrite

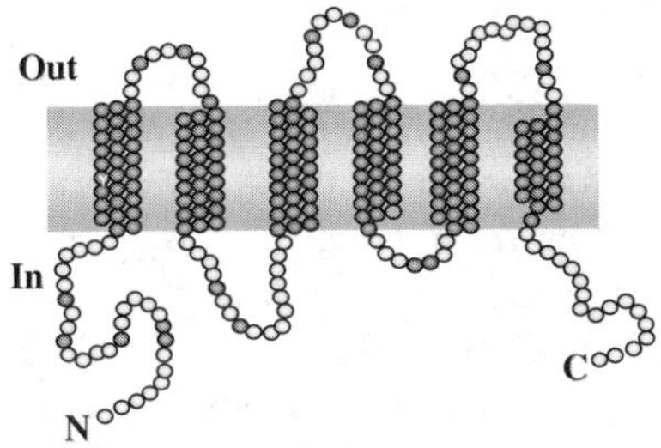

FIGURE 7.6 (**Color Figure 7.6 follows p. 144.**) Model for the NAR1.1 protein showing the six transmembrane domains. Residues mostly conserved among FNT proteins are shown in red. (Scheme modified from Galván, A. et al., *J. Exp. Bot.*, 53, 845, 2002.)

transporter allowing for efficient nitrate utilization, has been deduced on the basis of the following data:

Nar1.1 is located in the nitrate cluster and coregulated with the other clustered genes (Figure 7.2; [41]). Thus, *Nar1.1* is expressed in nitrate but not in ammonium media.

NAR1.1 corresponds to an integral membrane protein predicted to have six spanning-membrane domains, a plastidial localization, and significant identity to formate and nitrite transporters from bacteria (Figure 7.6) [34].

The nitrite uptake activity by intact chloroplasts isolated from Nar1;1$^+$ and Nar1;1$^-$ strains supports the notion that NAR1;1 is a plastidial nitrite transporter with an apparent *K*s for nitrite of about 5 μM [34].

Nar1.1 allows nitrate utilization when this nutrient is limiting for the cells. This limitation of nitrate takes place in strains lacking HANT systems I and II at nitrate millimolar concentrations, or in strains having HANT with nitrate micromolar concentrations in the medium [34].

Nar1.1 improves nitrate use efficiency for growth under light/dark cycles and low CO_2 environments. Under such conditions, strains lacking *Nar1.1* uncouple nitrate reduction from the cells' capability to assimilate the ammonium produced because of a significant deregulation in expression of enzymes and transporters for nitrate assimilation including GS1 [130].

7.6 AMMONIUM ASSIMILATION

Incorporation of ammonium is a basic process shared by nitrate assimilation and other alternative nitrogen source pathways [37]. The GS/GOGAT cycle is the major step for ammonium incorporation into carbon skeletons by photosynthetic organisms. GS catalyses the formation of glutamine from ammonium and glutamate in an ATP-dependent reaction. In plants, GS isozymes are encoded by multigene families and some of their members show cytosolic localization, while others have a plastidial or nodular localization [131]. The *Chlamydomonas* genome sequence shows three *GS* genes (Table 7.2); *GS1* encodes a cytosolic form and *GS2* and *GS3* are proposed to encode plastidial isozymes. Although a *GS3* expression pattern is not documented, the major transcript level corresponds to that of *GS2*, which is constitutively expressed with respect to nitrogen supply [39]. *GS1* transcripts increase in cells grown in nitrate and decrease in cells grown in ammonium [39]. GS1 is proposed to play an active role in nitrate assimilation [132].

GOGAT catalyses the transfer of the amide group from glutamine to 2-oxoglutarate in a reaction that requires reducing equivalents. Two molecules of glutamate, the substrate of GS, are produced. Two GOGAT isoenzymes—one specific for reduced ferredoxin as electron donor and another specific for NADH—have been characterized in plants. Fd-GOGAT is a 130- to 150-kDa monomer, with a [3Fe-4S] center located in plastids and roots; NADH-GOGAT is a 158- to 240-kDa monomer with the same prosthetic group, located in the plastids [133]. GOGAT gene number per genome differs between species [134]. In *Chlamydomonas,* single genes encoding the NADH- and the ferredoxin-GOGAT are present (Table 7.2). These enzymes have been characterized in detail [135,136].

7.7 AMMONIUM TRANSPORT GENES

As schematized in Figure 7.1, ammonium transporters are expected to operate in different cell localizations (plasma membrane, chloroplast, and mitochondria) mediating influx and efflux. Because ammonium is a strong negative signal of nitrate assimilation, it is important to know the different transporters that could mediate its effects. Knowledge of the mechanisms regulating ammonium transport systems is also essential for better understanding of nitrogen metabolism and plant growth.

At the physiological level, biphasic kinetics for ammonium uptake in several species have been assigned to two different systems of transport: low-affinity transport systems (LATS), which are related to passive K^+ channels [12,137,138], and high-affinity transport systems (HATS) that mediate active transport by coupling NH^+_4 influx to a H^+ gradient [139,140]. Similarly, at the physiological level, in *Chlamydomonas* cells, two transport systems subject to circadian rhythm [141] are thought to participate in ammonium uptake: a constitutively expressed, low-affinity version and a high-affinity version negatively regulated by ammonium [142].

Although LATSs have not yet been described at the molecular level, HATSs have been widely described in diverse organisms like plants, yeasts, bacteria, fungi, and animals, and form the AMT/MEP family [12,140,143]. This family has also been related by homology with the human rhesus (Rh) blood proteins. *Chlamydomonas* is one of the few organisms that have genes from both families: *Amt* and *Rh*. A possible role of the *Chlamydomonas* Rh proteins is as a bidirectional channel for ammonia and CO_2 [144,145].

In plants, most AMTs have been included in a large family designated AMT1 [12,140], with the exception of some AMTs identified in *Arabidopsis* and *Lotus japonicus* that are included in AMT2, a new family of transporters whose sequence identity is closer to yeast and bacteria than to plants [146–148]. Five members of the AMT1 family have been described in *Arabidopsis* [140,149,150] and three in tomato [151]; its existence is also known in other species such as rice [152], *Brassica napus* [153], and *Lotus japonicus* [154].

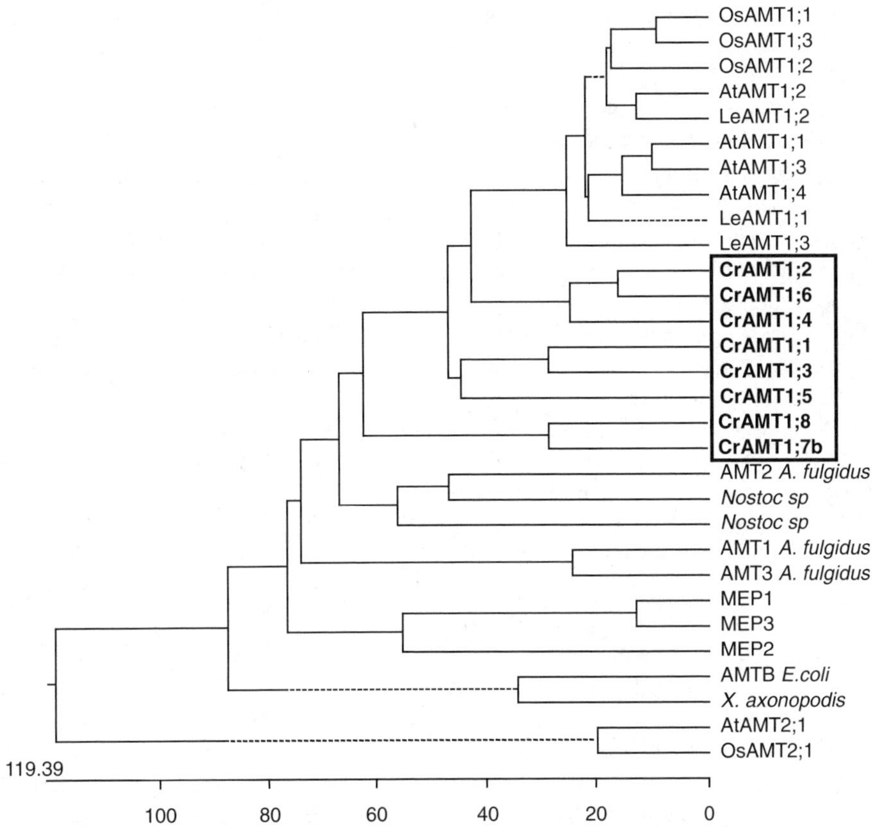

FIGURE 7.7 Phylogenetic tree of the AMT1 proteins. Alignment was performed with ClustalW and the tree with the DNAStar package. *Chlamydomonas* AMT1 are included in a box. Other details are given in the text.

Eight members of the *Amt1* gene family (*Amt1*.(1 to 8) genes) have been identified in *Chlamydomonas* (Table 7.2) [45], representing the largest family thus far described in any organism. On the other hand, no *Amt2* genes have been found in *Chlamydomonas*, suggesting that the proposed role of AMT2 transporters in plants could be undertaken by some of the AMT1 proteins in the alga. In Figure 7.7, the phylogenetic relationships among these transporters and those from other organisms are shown. AMT1.1 to 6 are closely related to the plant AMT proteins, whereas AMT7 and 8 are more distantly related to plants.

The *Chlamydomonas Amt1* genes show interesting evolutionary relationships because a particular feature present in all is the existence of some very small exons (less than 40 bp) and conservation of some intron positions that may be related to its transcriptional regulation. Alternative splicing has been found in the 5′ regions of *Amt1*.5 and *Amt1*.7. This complex regulation might even result in different protein localization, as suggested for the two isoforms deduced for AMT1.7 (Table 7.2) [45].

That a single cell such as *Chlamydomonas* has such a wide set of putative transport systems for ammonium and other nutrients such as nitrate may reflect the need for complementary affinities and activities to obtain different substrates efficiently under changing environmental conditions. However, in *Chlamydomonas,* the number of *Amt1* genes is higher than that of the *Nrt2* genes (Table 7.2), in contrast to plants where the opposite pertains [12,15]. This is surprising if *Chlamydomonas* is considered as a unicellular green alga and thus no tissue specialization exists.

In any case, ammonium, in contrast to nitrate, requires a regulated intracellular flux because it can be assimilated in the cytoplasm or the chloroplast, accumulated in the vacuole, or excreted to the medium, and ammonium produced during photorespiration needs to be exported from the mitochondria [15,139]. The localizations of the AMT1 proteins remain unknown, although several predictions suggest a clear chloroplast localization for AMT1.2 and especially for AMT1.4, whose hypothetical signal peptide has substantial similarity with that of the *Chlamydomonas* chloroplast protein NiR (Table 7.2).

Putative ammonium transporters AMT1 from *Chlamydomonas* show particular expression patterns in media containing different nitrogen sources (Figure 7.8): (1) *Amt1*.4 and *Amt1*.7 have the highest expression levels; and (2) by comparing different nitrogen conditions, putative ammonium transporters 1, 2, 4, and 5 have maximum expression in an N-free medium; the ammonium transporters 3 and 8 show maximum expression in nitrate medium, whereas 6 and 7 are enhanced in ammonium medium [45]. The expression of the *Chlamydomonas Amt1*;1 has a complex regulatory mechanism responding to repression by ammonium, ammonium derivatives, and nitrate in a process mediated by the regulatory gene *Nit2,* so NIT2 provides a signal for the presence of a usable nitrogen form—nitrate—connecting the pathway of ammonium uptake with that of nitrate assimilation. This means that NIT2 has a dual role in gene expression: the well-known positive one on nitrate assimilation and a novel negative one on *Amt1*;1 expression [45,60]. *Chlamydomonas Amt1*;1 has an expression pattern similar to that of the *Arabidopsis* gene *Amt1*;1, which encodes the major system responsible for ammonium uptake [150].

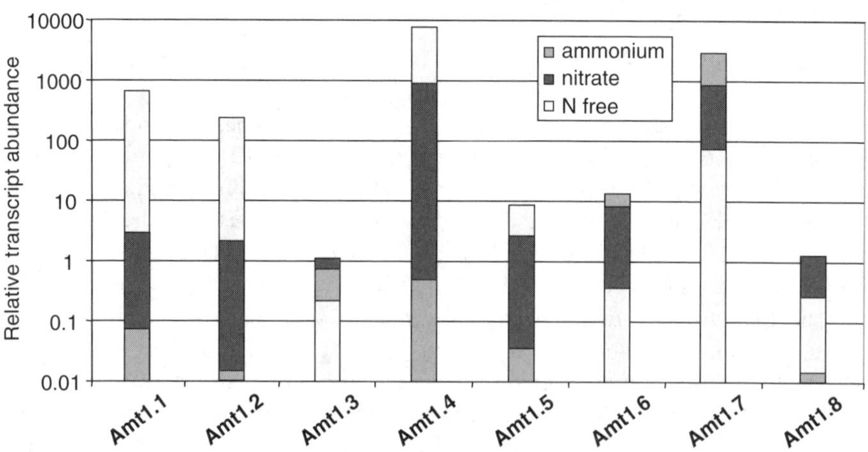

FIGURE 7.8 Expression of *Amt1* transcripts in different nitrogenous media. The relative abundance is referred to that of ubiquitin ligase transcript used as a control. (From González-Ballester, D. et al., *Plant Mol. Biol.*, 56, 863, 2004.)

7.8 NITRATE ASSIMILATION AND LIGHT/CIRCADIAN RHYTHM

Light positively regulates most of the nitrate assimilation pathway genes and activities. It is known that NR activity and mRNA levels of plants grown in a day/night cycle fluctuate between a maximal level at the beginning of the day period and an almost undetectable level at the end of the day [22]. Further evidence of this light regulation are the diurnal changes of nitrate and nitrite uptake that generally peak during the light period and reach a minimum in the dark [26,155].

An endogenous clock regulates the temporal expression of genes/mRNAs involved in the circadian pathway. In *Chlamydomonas,* a recently identified, clock-controlled RNA-binding protein (Chlamy 1) represents an analog of the circadian trans-acting factor, CCTR, from the phylogenetically diverse algal species *Gonyaulax polyedra* [156]. Chlamy1 protein could act as a translational repressor, preventing translation of UG-repeats-bearing mRNAs from the end of the light until the end of the dark phase. The strength of the interaction between Chlamy1 and transcripts may be influenced by the number of UG units in a region. Chlamy1 has been cloned and is composed of two subunits: C1 is involved in protein–protein interaction and C3 bears three RNA recognition motif domains [157].

Chlamy1 was reported to bind mRNAs whose products are key components of nitrogen and CO_2 metabolism [158]. Some of the genes under Chlamy1 control are involved in uptake of nitrite (*Nrt2.3*), its reduction to ammonium (*Nii1*), fixation of ammonium as glutamine (*GS2*), and arginine biosynthesis (*ARG7*). Other proteins encoded by these RNAs are related to photosynthesis and the CO_2 shuttle into the chloroplast (LIP36-G1), and CO_2 fixation as ribulose-1,5-bisphosphate carboxylase (RBCS1, the small subunit of RUBISCO). Another is the light-dependent NADPH-protochlorophyllide-oxidoreductase (L-POR), one of two PORs that catalyze the reduction of protochlorophyllide to chlorophyllide, a regulatory step in chlorophyll biosynthesis [159]. Finally, YPTC4 is a G-protein [160] whose function is not yet fully understood [158]. These genes, plus others containing UG units putatively controlled by Chlamy1, are shown in Figure 7.9.

The binding activity of Chlamy1 is controlled by the circadian clock. Thus, the regulatory properties of this factor are limited to a certain time window, which begins at the end of the day phase and ends in the middle of the night phase. During this period, binding activity of Chlamy1 is high, and at other time points it is low [161]. Clock-controlled RNA-binding proteins have also been identified in other species. In *Arabidopsis*, a glycine-rich RNA-binding protein, GRP7, was characterized. Both transcripts and the protein are components of a negative feedback circuit capable of generating a stable oscillation [162].

7.9 NITRATE ASSIMILATION AND REDOX REGULATION

It has been recently shown in *Chlamydomonas* that the redox state of the plastoquinone (PQ) pool plays a key role in *Nia1* gene expression [163]. A reduced PQ pool is needed as a positive signal to allow *Nia1* expression so that chemical inhibitors

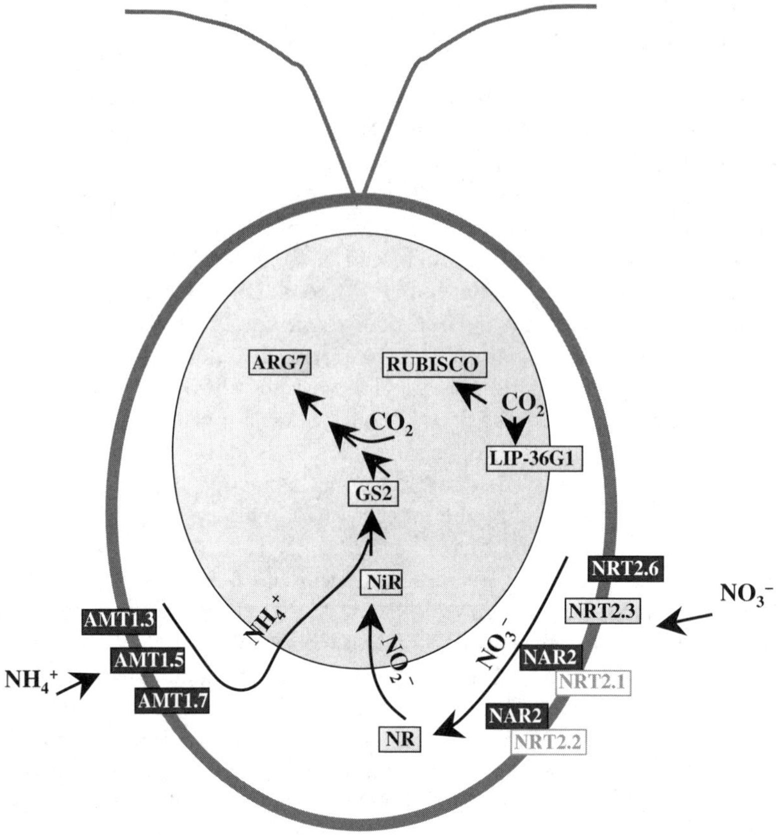

FIGURE 7.9 (**Color Figure 7.9 follows p. 144.**) Gene targets of Chlamy1 protein, modulating circadian rhythm. The scheme includes genes proposed to be a target of chlamy1 by Waltenberger et al. (pink boxes), plus other putative target genes proposed on the basis of their sequence containing UG repeats (blue boxes). (From Waltenberger, H. et al., *Mol. Genet. Genomics*, 265, 180, 2001.)

or mutations preventing loading of PQ with electrons have a negative effect on expression. These results are related to others reported in tobacco, *Arabidopsis*, and *Lemna*, where signals from the photosynthetic electron flow are controlling *Nia1* gene expression. Nonetheless, in plants, an oxidized state of an electron transport component appears to activate expression [164]. This apparent disagreement may reflect differences in the physiology and biochemistry of nitrate reduction in plants and algae.

Nitrate and nitrite reduction requires reducing power. In photosynthetic cells, this reducing power is supplied by the photosynthetic electron transfer chain (mediated by NADPH-ferredoxin oxido-reductase) generating reduced ferredoxin for nitrite reduction and redox valves, which feed the cytosol with NAD(P)H for nitrate reduction. Some participation of the Calvin cycle steps has been demonstrated: glucose-6-phosphate dehydrogenase and Calvin cycle activities are increased in nitrate media in *Chlamydomonas* and *Arabidopsis* [36,165].

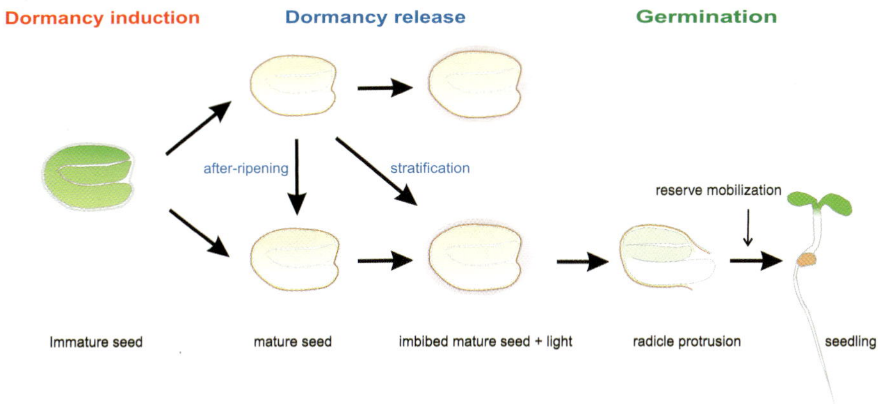

COLOR FIGURE 2.1 A general outline of dormancy and germination in *Arabidopsis* seeds.

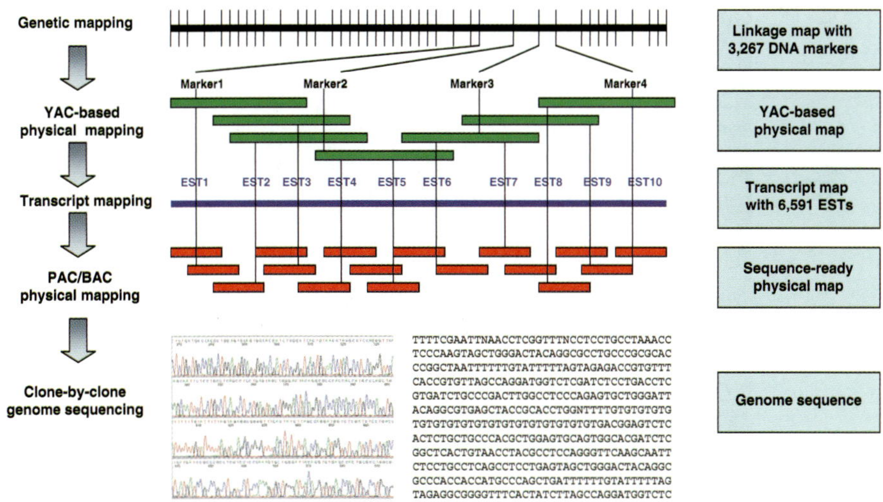

COLOR FIGURE 4.1 An overview of the strategy for generating a map-based sequence of the rice genome. A high-density linkage map was initially established and the mapped DNA markers were used to screen a YAC genomic library. The resulting YAC-based physical map was used as a template for transcript mapping. The mapped ESTs were then used to screen genomic libraries in PAC and BAC vectors to construct a sequence-ready physical map. The aligned clones were used for sequencing so that all sequenced clones could be assigned to specific positions in the chromosomes.

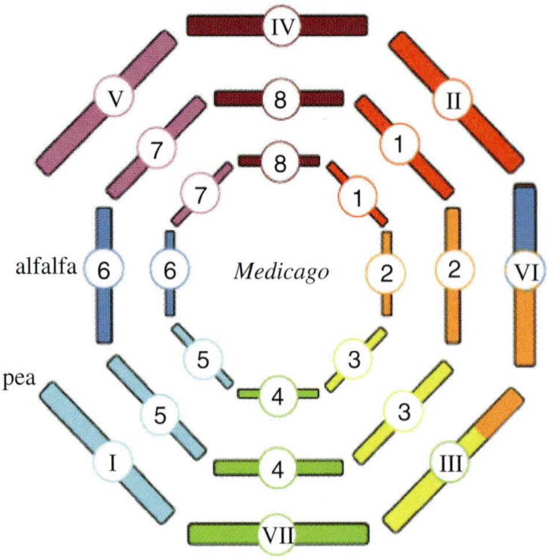

COLOR FIGURE 5.4 Nod factor signaling cascade as identified by forward genetics. Nod factors are perceived by LysM receptor kinases. These activate at least two downstream signaling pathways: one depending on the DMI proteins and a DMI independent pathway for which no specific genes have yet been identified. Both pathways can be discriminated based on distinct Ca^{2+} signals, external Ca^{2+} influx, and perinuclear Ca^{2+} spiking. The DMI pathway is shared mycorrhizal-secreted signal perceived by a hypothetical plant Myc receptor. For mycorrhizal-based signaling, a DMI independent pathway is predicted.

COLOR FIGURE 6.1 A diploid *Brachypodium distachyon* inbred line grown under long day-lengths can, depending upon the genotype, begin the transition to flowering within 3 weeks of planting. In this figure, a diploid genotype was grown under 20-h day-length restrictive conditions, as shown in the left panel, or under nutrient replete conditions, as shown in the right panel. The plant in the left panel is 4 weeks old, and the plant on the right is 8 weeks old, with seeds maturing. Alternatively, by growing this inbred diploid line under short-day conditions, it can be maintained in vegetative phase for long periods of time. (Source: David Garvin, USDA-Agricultural Research Service.)

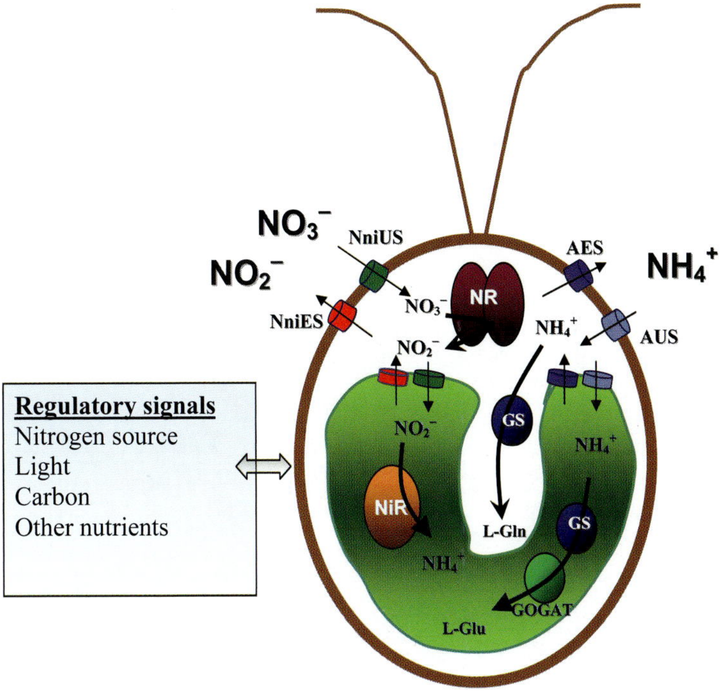

COLOR FIGURE 7.1 The nitrate assimilation pathway in *Chlamydomonas*. NniUS = nitrate/nitrite uptake influx systems; NniES = nitrate/nitrite efflux systems; AUS = ammonium uptake influx systems; AES = ammonium efflux systems. Other details are indicated in the text.

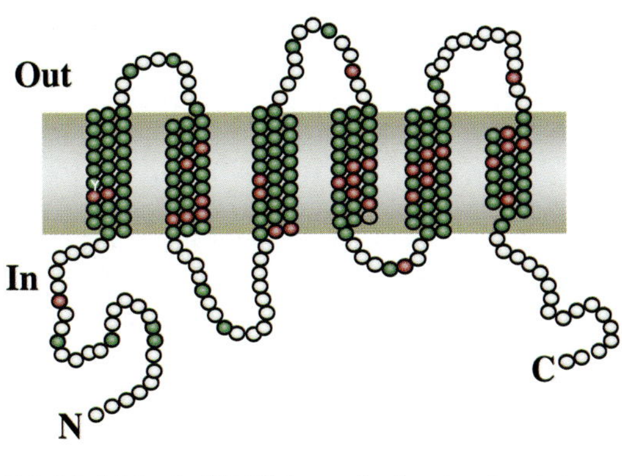

COLOR FIGURE 7.6 Model for the NAR1.1 protein showing the six transmembrane domains. Residues mostly conserved among FNT proteins are shown in red. (Scheme modified from Galván, A. et al., *J. Exp. Bot.*, 53, 845, 2002.)

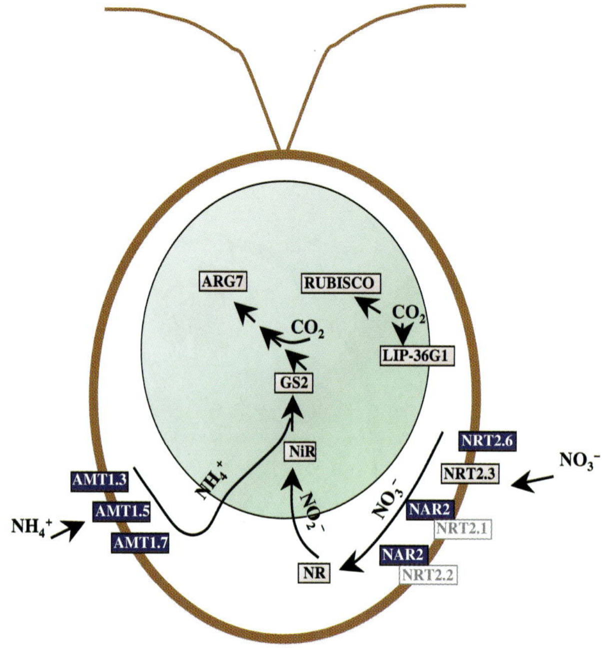

COLOR FIGURE 7.9 Gene targets of chlamy1 protein, modulating circadian rhythm. The scheme includes genes proposed to be a target of chlamy1 by Waltenberger et al. (pink boxes), plus other putative target genes proposed on the basis of their sequence containing UG repeats (blue boxes). (From Waltenberger, H. et al., *Mol. Genet. Genomics*, 265, 180, 2001.)

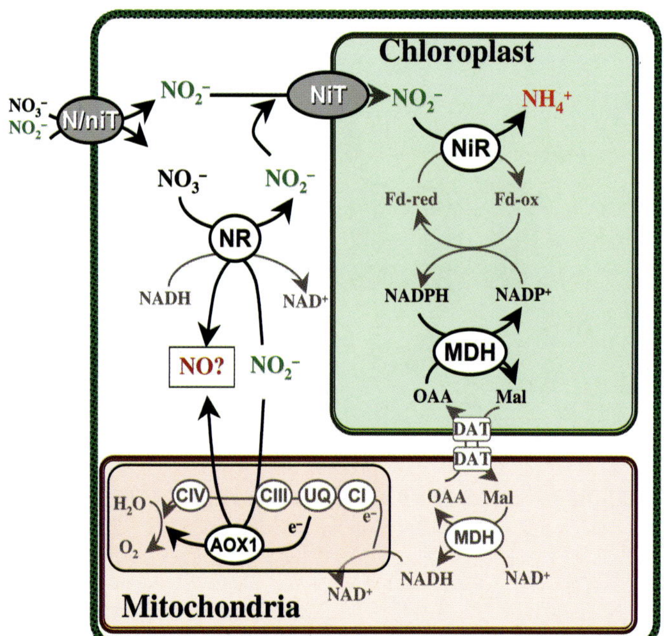

COLOR FIGURE 7.10 The redox valves and the nitrate assimilation pathway.

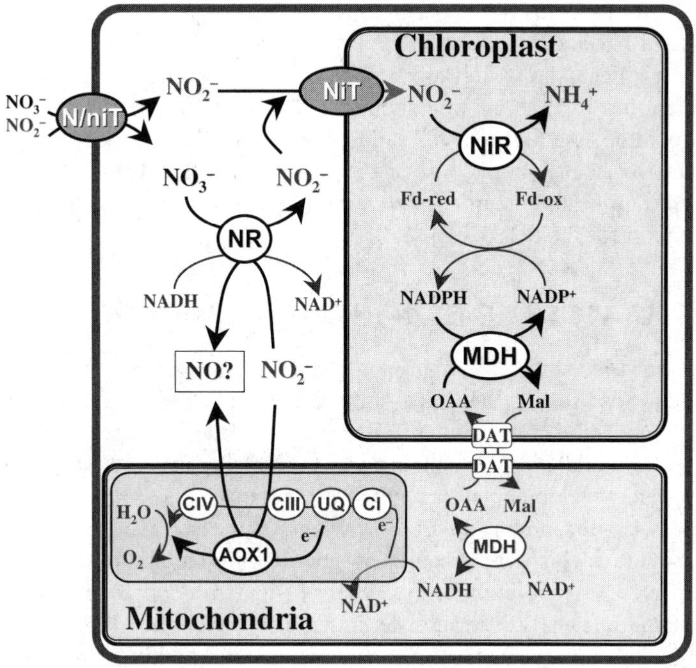

FIGURE 7.10 (**Color Figure 7.10 follows p. 144.**) The redox valves and the nitrate assimilation pathway.

In *Chlamydomonas,* NMDH and AOX1, which are closely related to their counterparts in plants, are involved in processes that control the fine-tuning of redox metabolism in chloroplast and mitochondrion [166] (Figure 7.10). NMDH participates in the export of reducing equivalents from the chloroplast to the cytosol by the malate–oxalacetate shuttle. Activity of this "malate valve" is regulated by the redox switch of NMDH, which is turned on and off by light signaling mediated by the thioredoxin–ferredoxin system [50,167].

AOX is a mitochondrial terminal oxidase that funnels electrons to oxygen from reduced ubiquinone. It bypasses the respiratory pathway through complexes III and IV of the mitochondrial electron transfer chain and the concomitant generation of electrochemical potential; thus, synthesis of the amount of ATP per electron pair is diminished. AOX could play a role as a redox valve to balance carbon metabolism and mitochondrial electron transport [168].

Clustering of *Nmdh1* and *Aox1* genes with nitrate assimilation genes in *Chlamydomonas* provides a direct molecular link between NADH and ATP pool regulation and nitrate assimilation (Figure 7.2 and Figure 7.10). NMDH provides reducing power for cytosolic nitrate reduction, and its activity level is compromised by competition for reduced ferredoxin, the electron donor for nitrite reductase [166].

Specific induction of *Aox1* by nitrate and not by ammonium might indicate the requirement to adapt mitochondrial electron flow and thus ATP synthesis during assimilation of these nitrogen sources. This adaptation is especially critical under stress conditions, such as high concentrations of nitrite in the cytosol that might act as substrate of NR and be converted to nitric oxide [169,170] or in the mitochondria where NO has been shown to be produced by the alternative oxidase in *Chlorella* [171]. NO in turn inhibits the mitochondrial electron transport chain, so alternative oxidase would be essential for ATP production under NO presence. Whether or not NO production is the result of stress or a metabolic condition that triggers a signaling cascade will require the identification of intermediates in this cascade and the final target of action for the cell response.

7.10 REGULATORY GENES IN NITRATE ASSIMILATION

Membrane proteins, such as pumps, ion channels, transporters, and receptors, are responsible for controlling the flow of nutrients and other solutes across the plasma membrane and cellular organelles. They also provide the appropriate signaling metabolites for modulating the cell response to changing needs according to nutrient availability and developmental or physiological status.

Nitrate is sensed, and this up- or down-regulates the expression of a large number of genes, some of which are specific for nitrate assimilation. Other genes link this route to different metabolic pathways. Studies with *Arabidopsis* have found that by using transcriptome analysis in short-term treatments with low levels of nitrate, more than 1000 genes are significantly induced or repressed within 20 min of treatment [172]. In roots, 270 genes were differentially expressed in media containing nitrate as opposed to ammonium nitrate [36,172,173].

In *Chlamydomonas,* nitrate signaling occurs intracellularly and directly depends on the activity of the nitrate transport systems [174,175]. Thus, in *Chlamydomonas* the presence of particular high-affinity nitrate transport systems has an important regulatory role in expression of nitrate assimilation genes. According to their affinity, specificity, and capacity for nitrate, these systems are responsible for differential nitrate signaling by regulating intracellular concentrations of nitrate.

Recently, serial analysis of gene expression (SAGE) transcriptome analysis in the nitrate transporter Chl1 mutant of *Arabidopsis* has revealed how a single transporter affects the expression of hundreds of genes. Because of the marked deregulation of *Nrt2.1* in the mutant, it was suggested that *Nrt1.1* plays a direct signaling role in regulating other nitrate transporters [176].

Ammonium transporters have been suggested to participate in the sensing of ammonium [177]. The yeast ammonium transporter MEP2, dispensable for cell growth, has been implicated in the sensing of ammonium [178,179]. In *Chlamydomonas,* the ammonium-uptake defective mutant strain 2170 is also relieved from the negative effect of ammonium or methylammonium on the nitrate pathway [180].

The regulatory genes for nitrate assimilation are well defined in fungi and yeast [30,181]. Two positively acting regulatory genes are required for expression of nitrate transporter and nitrate reduction genes in fungi [29,181]. *NirA/Nit4/YNA1* genes from *Aspergillus*, *Neurospora*, and *Hansenula*, respectively, are pathway-specific genes involved in nitrate induction and correspond to GAL4-like Cys6/Zn2 binuclear zinc cluster [181]. AREA/NIT2 from *Aspergillus* and *Neurospora* are major regulatory proteins mediating nitrogen repression from readily usable nitrogen sources such as ammonium and glutamine and correspond to GATA-binding transcription factors [182,183]. The mechanism proposed for the NIT2 negative function in *Neurospora* is that NMR1, a negatively acting protein, interacts with NIT2 in the presence of catabolic repressors and prevents the binding of NIT2 to DNA for transcription of target genes.

Various attempts have been made to identify regulatory genes from plants [22,184]. However, the relationship of the isolated genes to regulation of the nitrate pathway has not been shown, even though a GATA motif was found in the promoter of the spinach NiR gene [185] and *in silico* analysis revealed a large number of GATA-family transcription factors in *Arabidopsis* and rice [186].

The regulatory fungal model does not seem to fit in photosynthetic eukaryotes and clues for regulatory genes may therefore come from systems such as *Chlamydomonas*. It has been hypothesized that several genes act to mediate the positive effects of nitrate and the negative ones of ammonium. Their deficiency may lead to partial phenotypes that can explain the difficulties experienced in the genetic dissection of regulation in algae and plants [27].

In *Chlamydomonas*, the expression of nitrate assimilation genes is co-regulated (Figure 7.2). These genes are subject to repression by ammonium, induction by nitrate, and the control of the positive-acting regulatory gene *Nit2* [42,60]. This gene was cloned by transposon tagging from chlorate-resistant mutants [47]. Although structural analysis of the deduced NIT2 protein has not been performed, the *Chlamydomonas* genome database reveals that NIT2 is a transcription factor containing an RWPXRK box present in plant proteins involved in N control and the *Chlamydomonas* mid protein involved in minus mating type dominance [36,187,188]. *Nit2* is itself subject to ammonium repression, which implies an additional level of control [47].

A functional genomics approach to identify regulatory genes for nitrate assimilation has been recently performed in *Chlamydomonas*. By taking advantage of deletion events that occur during integration of a heterologous marker in *Chlamydomonas* transformation, along with the *Chlamydomonas* genome sequence, an ordered mutant library of 22,000 strains has recently been obtained [8]. Assuming that *Chlamydomonas* contains about 17,000 genes and deletions affect wide genomic regions, it has been proposed that such a number of mutants will be sufficient to cover most of the *Chlamydomonas* genome.

The arylsulfatase reporter gene under the control of the *Nia1* gene promoter has served as a sensor for identifying regulatory mutants (Figure 7.11). A forward screening of the library allowed for the selection of 145 mutants defective at putative genes related to the positive signal of nitrate or the negative signal of ammonium. The ammonium-insensitive mutants were found to be defective at genomic regions bearing putative new genes related to regulatory functions such as guanylate cyclase, protein kinase, peptidyl-prolyl isomerase, or DNA binding [8], although no direct evidence correlating these genes with ammonium repression exists. In addition, some insertions in ammonium-insensitive mutants map to unknown regions or genes in the genome. From these and previous results [189–191], it is becoming evident that a complex network of signaling proteins mediates the effects of ammonium and its derivatives on the nitrate assimilation pathway.

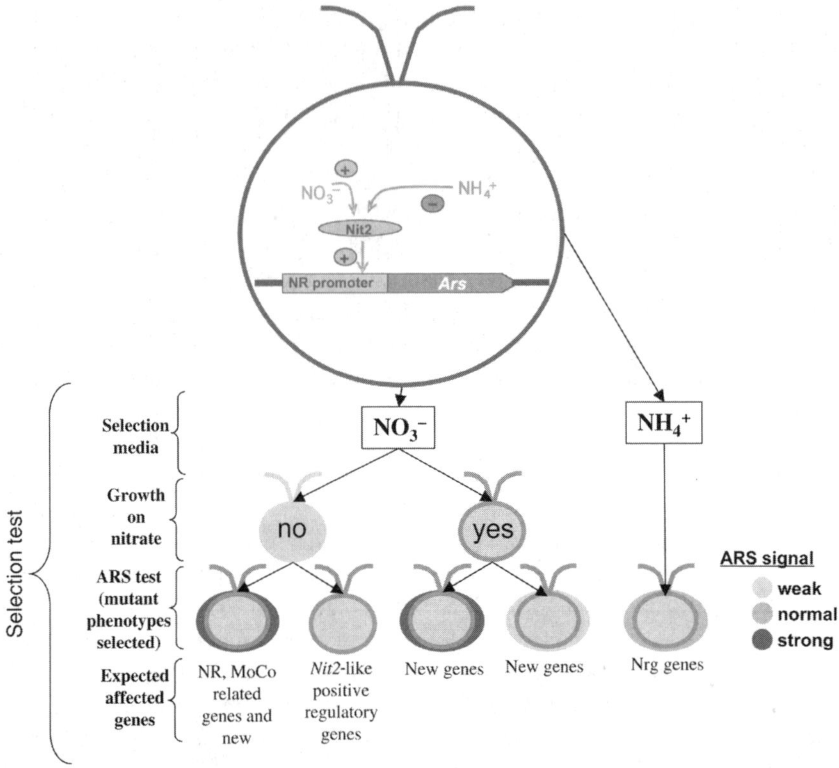

FIGURE 7.11 The isolation of nitrate assimilation mutants defective in positive and negative signaling. ARS = arylsulfatase activity. Other details are in the text.

Now, it will be possible to explore additional regulatory pathways by direct or reverse genetics—for example, those involving TOR or PII proteins reported in other organisms and considered to be integrators of nutrient availability (amino acids and energy) and as key players of nutrient-mediated signal transduction [182,192–196].

7.11 CONCLUSION

As the most important pathway for nitrogen acquisition by crop plants, nitrate assimilation needs to be understood in its most basic molecular aspects. In spite of its simplicity, this pathway has many intriguing questions still to solve. Model systems can provide invaluable information in the search for key genes and functions in plants and, finally, to the optimization of nitrogen use efficiency. In the age of genomics, comparative biology and mutant collections, the work carried out in *Chlamydomonas* should progress further in this direction.

ACKNOWLEDGMENTS

This work was supported by Ministerio de Ciencia y Tecnología, Spain (MCYT grant no. BFU2005-07521), European Commission (RTN grant HPRN-CT-2002-00247), and Junta de Andalucía, Spain (PAI, CVI-0128).

REFERENCES

1. Mandoli, D.F. and Olmstead, R., The importance of emerging model systems in plant biology, *J. Plant Growth Reg.*, 19, 249, 2000.
2. Silflow, C.D., Organization of the nuclear genome, in *The Molecular Biology of Chloroplasts and Mitochondria in Chlamydomonas,* Rochaix, J.D., Goldschmidt-Clermont, M., and Merchant, S., Eds., Kluwer Academic Publishers, Dordrecht, the Netherlands, 1998, 25–40.
3. Harris, E., *The Chlamydomonas Sourcebook*, Academic Press, New York, 1989.
4. Kindle, K.L., Nuclear transformation: technology and applications, in *The Molecular Biology of Chloroplasts and Mitochondria in Chlamydomonas*, Rochaix, J.D., Goldschmidt-Clermont, M., and Merchant, S., Eds., Kluwer Academic Publishers, Dordrecht, the Netherlands, 1998, 41–61.
5. León-Bañares, R. et al., Transgenic microalgae as green cell-factories, *Trends Biotechnol.*, 22, 45, 2004.
6. Fuhrmann, M. et al., The abundant retinal protein of the *Chlamydomonas* eye is not the photoreceptor for phototaxis and photophobic responses, *J. Cell Sci.*, 114, 3857, 2001.
7. Rohr, J. et al., Tandem inverted repeat system for selection of effective transgenic RNAi strains in *Chlamydomonas*, *Plant J.*, 40, 611, 2004.
8. González-Ballester, D. et al., Functional genomics of the regulation of the nitrate assimilation pathway in *Chlamydomonas*, *Plant Physiol.*, 137, 522, 2005.
9. Rochaix, J.D., Goldschmidt-Clermont, M., and Merchant, S., *The Molecular Biology of Chloroplasts and Mitochondria in Chlamydomonas*, Kluwer Academic Publishers, Dordrecht, the Netherlands, 1989.
10. Gutman, B.L. and Nigoyi, K.K., *Chlamydomonas* and *Arabidopsis*. A dynamic duo, *Plant Physiol.*, 135, 607, 2004.
11. Hart, S.C. and Stark, J.M., Nitrogen limitation of the microbial biomass in an old growth forest, *Ecoscience*, 4, 91, 1997.
12. Crawford, N.M., and Forde, B., Molecular and developmental biology of inorganic nitrogen nutrition, in *The Arabidopsis Book*, Somerville, C. and Meyerowitz, E., Eds., American Society of Plant Physiologists, Rockville, MD, 2002, doi/10.1199/tab.0011, http://www.aspb.org/publications/arabidopsis/.
13. Bloom, A.J., Sukrapanna, S.S., and Warner, R.L., Root respiration associated with ammonium and nitrate absorption and assimilation by barley, *Plant Physiol.*, 99, 1294, 1992.
14. Fernández, E. and Cárdenas, J., Regulation of the nitrate-reducing system enzymes in wild type and mutant strains of *Chlamydomonas reinhardtii*, *Mol. Genet. Genomics*, 186, 164, 1982.

15. Glass, A.D.M., Nitrogen use efficiency of crop plants: physiological constraints upon nitrogen absorption, *Crit. Rev. Plant. Sci.*, 22, 453, 2003.
16. Hirsch, R.E. and Sussman, M.R., Improving nutrient capture from soil by the genetic manipulation of crop plants, *Trends Biotechnol.*, 17, 356, 1999.
17. Britto, D.T. et al., Futile transmembrane $NH_4(^+)$ cycling: a cellular hypothesis to explain ammonium toxicity in plants, *Proc. Natl. Acad. Sci. USA*, 98, 4255, 2001.
18. Wray, J.L. and Kinghorn, J.R., *Molecular and Genetic Aspects of Nitrate Assimilation*, Oxford Science Publications, Oxford, 1989.
19. Caboche, M. and Rouzé, P., Nitrate reductase a target for molecular and cellular studies in higher plants, *Trends Genet.*, 6, 187, 1990.
20. Hoff, T., Truong, H.N., and Caboche, M., The use of mutants and transgenic plants to study nitrate assimilation, *Plant Cell Environ.*, 17, 489, 1994.
21. Crawford, N.M., Nitrate: nutrient and signal for plant growth, *Plant Cell*, 7, 859, 1995.
22. Daniel-Vedele, F. and Caboche, M., A tobacco cDNA clone encoding a GATA-1 zinc finger protein homologous to regulators of nitrogen metabolism in fungi, *Mol. Gen. Genet.*, 240, 365, 1993.
23. Trueman, L.J., Onyeocha, I., and Forde, B.G., Recent advances in the molecular biology of a family of eukaryotic high affinity nitrate transporters, *Plant Physiol. Biochem.*, 34, 621, 1996.
24. Crawford, N.M. and Glass, A.D.M., Molecular and physiological aspects of nitrate uptake in plants, *Trends Plant Sci.*, 3, 389, 1998.
25. Campbell, W., Nitrate reductase structure, function and regulation: binding the gap between biochemistry and physiology, *Annu. Rev. Plant. Physiol. Plant Mol. Biol.*, 50, 277, 1999.
26. Forde, B., Nitrate transporters in plants: structure, function and regulation. *Biochim. Biophys. Acta Biomembranes*, 1465, 219, 2000.
27. Galván, A. and Fernández, E., Eukaryotic nitrate and nitrite transporters, *Cell Mol. Life Sci.*, 58, 225, 2001.
28. Williams, L.E. and Miller, A., Transporters responsible for the uptake and partitioning of nitrogen solutes, *Annu. Rev. Plant Physiol. Plant Mol. Biol.*, 52, 659, 2001.
29. Crawford, N.M. and Arst, H.N., The molecular genetics of nitrate assimilation in fungi and plants, *Annu. Rev. Genet.*, 27, 115, 1993.
30. Siverio, J.M., Assimilation of nitrate by yeasts, *FEMS Microbiol. Rev.*, 26, 277, 2002.
31. Forde, B. and Cole, J., Nitrate finds a place in the sun, *Plant Physiol.*, 131, 395, 2003.
32. Azuara, M.P. and Aparicio, P., *In vivo* blue-light activation of *Chlamydomonas reinhardtii* nitrate reductase, *Plant Physiol.*, 71, 286, 1983.
33. Navarro, M.T. et al., Nitrite reductase mutants as an approach to understanding nitrate assimilation in *Chlamydomonas reinhardtii*, *Plant Physiol.*, 122, 283, 2000.
34. Rexach, J., Fernández, E., and Galván, A., The *Chlamydomonas reinhardtii Nar1* gene encodes a chloroplast membrane protein involved in nitrite transport, *Plant Cell*, 12, 1441, 2000.
35. Armengaud, P., Breitling, R., and Amtmann, A., The potassium-dependent transcriptome of *Arabidopsis* reveals a prominent role of jasmonic acid in nutrient signaling, *Plant Physiol.*, 136, 2556, 2004.
36. Scheible, W.R. et al., Genome-wide reprogramming of primary and secondary metabolism, protein synthesis, cellular growth processes, and the regulatory infrastructure of *Arabidopsis* in response to nitrogen, *Plant Physiol.*, 136, 2483, 2004.
37. Fernández, E. et al., Isolation and characterization of the nitrate reductase structural gene in *Chlamydomonas reinhardtii*, *Proc. Natl. Acad. Sci. USA*, 86, 6449, 1989.

38. Quesada, A., Gómez, I., and Fernández, E., Clustering of the nitrite reductase gene and a light-regulated gene with nitrate assimilation loci in *Chlamydomonas reinhardtii*, *Planta*, 206, 259 1998.
39. Chen, Q. and Silflow, C.D., Isolation and characterization of glutamine synthetase genes in *Chlamydomonas reinhardtii*, *Plant Physiol.*, 112, 987, 1996.
40. García-Sanchez, M.I. et al., Homology predicted structure and functional interaction of ferredoxin from the eukaryotic alga *Chlamydomonas reinhardtii* with nitrite reductase and glutamate synthase, *J. Biol. Inorg. Chem.*, 5, 713, 2000.
41. Quesada, A. et al., Five nitrate assimilation related loci are clustered in *Chlamydomonas reinhardtii*, *Mol. Genet. Genomics*, 240, 387, 1993.
42. Quesada, A., Galván, A., and Fernández, E., Identification of nitrate transporters in *Chlamydomonas reinhardtii*, *Plant J.*, 5, 407, 1994.
43. Galván, A. et al., Nitrite transport to the chloroplast in *Chlamydomonas*: molecular evidences for a regulated process, *J. Exp. Bot.*, 53, 845, 2002.
44. Miura, K. et al., Expression profiling-based identification of CO2-responsive genes regulated by CCM1 controlling a carbon-concentrating mechanism in *Chlamydomonas reinhardtii*, *Plant Physiol.*, 135, 1595, 2004.
45. González-Ballester, D., Camargo, A., and Fernández, E., Ammonium transporter genes in *Chlamydomonas*: the nitrate-specific regulatory gene *Nit2* is involved in *Amt1*1 expression, *Plant Mol. Biol.*, 56, 863, 2004.
46. Kim, K.S. et al., Spontaneous mutations in the ammonium transport gene AMT4 of *Chlamydomonas reinhardtii*, *Genetics*, 170, 631, 2005.
47. Schnell, R.A. and Lefebvre, P.A., Isolation of the *Chlamydomonas* regulatory gene *nit2* by transposon tagging, *Genetics*, 134, 737, 1993.
48. Ataya, F.S. et al., *Mcp1* encodes the molybdenum cofactor carrier protein in *Chlamydomonas reinhardtii* and participates in protection, binding, and storage functions of the cofactor, *J. Biol. Chem.*, 278, 10885, 2003.
49. Dinant, M. et al., Characterization of two genes encoding the mitochondrial alternative oxidase in *Chlamydomonas reinhardtii*, *Curr. Genet.*, 39, 101, 2001.
50. Gómez, I. et al., NADP-malate dehydrogenase from *Chlamydomonas*: prediction of new structural determinants for redox regulation by homology modeling, *Plant Mol. Biol.*, 48, 211, 2002.
51. Quesada, A., Hidalgo, J., and Fernández, E., Three *Ntr2* genes are differentially regulated in *Chlamydomonas reinhardtii*, *Mol. Genet. Genomics*, 258, 373, 1998.
52. Rexach, J. et al., Differential regulation of the high affinity nitrite transport systems III and IV in *Chlamydomonas reinhardtii*, *J. Biol. Chem.*, 274, 27801, 1999.
53. Johnston, I.L. et al., Isolation and characterization of the crnA-niiA-niaD gene cluster for nitrate assimilation in *Aspergillus nidulans*, *Gene*, 90, 181, 1990.
54. Perez, M.D. et al., The *YNT1* gene encoding the nitrate transporter in the yeast *Hansenula polymorpha* is clustered with genes *YNI1* and *YNR1* encoding nitrite reductase and nitrate reductase, and its disruption causes inability to grow in nitrate, *Biochem. J.*, 321, 397, 1997.
55. Punt, P.J. et al., The intergenic region between the divergently transcribed niiA and niaD genes of *Aspergillus nidulans* contains multiple NirA binding sites which act bidirectionally, *Mol. Cell. Biol.*, 15, 5688, 1995.
56. Oostindiër-Braaksma, F.J. and Feenstra, W.J., Isolation and characterization of chlorate-resistant mutants of *Arabidopsis thaliana*, *Mutation Res.*, 19, 175, 1973.
57. Cove, D.J., Chlorate toxicity in *Aspergillus nidulans*. Studies of mutants altered in nitrate assimilation, *Mol. Genet. Genomics*, 146, 147, 1976.

58. Prieto, R. and Fernández, E., Toxicity and mutagenesis by chlorate are independent of nitrate reductase activity in *Chlamydomonas reinhardtii*, *Mol. Genet. Genomics*, 237, 429, 1993.
59. Aberg, B., On the mechanism of the toxic action of chlorate and some related substances upon young wheat plants, *Annu. Rev. Agric. Coll. Sweden*, 15, 37, 1947.
60. Fernández, E., Galván, A., and Quesada, A., Nitrogen assimilation and its regulation, in *Molecular Biology of Chlamydomonas: Chloroplast and Mitochondria*, Rochaix, J.D., Goldschmidt-Clermont, M., and Merchant, S., Eds., Kluwer Academic Publishers, Dordrecht, the Netherlands, 1998, 637.
61. Galván, A., Quesada, A., and Fernández, E., Nitrate and nitrite are transported by different specific transport systems and by a bispecific transporter in *Chlamydomonas reinhardtii*, *J. Biol. Chem.*, 271, 2088, 1996.
62. Navarro, M.T. et al., *Chlamydomonas reinhardtii* strains expressing nitrate reductase under control of the *cabII-1* promoter: isolation of chlorate resistant mutants and identification of new loci for nitrate assimilation, *Photosynth. Res.*, 83, 151, 2005.
63. Cao, D. et al., The chlorate-resistant and photomorphogenesis-defective mutant cr88 encodes a chloroplast-targeted HSP90, *Plant J.*, 33, 107, 2003.
64. Campbell, W.H. and Kinghorn, J.R., Functional domains of assimilatory nitrate reductases and nitrite reductases, *Trends. Biochem. Sci.*, 15, 315, 1990.
65. Moura, J.J. et al., Mo and W bis-MGD enzymes: nitrate reductases and formate dehydrogenases, *J. Biol. Inorg. Chem.*, 9, 791, 2004.
66. Moreno-Vivian, C. et al., Prokaryotic nitrate reduction: molecular properties and functional distinction among bacterial nitrate reductases, *J. Bacteriol.*, 181, 6573, 1999.
67. Eilers, T. et al., Identification and biochemical characterization of *Arabidopsis thaliana* sulfite oxidase. A new player in plant sulfur metabolism, *J. Biol. Chem.*, 14, 46989, 2001.
68. Mendel, R.R. and Hansch, R., Molybdoenzymes and molybdenum cofactor in plants, *J. Exp. Bot.*, 53, 1689, 2002.
69. Mendel, R. and Schwarz, G., Molybdoenzymes and molybdenum cofactor in plants, *Crit. Rev. Plant Sci.*, 18, 33, 1999.
70. Fernández, E. and Matagne, R.F., Genetic analysis of nitrate reductase-deficient mutants in *Chlamydomonas reinhardtii*, *Curr. Genet.*, 8, 635, 1984.
71. Aguilar, M. et al., Direct transfer of molybdopterin cofactor to aponitrate reductase from a carrier protein in *Chlamydomonas reinhardtii*, *FEBS Lett.*, 307, 162, 1992.
72. Llamas, A., Kalakoutskii, K.L., and Fernández, E., Molybdenum cofactor amounts in *Chlamydomonas reinhardtii* depend on the *Nit5* gene function related to molybdate transport, *Plant Cell Environ.*, 23, 1247, 2000.
73. Kalakoutskii, K. and Fernández, E., *Chlamydomonas reinhardtii* nitrate reductase complex has 105-kDa subunits in the wild-type strain and a structural mutant, *Plant Sci.*, 105, 195, 1995.
74. Witte, C.P. et al., The *Chlamydomonas reinhardtii* MoCo carrier protein is multimeric and stabilizes molybdopterin cofactor in a molybdate charged form, *FEBS Lett.*, 431, 205, 1998.
75. Bachmann, M. et al., 14-3-3 proteins associate with the regulatory phosphorylation site of spinach leaf nitrate reductase in an isoform-specific manner and reduce dephosphorylation of Ser-543 by endogenous protein phosphatases, *FEBS Lett.*, 398, 26, 1996.
76. Moorhead, G. et al., Phosphorylated nitrate reductase from spinach leaves is inhibited by 14-3-3 proteins and activated by fusicoccin, *Curr. Biol.*, 6, 1104, 1996.

77. Kaiser, W.M. et al., Modulation of nitrate reductase: some new insights, an unusual case and a potentially important side reaction, *J. Exp. Bot.*, 53, 875, 2002.
78. Moorhead, G. et al., Phosphorylation-dependent interactions between enzymes of plant metabolism and 14-3-3 proteins, *Plant J.*, 18, 1, 1999.
79. Lillo, C. et al., Characterization of nitrate reductase from light- and dark-exposed leaves. Comparison of different species and effects of 14-3-3 inhibitor proteins, *Plant Physiol.*, 114, 1377, 1997.
80. Mackintosh, C., Dynamic interactions between 14-3-3 proteins and phosphoproteins regulate diverse cellular processes, *Biochem. J.*, 381, 329, 2004.
81. Comparot, S. et al., Function and specificity of 14-3-3 proteins in the regulation of carbohydrate and nitrogen metabolism, *J. Exp. Bot.*, 54, 595, 2003.
82. Pozuelo, M. et al., Cytosolic glutamine synthetase and not nitrate reductase from the green alga *Chlamydomonas reinhardtii* is phosphorylated and binds 14-3-3 proteins, *Planta*, 212, 264, 2001.
83. Huber, S.C., MacKintosh, C., and Kaiser, W.M., Metabolic enzymes as targets for 14-3-3 proteins, *Plant Mol. Biol.*, 50, 1053, 2002.
84. Guerrero, M.G., Vega, J.M., and Losada, M. The assimilatory nitrate-reducing system and its regulation, *Annu. Rev. Plant Physiol.*, 32, 169, 1981.
85. Franco, A.R., Cárdenas, J., and Fernández, E., Involvement of reversible inactivation in the regulation of nitrate reductase enzyme levels in *Chlamydomonas reinhardtii*, *Plant Physiol.*, 84, 665, 1987.
86. Vega, J.M. and Kamin, H., Spinach nitrite reductase. Purification and properties of a siroheme-containing iron-sulfur enzyme, *J. Biol. Chem.*, 252, 896, 1977.
87. Kato, C. et al., Differential expression of the nitrite reductase gene family in tobacco as revealed by quantitative competitive RT-PCR, *J. Exp. Bot.*, 55, 1761, 2004.
88. Martínez-Rivas, J.M., Vega, J.M., and Márquez, A., Differential regulation of the nitrate-reducing and ammonium-assimilatory systems in synchronous cultures of *Chlamydomonas reinhardtii*, *FEMS Microbiol. Lett.*, 78, 85, 1991.
89. Crete, P., Caboche, M., and Meyer, C., Nitrite reductase expression is regulated at the post-transcriptional level by the nitrogen source in *Nicotiana plumbaginifolia* and *Arabidopsis thaliana*, *Plant J.*, 11, 625, 1997.
90. Glass, A.D.M., Shaff, J.E., and Kochian, L.V., Studies of the uptake of nitrate in barley. IV. Electrophysiology, *Plant Physiol.*, 99, 456, 1992.
91. Tsay, Y.F. et al., An herbicide sensitive gene *CHL1* of *Arabidopsis* encodes a nitrate-inducible nitrate transporter, *Cell*, 72, 705, 1993.
92. Zhou, J.J. et al., A high affinity nitrate transport system from *Chlamydomonas* requires two gene products, *FEBS Lett.*, 28, 225, 2000.
93. Miller, A.J. et al., The use of microelectrodes to investigate compartmentation and the transport of metabolized inorganic ions in plants, *J. Exp. Bot.*, 52, 541, 2001.
94. Koltermann, M. et al., Cloning, functional expression and expression studies of the nitrate transporter gene from *Chlorella sorokiniana* (strain 211-8k), *Plant Mol. Biol.*, 54, 855, 2003.
95. McClure, P.R. et al., Evidence for cotransport of nitrate and proton in maize roots. I. Effect of nitrate on the membrane potential, *Plant Physiol.*, 93, 281, 1990.
96. McClure, P.R. et al., Evidence for cotransport of nitrate and protons in maize roots. II. Measurement of NO3- and H+ fluxes with ion-selective microelectrodes, *Plant Physiol.*, 93, 281, 1990.
97. Zhou, J.J. et al., A high affinity fungal nitrate carrier with two transport mechanisms, *J. Biol. Chem.*, 275, 39894, 2000.

98. Rodríguez, R., Lara, C., and Guerrero, M.G., Nitrate transport in the cyanobacterium *Anacystis nidulans* R2. Kinetic and energetic aspects, *Biochem. J.*, 15, 639, 1992.
99. Omata, T., Structure, function and regulation of the nitrate transport system of the cyanobacterium *Synechococcus* sp. PCC7942, *Plant Cell Physiol.*, 36, 207, 1995.
100. DeMoss, J.A. and Hsu, P.Y., NarK enhances nitrate uptake and nitrite excretion in *Escherichia coli*, *J. Bacteriol.*, 173, 3303, 1991.
101. Clegg, S. et al., The roles of the polytopic membrane proteins NarK, NarU and NirC in Escherichia coli K-12: two nitrate and three nitrite transporters, *Mol. Microbiol.*, 44, 143, 2002.
102. Geelen, D. et al., Disruption of putative anion channel gene AtCLC-a in *Arabidopsis* suggests a role in the regulation of nitrate content, *Plant J.*, 21, 259, 2000.
103. Fraisier, V. et al., Constitutive expression of a putative high-affinity nitrate transporter in *Nicotiana plumbaginifolia*: evidence for post-transcriptional regulation by a reduced nitrogen source, *Plant J.*, 23, 489, 2000.
104. Aslam, M., Travis, R.L., and Huffaker, R.C., Comparative induction inhibition of nitrate and nitrite uptake and reduction systems by ambient nitrate and nitrite in intact roots of barley seedlings, *Plant Physiol.*, 102, 811, 1993.
105. Wang, R. and Crawford, N.M., Genetic identification of a gene involved in constitutive high-affinity nitrate transport in higher plants, *Proc. Natl. Acad. Sci. USA*, 93, 9297, 1996.
106. Quesada, A. et al., PCR-identification of a *Nicotiana plumbaginifolia* cDNA homologous to the high-affinity nitrate transporters of the crnA family, *Plant Mol. Biol.*, 34, 265, 1997.
107. Machín, F. et al., The role of Ynt1 in nitrate and nitrite transport in the yeast *Hansenula polymorpha*, *Yeast*, 21, 265, 2004.
108. Orsel, M., Krapp, A., and Daniel-Vedele, F., Analysis of the NRT2 nitrate transporter family in *Arabidopsis*. Structure and gene expression, *Plant Physiol.*, 129, 886, 2002.
109. Okamoto, M., Vidmar, J.J., and Glass, A.D., Regulation of NRT1 and NRT2 gene families of *Arabidopsis thaliana*: responses to nitrate provision, *Plant Cell Physiol.*, 44, 304, 2003.
110. Zhou, J.J. et al., Cloning and functional characterization of a *Brassica napus* transporter that is able to transport nitrate and histidine, *J. Biol. Chem.*, 273, 12017, 1998.
111. Paulsen, I.T. and Skurray, R.A., The POT family of transport proteins, *Trends Biochem. Sci.*, 19, 404, 1994.
112. Pao, S.S., Paulsen, I.T., and Saier, M.H., The major facilitator superfamily, *Microbiol. Mol. Biol. Rev.*, 62, 1, 1998.
113. Doddema, H. and Telkamp, G.P., Uptake of nitrate by mutants of *Arabidopsis thaliana*, disturbed in uptake or reduction of nitrate: II. Kinetics, *Physiol. Plant.*, 45, 332, 1979.
114. Huang, N.C., Cloning and functional characterization of an *Arabidopsis* nitrate transporter gene that encodes a constitutive component of low-affinity uptake, *Plant Cell*, 11, 1381, 1999.
115. Huang, N.C. et al., CHL1 encodes a component of the low-affinity nitrate uptake system in *Arabidopsis* and shows cell type-specific expression in roots, *Plant Cell*, 8, 2183, 1996.
116. Touraine, B. and Glass, A.D., NO_3^- and ClO_3^- fluxes in the chl1-5 mutant of *Arabidopsis thaliana*, Does the CHL!-5 gene encode a low-affinity NO_3^- transporter? *Plant Physiol.*, 114, 137, 1997.
117. Wang, R., Liu, D., and Crawford, N.M., The *Arabidopsis* CHL1 protein plays a major role in high-affinity nitrate uptake, *Proc. Natl. Acad. Sci. USA*, 95, 15134, 1998.

118. Liu, K.H., Huang, C.Y., and Tsay, Y.F., CHL1 is a dual-affinity nitrate transporter of *Arabidopsis* involved in multiple phases of nitrate uptake, *Plant Cell*, 11, 865, 1999.
119. Liu, K.H. and Tsay, Y.F., Switching between the two action modes of the dual-affinity nitrate transporter CHL1 by phosphorylation, *EMBO J.*, 22, 1005, 2003.
120. Guo, F.Q. et al., The *Arabidopsis* dual-affinity nitrate transporter gene AtNRT1.1 (CHL1) is activated and functions in nascent organ development during vegetative and reproductive growth, *Plant Cell,* 13, 1761, 2001.
121a. Chiu, C.C. et al., Mutation of a nitrate transporter, AtNRT1:4, results in a reduced petiole nitrate content and altered leaf development, *Plant Cell. Physiol.*, 45, 1139, 2004.
121b. Guo, F.Q., Young, J., and Crawford, N.M., The nitrate transporter AtNRT1.1 (CHL1) functions in stomatal opening and contributes to drought susceptibility in *Arabidopsis*, *Plant Cell*, 15, 107, 2003.
122. Unkles, S.E. et al., *Crna* encodes a nitrate transporter in *Aspergillus nidulans*, *Proc. Natl. Acad. Sci. USA*, 88, 204, 1991.
123. Unkles, S.E. et al., Two perfectly conserved arginine residues are required for substrate binding in a high-affinity nitrate transporter, *Proc. Natl. Acad. Sci. USA*, 101, 17549, 2004.
124. Tong, Y. et al., A two-component high affinity nitrate uptake system in barley, *Plant J.*, 41, 442, 2005.
125. Enser, U. and Heber, U, Metabolic regulation by pH gradients. Inhibition of photosynthesis by indirect proton transfer across the chloroplast envelope, *Biochim. Biophys. Acta*, 159, 577, 1980.
126. Brunswick, P. and Cresswell, C.F., Nitrite uptake into intact pea chloroplasts. I. Kinetics and relationship with nitrite assimilation, *Plant Physiol.*, 86, 378, 1988.
127. Brunswick, P. and Cresswell, C.F., Nitrite uptake into intact pea chloroplasts. II. Influence of electron transport regulators, uncouplers, ATPase and anion uptake inhibitors and protein binding reagents, *Plant Physiol.*, 86, 384, 1988.
128. Shingles, R., Roh, M.H., and McCarty, R.E., Nitrite transport in chloroplast inner envelope vesicles. I. Direct measurement of proton-linked transport, *Plant Physiol.*, 112, 1375, 1996.
129. Riens, B. and Heldt, W., Decrease of nitrate reductase activity in spinach leaves during a light–dark transition, *Plant Physiol.*, 98, 573, 1992.
130. Mariscal, V. et al., The plastidic nitrite transporter NAR1;1 improves nitrate use efficiency for growth in *Chlamydomonas*, *Plant Cell Environ.*, 27, 1321, 2004.
131. Miflin, J. and Habash, D.J. The role of glutamine synthetase and glutamate dehydrogenase in nitrogen assimilation and possibilities for improvement in the nitrogen utilization of crops, *J. Exp. Bot.*, 370, 979, 2002.
132. Cullimore, J. and Sims, P., Glutamine synthetase of *Chlamydomonas*: its role in the control of nitrate assimilation, *Planta*, 153, 1824, 1981.
133. Tobin, A.K. and Yamaya, T., Cellular compartmentation of ammonium assimilation in rice and barley, *J. Exp. Bot.*, 52:591, 2001.
134. Temple, S.J., Vance, C.P., and Gantt, J.S., Glutamate synthase and nitrogen assimilation, *Trends Plant Sci.*, 3, 51, 1998.
135. Márquez, A.J., Galván, F., and Vega, J.M., Purification and characterization of the NADH-glutamate synthase from *Chlamydomonas reinhardtii*, *Plant Sci. Lett.*, 34, 305, 1984.
136. Márquez, A.J. et al., Ferredoxin–glutamate synthase from *Chlamydomonas reinhardtii*. Prosthetic groups and preliminary studies of mechanism, *Int. J. Biochem.*, 18, 531, 1986.

137. Ullrich, W.R. et al., Ammonium uptake in *Lemna gibba* G1, related membrane potential change and inhibition of anion uptake, *Physiol. Plant.*, 61, 369, 1984.
138. Wang, M.Y. et al., Ammonium uptake by rice roots (I. Fluxes and subcellular distribution of $^{13}NH_4^+$), *Plant Physiol.*, 103, 1249, 1993.
139. Howitt, S.M. and Udvardi, M.K., Structure, function and regulation of ammonium transporters in plants, *Biochim. Biophys. Acta*, 1465, 152, 2000.
140. von Wirén, N. et al., The molecular physiology of ammonium uptake and retrieval, *Curr. Opin. Plant Biol.*, 3, 254, 2000.
141. Byrne, T.E., Wells, M.R., and Johnson, C.H., Circadian rhythms of chemotaxis to ammonium and of methylammonium uptake in *Chlamydomonas*, *Plant Physiol.*, 93, 879, 1992.
142. Franco, A.R., Cárdenas, J., and Fernández, E., Two different carriers transport both ammonium and methylammonium in *Chlamydomonas reinhardti*, *J. Biol. Chem.*, 263, 14039, 1988.
143. Marini, A.M. et al., A family of ammonium transporters in *Saccharomyces cerevisiae*, *Mol. Cell Biol.*, 17, 4282, 1997.
144. Soupene, E. et al., Rhesus expression in a green alga is regulated by CO_2, *Proc. Natl. Acad. Sci. USA*, 99, 7769, 2002.
145. Soupene, E., Inwood, W., and Kustu, S., Lack of the rhesus protein Rh1 impairs growth of the green alga *Chlamydomonas reinhardtii* at high CO_2, *Proc. Natl. Acad. Sci. USA*, 101, 7787, 2004.
146. Sohlenkamp, C. et al., Characterization of *Arabidopsis* AtAMT2, a novel ammonium transporter in plants, *FEBS Lett.*, 467, 271, 2000.
147. Sohlenkamp, C. et al., Characterization of *Arabidopsis* AtAMT2, a high-affinity ammonium transporter of the plasma membrane, *Plant Physiol.*, 130, 1, 2002.
148. Simon-Rosin, U., Wood, C., Udvardi, M.K., Molecular and cellular characterization of LjAMT2;1, an ammonium transporter form the model legume *Lotus japonicus*, *Plant Mol. Biol.*, 51, 99, 2003.
149. Ninnemann, O., Jauniaux, J.C., and Frommer, W.B., Identification of a high affinity ammonium transporter from plants, *EMBO J.*, 13, 3464, 1994.
150. Gazzarrini, S. et al., Three functional transporters for constitutive, diurnally regulated and starvation-induced uptake of ammonium into *Arabidopsis* roots, *Plant Cell*, 11, 937, 1999.
151. von Wirén, N. et al., Differential regulation of three functional ammonium transporters genes by nitrogen in root hairs and by light in leaves of tomato, *Plant J.*, 21, 167, 2000.
152. von Wirén, N. et al., OsAMT1-1: a high-affinity ammonium transporter from rice (*Oryza sativa* cv. Nipponbare), *Plant Mol. Biol.*, 35, 681, 1997.
153. Pearson, J.N., Finnemann, J., and Schjoerring, J.K., Regulation of the high-affinity ammonium transporter (BnAMT1;2) in the leaves of *Brassica napus* by nitrogen status, *Plant Mol. Biol.*, 49, 483, 2002.
154. Salvemini, F. et al., Functional characterization of an ammonium transporter gene from *Lotus japonicus*, *Gene*, 270, 237, 2001.
155. Pajuelo, E. et al., Regulation of the expression of ferredoxin-nitrite reductase in synchronous cultures of *Chlamydomonas reinhardtii*, *Biochim. Biophys. Acta*, 1249, 72, 1995.
156. Mittag, M. and Waltenberger, H., *In vitro* mutagenesis of binding site elements for the clock-controlled proteins CCTR and Chlamy 1, *J. Biol. Chem.*, 378, 1167, 2001.
157. Zhao, L. et al., Overexpression of LSH1, a member of an uncharacterized gene family, causes enhanced light regulation of seedling development, *Plant J.*, 37, 694, 2004.

158. Waltenberger, H. et al., Identification of target mRNAs for the clock-controlled RNA-binding protein Chlamy 1 from *Chlamydomonas reinhardtii*, *Mol. Genet. Genomics*, 265, 180, 2001.
159. Li, J. and Timko, M.P., The pc-1 phenotype of *Chlamydomonas reinhardtii* results from a deletion mutation in the nuclear gene for NADPH:protochlorophyllide oxidoreductase, *Plant Mol. Biol.*, 30,15, 1996.
160. Dietmaier, W. et al., Analysis of a family of ypt genes and their products from *Chlamydomonas reinhardtii*, *Gene*, 158, 41, 1995.
161. Mittag, M., Conserved circadian elements in phylogenetically diverse algae, *Proc. Natl. Acad. Sci. USA*, 93, 14401, 1996.
162. Heintzen, C. et al., AtGRP7, a nuclear RNA-binding protein as a component of a circadian-regulated negative feedback loop in *Arabidopsis thaliana*, *Proc. Natl. Acad. Sci. USA*, 94, 8515, 1997.
163. Giordano, M. et al., Regulation of nitrate reductase in *Chlamydomonas reinhardtii* by the redox state of the plastoquinone pool, *Eur. J. Phycol.*, 40, 345, 2005.
164. Sherameti, I. et al., Photosynthetic electron transport determines nitrate reductase gene expression and activity in higher plants, *J. Biol. Chem.*, 277, 46594, 2002.
165. Huppe, H.C. and Turpin, D.H., Appearance of novel glucose-6-phosphate dehydrogenase isoforms in *Chlamydomonas reinhardtii* during growth on nitrate, *Plant Physiol.*, 110, 1431, 1996.
166. Quesada, A., Gómez-García, L.A., and Fernández, E., Involvement of chloroplast and mitochondria redox valves in nitrate assimilation, *Trends Plant Sci.*, 5, 463, 2000.
167. Lemaire, S.D. et al., NADP-malate dehydrogenase from unicellular green alga *Chlamydomonas reinhardtii*, a first step towards redox regulation? *Plant Physiol.*, 137, 514, 2005.
168. Vanlerberghe, G.C. and McIntosh, L., Alternative oxidase: from gene to function, *Annu. Rev. Plant Physiol. Plant Mol. Biol.*, 48, 703, 1997.
169. Morot-Gaudry-Talarmain, Y. et al., Nitrite accumulation and nitric oxide emission in relation to cellular signaling in nitrite reductase antisense tobacco, *Planta*, 215, 708, 2002.
170. Sakihama, Y., Nakamura, S., and Yamasaki, H., Nitric oxide production mediated by nitrate reductase in the green alga *Chlamydomonas reinhardtii*: an alternative NO production pathway in photosynthetic organisms, *Plant Cell Physiol.*, 43, 290, 2002.
171. Tischner, R., Planchet, E., and Kaiser, W.M., Mitochondrial electron transport as a source for nitric oxide in the unicellular green alga *Chlorella sorokiniana*, *FEBS Lett.*, 576, 151, 2004.
172. Wang, R. et al., Genomic analysis of a nutrient response in *Arabidopsis* reveals diverse expression patterns and novel metabolic and potential regulatory genes induced by nitrate, *Plant Cell*, 12, 1491, 2000.
173. Fizames, C. et al., The *Arabidopsis* root transcriptome by serial analysis of gene expression. Gene identification using the genome sequence, *Plant Physiol.*, 134, 67, 2004.
174. Llamas, A. et al., Nitrate signaling on the nitrate reductase gene promoter depends directly on the activity of the nitrate transport systems in *Chlamydomonas*, *Plant J.*, 30, 261, 2002.
175. Rexach, J. et al., The activity of the high-affinity nitrate transport system I (NRT2;1, NAR2) is responsible for the efficient signaling of nitrate assimilation genes in *Chlamydomonas reinhardtii*, *Planta*, 215, 606, 2002.
176. Munos, S. et al., Transcript profiling in the chl1-5 mutant of *Arabidopsis* reveals a role of the nitrate transporter NRT1.1 in the regulation of another nitrate transporter, NRT2.1, *Plant Cell*, 16, 2433, 2004.

177. Kronzucker, H.J. et al., Inhibition of nitrate uptake by ammonium in barley: analysis of component fluxes, *Plant Physiol.*, 120, 283, 1999.
178. Lorenz, M.C. and Heitman, J., Regulators of pseudohyphal differentiation in *Saccharomyces cerevisiae* identified through multicopy suppressor analysis in ammonium permease mutant strains, *Genetics*, 150, 1443, 1998.
179. Marini, A.-M. and André, B., *In vivo* N-glycosylation of the Mep2 high-affinity ammonium transporter of *Saccharomyces cerevisiae* reveals an extracytosolic N-terminus, *Mol. Microbiol.*, 38, 552, 2000.
180. Franco, A.R., Cárdenas, J., and Fernández, E., A mutant of *Chlamydomonas reinhardtii* altered in the transport of ammonium and methylammonium, *Mol. Gen. Genet.*, 206, 414, 1987.
181. Marzluf, G.A., Genetic regulation of nitrogen metabolism in the fungi, *Microbiol. Mol. Biol. Rev.*, 61, 17, 1997.
182. Cooper, T.G., Transmitting the signal of excess nitrogen in *Saccharomyces cerevisiae* from Tor proteins to the *Saccharomyces cerevisiae* GATA-factors: connecting the dots, *FEMS Microbiol. Rev.*, 26, 223, 2002.
183. Scazzocchio, C., The fungal GATA factors, *Curr. Opin. Microbiol.*, 3, 126, 2000.
184. Troung, H.N., Caboche, M., and Daniel-Vedele, F., Sequence and characterization of two *Arabidopsis thaliana* cDNAs isolated by functional complementation of yeast *gln3 gdh1* mutant, *FEBS Lett.*, 410, 213, 1997.
185. Rastogi, R. et al., Footprinting of the spinach nitrite reductase gene promotor reveals the preservation of nitrate regulatory elements between fungi and higher plants, *Plant Mol. Biol.*, 34, 465, 1997.
186. Reyes, J.C., Muro-Pastor, M.I., and Florencio, F.J., The GATA family of transcription factors in *Arabidopsis* and rice, *Plant Physiol.*, 134, 1718, 2004.
187. Ferris, P.J. and Goodenough, U.W., Mating type in *Chlamydomonas* is specified by mid, the minus-dominance gene, *Genetics*, 146, 859, 1997.
188. Schauser, L. et al., A plant regulator controlling development of symbiotic root nodules, *Nature*, 402, 191, 1999.
189. Prieto, R. et al., Isolation and characterization of two regulatory mutants for nitrate assimilation in *Chlamydomonas reinhardtii*, *Mol. Genet. Genomics*, 251, 461, 1996.
190. Zhang, D. and Lefebvre, P.A., *FAR1*, a new negative regulatory locus required for the repression of the nitrate reductase gene in *Chlamydomonas reinhardtii*, *Genetics*, 146, 121, 1997.
191. Pérez-Alegre, M., Dubus, A., and Fernández, E., REM1, a new type of lng terminal repeat retrotransposon in *Chlamydomonas*, *Mol. Cell Biol.*, 25,10628, 2005.
192. Lee, H.M. et al., Phosphorylation of the signal transducer PII protein and an additional effector are required for the PII-mediated regulation of nitrate and nitrite uptake in the *Cyanobacterium synechococcus* sp. PCC 7942, *Eur. J. Biochem.*, 267, 591, 2000.
193. Arcondeguy, T., Jack, R., and Merrick, M., P(II) signal transduction proteins, pivotal players in microbial nitrogen control, *Microbiol. Mol. Biol. Rev.*, 65, 80, 2001.
194. Bertram, P.G. et al., Convergence of TOR-nitrogen and Snf1-glucose signaling pathways onto Gln3, *Mol. Cell Biol.*, 22, 1246, 2002.
195. Moorhead, G.B. and Smith, C.S., Interpreting the plastid carbon, nitrogen, and energy status. A role for PII? *Plant Physiol.*, 133, 492, 2003.
196. Herrero, A., New targets of the PII signal transduction protein identified in cyanobacteria, *Mol. Microbiol.*, 52, 1225, 2004.

8 Transcription Factors Regulating Plant Defense Responses

Pierre R. Fobert

CONTENTS

8.1 INTRODUCTION

8.1.1 Transcriptional Regulation of Plant Defense Responses

Exposure to microbial pathogens results in massive transcriptional reprogramming of the plant genome [1]. Historically, genes and proteins induced in response to pathogen challenge were labeled as pathogenesis-related (PR). At least 14 classes of PR proteins are recognized, several of which have the potential for direct antimicrobial activity, including chitinases, glucanases, and cationic peptides [2]. In general, ectopic (over) expression of individual or pairs of *PR* genes in transgenic plants does not substantially augment disease resistance [3,4], and there has been no reported characterization of *PR* gene mutants. Accordingly, the specific contribution of individual PR proteins to disease resistance remains elusive.

The magnitude, complexity, and dynamics of pathogen challenge on plant gene expression are currently being revealed by genome-wide transcript profiling studies [1,4]. In addition to classical *PR* genes, several hundred, if not thousands of, genes encoding products implicated in almost every aspect of plant physiology have been shown to be affected. Such changes in gene expression likely represent a combination of plant defense and disease susceptibility responses [4]. Although the precise function of most genes modulated by pathogen challenge remains unknown, it is clear that the timely, coordinated transcriptional control of large sets of genes is crucial for plant disease resistance [5–7].

Transcription of a gene is ultimately determined by the combination of the *cis*-acting transcriptional regulatory elements that it possesses and the repertoire of active *trans*-acting transcription factors (TFs) present in the cell. Whereas a gene's complement of *cis*-acting elements is "hardwired" in the genome, the abundance and activity of numerous TFs are modulated by signal transduction events initiated following pathogen recognition by the plant cell. Such TFs will include key regulators of the plant's inducible defense responses against pathogens and are the subject of this chapter. I will review how these TFs were identified and are functionally analyzed and regulated following pathogen challenge. To place the discussion into proper context, a brief overview of different plant defense systems and the importance of model systems in the study of TFs mediating plant defense responses are first provided.

Three broad types of *trans*-acting TFs can be distinguished: general (or basal) TFs, sequence-specific TFs, and cofactors [8,9]. With the exception of the large body of literature on the cofactor NONEXPRESSOR OF PATHOGENESIS-RELATED GENES1 (NPR1) [10,11], most studies have focused on the sequence-specific DNA-binding TFs. Accordingly and unless otherwise stated, these will be referred to here simply as TFs. Whenever possible, I reference recent reviews or representative original publications in an effort to minimize the number of references.

8.1.2 Inducible Plant Defense Systems

Plant defense against pathogen attack involves recognition of pathogen-derived molecules called elicitors. One of the best characterized defense systems, race-specific resistance, is governed by genetic interactions between genes encoding a class of elicitors called avirulence factors and plant resistance (*R*) genes (see Chapter 9 in this volume for review of *R*-genes). Avirulence factors are polymorphic among isolates of a single pathogen species and trigger a very rapid and specific response in plants expressing the corresponding *R* gene product [12]. In these cases, host–pathogen interaction is said to be incompatible, the pathogen is avirulent, and the plant host is resistant. Absence of specific genetic recognition results in a compatible interaction, in which the pathogen is said to be virulent and the host susceptible, and disease ensues.

Under these circumstances, plants rely on a basal defense response triggered by the recognition of pathogen-associated molecular patterns (PAMPs) that are conserved among several microbial species [12]. Basal defense responses are not as specific or rapid as those mediated by *R*-gene–avirulence gene recognition. They do not prevent disease but restrict pathogen spread. Accordingly, mutations in components of the basal defense system result in hypersusceptibility to virulent pathogens [13].

Regardless of whether pathogens are detected through avirulence determinants or PAMPS, the signaling events that are triggered rapidly converge into a limited number of interacting pathways, or networks, that rely on small molecules, including salicylic acid (SA), jasmonic acid (JA), and ethylene (ET), as secondary messengers [13,14]. SA-dependent signaling is required for resistance to certain pathogens that derive energy from living host cells (biotrophs; see Oliver and Ipcho [15] for definitions) and for mediating a type of broad-spectrum, inducible disease resistance known as systemic acquired resistance (SAR) [16]. Pathogen activation of the *PR* genes *PR-1*, *PR-2*, and *PR-5* depends on SA, and these genes serve as marker genes for SA-dependent signaling events. JA and ET signaling are generally required for resistance to necrotrophic pathogens (i.e., derive energy from killed cells) and for an SA-independent form of induced systemic resistance called ISR [13,14].

Expression of *PDF1.2*, which codes for a cationic antimicrobial peptide, is often used as a marker for JA signaling in *Arabidopsis*, while expressions of *PR-3* and *PR-4* are popular markers for ET-mediated signaling in tobacco. The SA- and JA/ET-dependent signaling pathways appear to interact in a complex fashion; the primary mode of interaction is mutual antagonism [17]. However, examples of positive interactions have also been reported. Interpathway communication has been speculated to help plants fine-tune and prioritize defense responses upon encountering multiple signals. As is discussed later, it is common for TFs to be regulated by multiple signaling molecules, suggesting that they play an important role in integrating different signaling pathways.

8.1.3 Model Systems

Early model systems for studying TFs mediating defense responses included tobacco, potato, and cell suspension cultures from parsley and soybean. The rapid and efficient genetic transformation systems in tobacco were ideal for functional testing of candidate TF genes. Protoplasts and cell cultures were also convenient systems for functional assays based on transient or stable expression of foreign genes. Furthermore, they provided large amounts of relatively uniform starting material well suited for studying physiological and biochemical responses induced by the addition of defined chemical elicitors of defense responses. The availability of a well-defined chemical elicitor was a contributing factor in stimulating research on TFs in potato tubers. These early model systems also provided ample starting material for gene isolation and biochemical analysis of TFs.

Well-defined host–pathogen interactions were also critical for the establishment of model systems. The characterization, in the 1980s and early 1990s, of several host–pathogen systems involving *Arabidopsis thaliana* permitted application of this popular model species to the study of molecular plant pathology [18]. The genetic and genomic resources available in *Arabidopsis*, including the full genome sequence, are particularly relevant to studies of TFs. Because TFs regulate gene expression on a genome-wide basis, *Arabidopsis* resources maximize the potential for identifying and characterizing target genes [19]. Many TFs are encoded by large multigene families (see Table 8.1). Accordingly, the availability of annotated full-genome sequence information greatly facilitates identification of closely related

TABLE 8.1
Transcription Factors Implicated in Mediating Plant Defense Responses

Family name	Family size (*Arabidopsis*)	Cognate *cis*-element	Comments
ERF	56	GCCGCC (CGG-box)	Part of AP2/ERF superfamily; contains single ~60 a.a. DNA-binding domain consisting of three-stranded β-sheet with parallel running α-helix; binds DNA as monomers
WRKY	74	(T)GACC/T (W-box)	Contains one or two ~60 a.a. DNA binding domain consisting of four-stranded β-sheet and novel zinc binding pocket; binds DNA as monomers
Whirly	3	GTCAAAAA/T (elicitor response element [ERE])	Highly conserved β-sheet surface within family members (Whirly domain) involved in DNA binding; acts as tetramers; bind single-stranded DNA
TGA	10	TGACGT[a] (*as-1* element)	Part of bZIP superfamily; extended α-helix contains basic DNA-binding domain and leucine-zipper dimerization motif; act as homo- and heterodimers
R2R3MYB	125	Various	Largest group of plant MYB superfamily; contains two ~52 a.a. domains with helix–turn–helix structure

[a] This motif appears as a direct repeat in the *as-1* element. TGA factors are also capable of binding *cis*-elements containing only one copy of the motif. (Modified from Eulgem, T., *Trends Plant Sci.*, 10, 71, 2005.)

family members with potentially redundant functionalities [20], while resources for reverse genetics allow identification and analysis of mutations in such genes [21]. The large collection of *Arabidopsis* mutants affected in defense responses also represents a formidable analytical tool [13].

8.2 SEQUENCE-SPECIFIC TRANSCRIPTION FACTORS

8.2.1 ISOLATION

The major classes of TFs implicated in mediating plant defense gene expression are listed in Table 8.1. This section summarizes some of the approaches exploited to isolate them. Detailed information on structure and evolution of the TF families is provided in the following section.

Each of the methods described here has specific advantages and disadvantages. Those based on DNA-binding, protein–protein interactions, and mutant phenotypes do not require a priori knowledge of gene or protein sequence and do not rely on differential gene expression. With the exception of genetic approaches, none of the methods listed relies on the recovery of recognizable mutant phenotypes. However, none of the methods, by itself, provides as convincing evidence for implicating a TF in mediating defense responses as does the genetic approach. Several of the methods can be applied to model species as well as crops. Genomic approaches and those based on sequence similarity are well suited for identifying from crop species putative homologs of TFs initially characterized in model systems.

8.2.1.1 DNA Binding

The ability to bind short stretches of DNA of defined sequence with high affinity is a characteristic feature of TFs [9] and has been exploited extensively as a means of isolating these proteins and their corresponding genes. The founding members of all the families of TFs listed in Table 8.1, with the exception of MYB proteins, were isolated based on their ability to bind to *cis*-acting elements required for gene expression in response to microbial pathogens or signaling molecules such as SA, JA, or ET.

Southwestern hybridization, which involves screening expression libraries with radiolabeled oligonucleotide probes containing consensus *cis*-acting DNA sequences, was exploited to isolate the ethylene response factors (ERFs) [22], WRKY proteins [23], and different classes of basic leucine zipper (bZIP) proteins, including TGA factors [24] and G/HBF-1 [25], as well as homeobox proteins [26]. Biochemical purification of proteins capable of binding elicitor responsive or silencing elements found in the potato *PR-10* promoter was used to isolate a Whirly protein [27] and the novel *s*ilencing *e*lement *b*inding *f*actor (SEBF) [28], respectively. Partial peptide sequencing of purified proteins was subsequently used to clone the corresponding genes by degenerate PCR [27,28]. The yeast one-hybrid system has also been exploited to isolate TFs, including MYB [29] and MYC [30] type proteins, based on their ability to bind specific *cis*-acting elements.

8.2.1.2 Genetic Approaches

The isolation of mutants compromised in their response to pathogens or the growth regulators SA, JA, and ET represents the most productive strategy for identifying genes that regulate plant defense responses [13]. However, only a small fraction of the genes recovered using this approach encodes TFs. Examples include the *Arabidopsis LESION SIMULATING DISEASE1* (*LSD1*), which encodes a novel zinc finger protein [31]; *BOTRYTIS SUSCEPTIBLE1* (*BOS1*), which encodes an R2R3MYB protein [32]; and *JASMONATE INSENSITIVE1* (*JIN1*), which encodes an MYC protein [33].

Many plant TFs belong to large multigene families [20,34]. Consequently, it is very probable that functional redundancy between related members of a family limits the applicability of conventional genetic screens as a means of isolating TF genes

[35]. This is well illustrated by a recent study of *Arabidopsis* TGA TFs, which revealed that knockout of three related genes (*TGA2*, *TGA5*, and *TGA6*) was required before differences in *PR* gene expression and disease resistance were apparent [36]. Alternatively, loss of TF function may be lethal to the plant in certain instances [37,38]. Mutagenic strategies based on ectopic gene expression, such as activation tagging [39], may overcome the limitations of conventional, loss-of-function genetic screens for identifying TF genes mediating plant defense responses.

8.2.1.3 Protein–Protein Interaction

Protein–protein interaction screens, such as the yeast two-hybrid system, using components of signal transduction pathways mediating plant defense responses have also led to identification of TFs. In many cases, the TFs recovered (or closely related factors) had been previously identified by other means. Nevertheless, their physical interaction with proteins implicated in mediating defense responses provided valuable functional information. Examples include TGA factors found to interact with NPR1 [40] and the ERF proteins Pti4/5/6 recovered on the virtue of their ability to interact with the tomato R protein Pto [41].

8.2.1.4 Gene Expression

Methods used to identify genes based on their differential expression following pathogen infection or elicitation, including differential screening [42], suppressive subtractive hybridization [43], differential-display reverse transcription-PCR [44], and cDNA amplified fragment length polymorphism (cDNA-AFLP) [45], have also led to identification of TF genes. Furthermore, large-scale transcript profiling studies, including those making use of microarrays, *s*erial *a*nalysis of *g*ene *e*xpression (SAGE) or *m*assively *p*arallel *s*ignature *s*equencing (MPSS) [46], are revealing that large numbers of TF genes are differentially expressed under these conditions (e.g., Chen et al. [47], reviewed in Wan et al. [4]). Data from many of these studies are available in public databases (e.g., several links can be found at www.arabidopsis.org/info/expression/index.jsp) and tools facilitating the search of these databases are emerging [48]. In addition to the TFs listed in Table 8.1, results from large-scale transcript profiling are implicating many other classes of TFs in plant defense responses.

8.2.1.5 Sequence Similarity

The isolation of relatively few TF genes using the preceding methods enabled rapid cloning of numerous genes encoding related factors on the basis of sequence similarity. Methods exploited for these purposes include screening libraries at reduced stringency with DNA probes [49] or with degenerate oligonucleotides [50], as well as PCR with degenerate oligonucleotides [51].

8.2.1.6 Genomic Approaches

Large-scale *e*xpressed *s*equence *t*ag (EST) projects under way in most major crops are now identifying large numbers of TF genes expressed following infection with several pathogens of economic importance (e.g., www.plantgdb.org). Furthermore, whole genome sequencing efforts in *Arabidopsis*, rice, and, more recently, poplar, are revealing the entire repertoire of TF genes in these organisms [20,52]. Although the rice genome is close to three times the size of the *Arabidopsis* genome, both encode similar numbers of TFs (1300 to 1500). Both species also contain large families of genes encoding classes of TFs implicated in mediating defense responses, such as ERF, WRKY, and R2R3 MYB proteins (Table 8.1).

The sizes of other gene families, such as those encoding Whirly and TGA factors, are considerably smaller. It is noteworthy that, even in *Arabidopsis*, only a small fraction of the total number of TF genes has been functionally characterized. Some families of TFs, such as the R2R3MYB proteins, contain members with very diverse functions [53]; only a fraction of genes within these families may be involved in mediating defense gene expression. Conversely, some classes of TFs may be specialized for modulating gene expression in response to pathogen challenge. This may be the case for the WRKY family, in which 49 of 72 *WRKY* genes tested were found to be regulated in response to pathogen infection or SA treatment [54].

8.2.2 Structure and Evolution of Transcription Factor Families Implicated in Mediating Defense Responses

Transcription factors are modular, consisting of one or more separate DNA-binding and effector (i.e., transcriptional activation or repression) domains [9]. DNA-binding domains are by far the most conserved portion of TFs and are commonly used as the basis for classifying these proteins [20]. Outside the DNA binding domains, family members may have little or no sequence similarity.

8.2.2.1 The ERF Family

ERF proteins contain a novel DNA-binding domain of about 60 amino acids previously identified in the product of the floral homeotic gene *APETALLA 2* (*AP2*) [22]. Accordingly, AP2 and ERF TFs are commonly grouped into a single family (AP2/ERF family) that includes three other subfamilies: the *d*ehydration-*r*esponsive *e*lement-*b*inding (DREB) factors, related to ABI3/VP1 (RAV), and others [20,55,56]. To date, only members of the ERF subfamily have been implicated in plant defense responses [56]. Based on sequence similarity, the ERF family can be further divided into a number of subclasses [56].

The structure of the AP2/ERF domain from the *Arabidopsis* ERF1 protein in complex with its target DNA was solved by NMR [57]. It consists of a three-stranded antiparallel β-sheet and one α-helix running almost parallel to the β-sheet. DNA contact is achieved through arginine and tryptophan residues in the β-sheet of the ERF1 monomer. Whereas amino acid residues making contact with the DNA are highly conserved among ERF proteins, they are not present in AP2 proteins [49], reinforcing the view that AP2 and ERF proteins are distinct. The three-dimensional structure of the ERF domain is related to those of the Tn916 and λ integrases and the human methyl-CpG binding domain MBD, even though these proteins do not appear to share any amino acid sequence similarity [58].

The ERF/AP2 domain was initially thought to be plant specific [20]. However, genes capable of encoding this domain have recently been identified in the ciliate *Tetrahymena* [58,59], the cyanobacteria *Trichodesmium erythraeum*, and two bacteriophages [58]. None of the nonplant AP2/ERF domain proteins are TFs. Instead, they are predicted HNH homing endonucleases, a class of proteins with catalytic and DNA-binding activities responsible for the lateral transfer of intervening sequences from genes into cognate alleles lacking them. It has been proposed that AP2/ERF TFs originated through lateral transfer of an HNH-AP2/ERF homing endonuclease gene from bacteria or bacteriophages into plants [58].

8.2.2.2 The WRKY Family

WRKY proteins contain one or two conserved domains of approximately 60 amino acids harboring the conserved sequence WRKYGQK at its N-terminal end and a novel zinc finger-like motif (reviewed in Eulgem et al. [60] and Zhang and Wang [61]; see Section 8.2.2.6 for information on zinc fingers). This family of TFs was originally classified into three groups [60]; however, reclassification into five groups has recently been proposed based on comparison of a more extensive set of proteins [61].

As revealed by NMR, the solution structure of the C-terminal WRKY domain of the *Arabidopsis* WRKY4 protein consists of a four-stranded antiparallel β-sheet with a novel zinc-binding pocket at one end [62]. Proper folding of the domain depends on the presence of zinc ions, which had previously been shown to be required for the DNA-binding activity of WRKY proteins *in vitro*. The hallmark WRKYGQK motif is localized within the N-terminal-most β-strand and has been proposed to be involved directly in DNA binding. Based on partial structural similarity, an evolutionary relationship was proposed to exist between WRKY domains and the drosophila GCM TFs [62].

Similar to ERF proteins, WRKY proteins were originally thought to be plant specific, but have recently been identified in other eukaryotes. More specifically, WRKY-like domains have been identified in the slime mold *Dictyostelium discoideum* and the unicellular protist *Giardia lamblia* [62,63]. They have also been reported in ferns, mosses, and green algae. No functional information is available on *WRKY* genes from nonplant and lower plant sources. Thus, it appears that the structural framework of DNA-binding domains from so-called plant-specific TFs were in fact established before divergence of the plant kingdom [62].

8.2.2.3 The Whirly Family

The potato Whirly factor StWhy1 is the only sequence-specific TF from plant for which the crystal structure has been solved [64] (reviewed in Desveaux et al. [65]). The active TF consists of four protomers packed perpendicularly against each other. Each protomer is made up of two antiparallel β-sheets and a helix–loop–helix motif. Interaction of protomers occurs through this motif, with the β-strands protruding outwards and resulting in the whirligig appearance that inspired the family name.

Sequence comparison between Whirly-like proteins from different plant species indicates that the region forming the β-sheet surface of StWhy1 is the most highly conserved and has been named the Whirly domain [65]. This region is thought to be the major surface involved in binding DNA. Interestingly, the StWhy1 tetramer binds to single-stranded DNA [27] and StWhy1 shares limited sequence similarity with single-stranded binding proteins from a number of sources [65]. However, genes encoding Whirly proteins have only been found in higher plants (angiosperms and gymnosperms) and the unicellular green algae *Chlamydomonas reinhardtii* [65].

8.2.2.4 The TGA Factor Family

TGA factors are a class of bZIP TFs [66] originally isolated based on their ability to bind to the SA-, JA-, and auxin-inducible a*ctivating* s*equence-1* (*as-1*) element found in the cauliflower mosaic virus 35S promoter or the related *ocs* element in the *oc*topine *s*ynthase promoter. As such, this class of factors is occasionally referred to as *o*cs-element *b*inding *f*actors (OBFs). Although no proteins closely related to TGA factors are found outside the plant kingdom, the bZIP domain is found in TFs from all eukaryotic kingdoms.

bZIP proteins typically function as homodimers and/or heterodimers. When bound to DNA, the bZIP domain of each monomer exists as a contiguous α-helix. The N-terminal basic region consists of approximately 16 amino acids that bind in the major groove of double-stranded DNA. The C-terminal consists of heptad repeats of leucines or other bulky hydrophobic amino acids and is amphipathic. This region mediates dimerization, forming a parallel coiled coil called the leucine zipper. TGA factors also contain an additional, novel domain important for dimerization [67]. Three groups of TGA factors (I, II, III) can be distinguished based on sequence similarity.

8.2.2.5 The R2R3MYB Family

The MYB domain is a conserved region of about 52 amino acids that displays a helix–turn–helix structure capable of intercalating into the major groove of DNA [68]. One to three copies of this domain (R1, R2, R3) are typically present in MYB proteins. Compared to other eukaryotes, plant genomes encode a large number of MYB proteins; most contain two MYB repeats (R2R3) [34].

8.2.2.6 Zinc Finger-Related Proteins

Zinc fingers are protein domains that use conserved cysteine and/or histidine residues to coordinate a zinc iron, yielding a compact "finger"-like structure [69]. The specific arrangements of cysteines and histidines define different types of zinc finger domains, some of which have the potential to bind DNA and others that mediate protein–protein interactions. Plant TFs implicated in mediating defense responses appear to contain novel zinc finger domains. The pepper CAZFP1 contains two novel C_2H_2-type zinc fingers [70], while LSD1 defines a novel type of plant-specific C_2C_2 zinc finger [31,38]. Neither LSD1 nor the related protein *L*SD *O*ne-*L*ike1 (LOL1) has been shown to bind DNA, and it is possible that they function as scaffolds instead of TFs [38].

8.2.2.7 Others

The structures of TF domains conserved across eukaryotic kingdoms, such as the homeobox and MYC basic helix–loop–helix (bHLH), have been summarized elsewhere [9,69]. Some homeobox TFs implicated in regulating defense response genes also contain a leucine zipper [26]. Interestingly, the plant-specific SEBF shares sequence similarity with nuclear-encoded chloroplast RNA-binding proteins, suggesting that it may also be involved in RNA processing [28].

8.2.3 Functional Characterization

Following their isolation, additional analyses are typically required to further elucidate the role of TFs in mediating plant defense responses. This section briefly summarizes information obtained on characterizing their DNA-binding sequences, transactivation and transrepression properties, gene expression patterns, and subcellular localization. Research aimed at identifying target genes, studying post-translational regulation, and assessing the role of TFs for disease resistance by genetic or transgenic approaches is discussed in subsequent sections.

8.2.3.1 DNA Binding Preferences

In general, the DNA-binding targets of TFs are initially determined *in vitro*, using approaches such as the electrophoretic mobility shift assay (EMSA) or PCR-based oligo selection (see Carey and Smale [71] for a detailed account of methods used to study TFs). Transcription factors to be tested are typically produced *in vitro* or highly purified from plant tissues. It is also possible to make use of cell extracts in combination with antibodies against specific TFs to "supershift" the DNA-binding activity (see, for example, Lam and Lam [72] and Niggeweg [73]). At a minimum, studies should demonstrate specificity of DNA binding. This is typically achieved using DNA probes with mutations at key residues or by addition of excess nonlabeled competitor DNA.

All ERF, WRKY, Whirly, and TGA factor proteins tested to date have the ability, *in vitro*, to bind to their cognate *cis*-elements listed in Table 8.1. Of note, two ERF proteins, Tsi1 from tobacco and CaPF1 from pepper, were shown to have dual specificity in binding to the GCC-box and the DRE-box [44,74]. Binding of Tsi1 and CaPF1 to the DRE-box appears to be biologically relevant because overexpression of these TFs results in the constitutive expression of DRE-box-containing genes and enhanced tolerance to abiotic stress (osmotic [74] and freezing [44], respectively). Most ERFs have not been tested for binding to the DRE-box and, accordingly, it is not known how common this dual specificity may be. However, Tsi1 and CaFP1 are not closely related family members [44], suggesting that other ERF proteins may share this property.

As a group, MYB proteins appear to have broader DNA-binding specificities [68]. The tobacco Myb1 protein was shown to bind MBSII (GTTTGGT)- and MBSI (TAACTG)-related elements in the promoter of the *PR-1a* gene [75]; NtMYB2 was found to bind to a wounding and elicitor-responsive L-box element (TCTCACCTACC) present in the promoters of genes involved in phenylpropanoid biosynthesis [29].

The above groups of TFs bind to double-stranded DNA. In contrast, the Whirly proteins StWhy1 [27] and AtWhy1 [37] (and probably all family members [65]), and SEBF [28] preferentially bind to single-stranded DNA. Unwinding of DNA to generate single strands may be a means of relieving the torsional stress associated with gene transcription [27]. It has been proposed that the Whirly proteins and SEBF may stabilize unstable regions of melted DNA *in vivo* [27].

Relatively few studies have attempted to determine the preferred binding sequences of TFs implicated in mediating plant defense responses (see, for example, Desveaux et al. [27] and Krawczyk et al. [76]). When compared, the binding specificity of individual members of a TF family displays slight differences [49,72], which are likely of regulatory significance in determining target site selection and gene expression parameters *in vivo*.

8.2.3.2 Transeffector Properties

A key function of TFs is to recruit the transcriptional machinery to specific gene promoters [9]. This is achieved through direct interactions with one or more general TFs or indirectly through cofactors (coactivators) that do not bind DNA. Several classes of coactivator have been described [77,78]. Many comprise large multiprotein complexes that possess chromatin remodeling and/or modifying activities. These activities facilitate access of TFs and the basal transcriptional machinery to specific gene promoters. Conversely, transrepression domains may interact with co-repressors possessing chromatin remodeling and/or modifying activities that impede access of the basal transcriptional machinery.

Transcription factor interfaces responsible for recruiting the preceding classes of proteins are called effector (transactivation or transrepression) domains. The specific amino acid sequences and structural features required to create effector domains are not as well defined as those responsible for DNA binding. Many transactivation domains are rich in acidic amino acids; others are rich in glutamine or isoleucine [9]. Repression domains have been characterized as being charged, rich in alanine or in alanine and proline [79]. Proline-rich domains have been implicated in transcriptional activation and repression. It is also possible for individual TFs to display transactivation and repression properties. The cell-specific concentration of a TF as well as the presence and concentration of interacting proteins are factors that may determine its ability to transactivate vs. transrepress [80].

Sequence analysis of TFs implicated in plant defense responses reveals that most contain features characteristic of effector domains (see, for example, references 56, 60, 65, 81, and 82). Several of these proteins have been experimentally shown to possess transactivation or transrepression properties. To facilitate analysis, tests are usually performed in yeast cells [82] or transient plant-based assays [49], although stably transformed plant tissues have also been used [83]. It is also common in these tests to fuse the TF to heterologous DNA-binding domains, such as the one from the yeast GAL4 protein, which have well characterized DNA-binding elements.

Each of the ERF, WRKY, and TGA families contain members that transactivate and others that transrepress [7,56,80]. The potato StWhy1 has been shown to transactivate [37]. However, several Whirly factors lack obvious effector domains, a feature that was considered as a main source of divergence within the family [65]. SEBF possesses transrepression activity [28]. The intact CaZFP1 protein did not display transactivation potential in yeast [70]. Otherwise, the transactivation properties of R2R3MYB and zinc finger proteins implicated in mediating defense responses have not been assessed.

A number of studies have attempted to localize protein regions important for the transactivation or transrepression activity of TFs implicated in plant defense responses. In some cases, sequences commonly associated with effector domains were found to be important (see, for example, Desveaux et al. [37]). In other cases, mutational analyses were not sufficiently detailed to resolve individual amino acids required or indicated a role for multiple parts of the proteins [84,85]. Comparisons of effector domains between family members suggest distinct modes of transactivation within individual TF families.

The transrepression properties of ERF proteins have been studied in detail. In addition to repressing basal transcription in transient assays, NtERF3, AtERF3, and AtERF4 also repress transactivation of other TFs [49,85,86]. Repression occurred in a dose-dependent fashion and was effective against ERF and non-ERF transcriptional activators [49]. Repression also required DNA binding of the repressor protein, but did not rely on competition for DNA binding sites with the transcriptional activator [49]. The transrepression activity of ERF proteins was localized to a conserved amphiphilic motif $^{L}/_{F}DLN^{L}/_{F}(X)P$ (the EAR motif) with the capacity to suppress transactivation when fused to heterologous DNA-binding domains [86,87]. Several ERF proteins containing the EAR motif cluster into the same subfamily [56].

8.2.3.3 Subcellular Localization

As would be expected for proteins with a role in regulating gene expression, TFs implicated in mediating plant defense responses that have been analyzed to date localize to the nucleus (see, for example, References 70, 85, 88, and 89). Functional nuclear localizing signals have been identified in WRKY [88] and TGA factors [90]. SEBF, which shares sequence similarity to chloroplast RNA-binding proteins, was found to be localized to the nucleus and chloroplasts [28]. The SEBF cDNA has the capacity to encode a putative transit peptide. Interestingly, the N-terminus of several Whirly proteins is also predicted to be chloroplast transit peptides [65]. These results suggest that SEBF and Whirly factors could play a role in coordinating defense responses in both compartments [28,65].

Control of subcellular protein localization, in particular nuclear import and exclusion of TFs, can be of considerable regulatory importance [91]. However, there have been no reports indicating that nuclear localization of TFs is regulated in response to pathogen challenge.

8.2.3.4 Gene Expression

Many, if not most, of the TF genes implicated in regulating gene expression during defense responses are differentially regulated following pathogen challenge [47,56,60]. Collectively, and even within individual TF gene families, the range of expression patterns observed is complex, showing differences in directionality (up- or downregulated), kinetics (immediate early, early, late expression), amplitude (strong, weak) and duration (transient, sustained). For any given TF gene, these parameters may be influenced as a function of the infecting pathogen and the plant host's ability to resist the pathogen (see, for example, References 44, 70, 92, and 93).

In general, there does not appear to be any correlation between the directionality of gene expression and the transactivation properties of the encoded protein (i.e., genes encoding transactivators are not necessarily upregulated and those encoding transrepressors are not all downregulated; see, for example, Fujimoto et al. [49]). Regulation of TF gene expression may occur at the site of infection [84] as well as in systemic, noninfected tissues [94]. Gene expression may also be affected by abiotic stresses [44,63,70,92] and can show developmental regulation in the absence of biotic or abiotic stress [60,70,92].

It is also common for expression of these TF genes to be modulated by one or more of the defense-related growth regulators, with simultaneous exposure to combinations of growth regulators having synergistic effects on gene expression. For example, many *ERF* genes are regulated by ET, with ET and JA having synergistic effects [56]. However, SA has also been shown to induce the expression of some *ERF* genes and SA/ET synergism has been reported [92]. Transcription of certain *ERF* genes [74,95] as well as the *CAZFP1* gene encoding a novel zinc finger [70] is induced by SA, JA, and ET. Many *WRKY* genes are also regulated by SA or JA. Two *MYC* genes implicated in JA signaling were found to be induced by this growth regulator [30], while the tobacco *Myb1* gene associated with tobacco mosaic virus (TMV) infection is induced by SA [96]. Some genes encoding TFs implicated in mediating defense responses do not appear to be modulated in response to pathogen challenge or defense-related growth regulators (see, for example, References 49, 92, 97, and 98).

To further elucidate the signaling pathways mediating TF gene expression, several groups have exploited mutants compromised in defense-related signaling [93,99,100]. Of note, the expression of several *WRKY* genes depends on NPR1 [43,101], and an *MYB* gene (*AtMYB30*) associated with the hypersensitive response is constitutively expressed in several *lsd* mutants that spontaneously form disease-like lesions [102].

Little is known about the mechanism regulating expression of TF genes implicated in defense responses. Several immediate-early *WRKY* genes contain W-boxes in their promoters, and it was recently shown using chromatin immunoprecipitation (ChIP; see Section 8.2.6) that the parsley *WRKY3* and *WRKY1* genes are *in vivo* targets of WRKY1 [103]. This suggests possible autoregulatory control of *WRKY* gene expression.

Transcriptional regulation of the *Arabidopsis ERF1* gene has been studied in some detail. Expression of *ERF1* requires the ETHYLENE INSENSITIVE 3 (EIN3)

TF, which specifically binds a primary ethylene response element (PERE) distinct from the GCC-box, in the *ERF1* promoter *in vitro* [104]. In the absence of ET, EIN3 is continuously degraded through the action of two closely related F-box proteins, EBF1 and EBF2 (*E*IN3-*b*inding *F* box protein 1 and 2) [105–107]. Exposure to ET or mutation of *EBF1* and *EBF2* leads to increased levels of EIN3 [105–107] and increased levels of *ERF1* [106].

It is not known how ET modifies EIN3 or EBF proteins to prevent EIN3 degradation, but it has been speculated that ubiquitination and degradation of EIN3 may be triggered by its phosphorylation status [105]. In support of this hypothesis, a *m*itogen-*a*ctivated *p*rotein *k*inase (MAPK) signaling cascade proposed to operate upstream of EIN3 has recently been identified. Constitutive activation of the MAPK kinase involved (SIMKK) leads to constitutive expression of *ERF1* [108]. The consequences on disease resistance of elevated *ERF1* expression in the *EBF1/EBF2* mutant and SIMKK plants have yet to be determined.

Treatment of tobacco calli with inhibitors of protein kinases and phosphatases substantially inhibited the levels of *ERF2-4* transcripts [51], suggesting that protein phosphorylation is also important for expression of these genes. Curiously, expression of these genes [51] as well as several *Arabidopsis ERFs* [49] was induced by the protein synthesis inhibitor cycloheximide (CHX). It was proposed that CHX may be preventing the *de novo* synthesis of a labile transcriptional repressor or mRNA degrading enzymes [49]. It will be interesting to determine whether the putative labile repressor is associated with EIN3 degradation.

8.2.4 Functional Analysis of Mutant and Transgenic Plants with Altered Levels of Transcription Factor Genes

8.2.4.1 Definitions and Limitations

Convincing proof of a TF's participation in mediating defense responses usually requires the characterization of plants containing altered levels of it (see Chandler and Werr [109] for a recent review on different approaches possible). The gold standard for such analyses is the study of loss-of-function mutations. The more traditional, forward genetic approach involves screening populations of mutagenized plants for a given phenotype (e.g., altered *PR* gene expression or disease resistance). The alternative, reverse genetics, involves identifying mutations in genes of known sequence and subsequently determining the resulting phenotypes.

Such strategies have become increasingly popular with onset of the genomic era. Reverse genetics offer the researcher more control in targeting genes of interest and facilitate creation of plants harboring mutations in multiple (related) genes. However, there are no guarantees that alleles recovered by reverse genetics will yield visible phenotypes. In fact, data from *Arabidopsis* and nonplant model systems suggest that few do [35]. It is also important to recognize that unrelated, second-site mutations may be generated by mutagenic treatments [110].

Several transgenic strategies have also been exploited to reduce levels of TF gene expression. These include antisense and RNA interference (RNAi) technology. Compared to antisense, RNAi is usually more effective at reducing levels of target gene transcripts. The creation of dominant-negative TFs has been a popular means of interfering with TGA factor function. Basically, transgenes are designed that will produce proteins lacking DNA binding activity but retain the ability to dimerize with biologically active, endogenous factors. Expression of the dominant-negative TF sequesters the endogenous dimerization partners, preventing them from binding DNA and regulating gene expression. One advantage of the dominant-negative approach is that it may overcome issues of functional redundancy in cases where related TFs have the potential to dimerize.

Transgenic ectopic (over) expression has represented the most popular means of functionally testing plant TFs [35]. It is not affected by functional redundancy and can be applied to crop species in which tools for reverse genetics are not available. Even when loss-of-function phenotypes are available, ectopic expression has been shown to yield valuable new information [35]. A twist to the ectopic expression approach involves use of TF fused to strong, constitutive transactivation domains, such as the one from viral particle 16 (VP16). This strategy may be particularly informative in cases in which the transactivation potential of a TF is weak or tightly regulated.

It is noteworthy that overexpression of TFs can trigger artifacts, so results need to be interpreted with caution. One potential problem is that of squelching [35]. As discussed in Section 8.2.3.2, effector domains represent interfaces for protein–protein interactions. Accordingly, increasing the quantity of a TF through transgenic ectopic expression can lead to sequestering of interacting proteins, such as cofactors, which may be limiting and required for the function of unrelated TFs. Expression of TFs at unphysiologically high levels may also result in binding of low-affinity DNA elements not normally bound by the TFs and the subsequent nonspecific activation or repression of gene sets. As a result of nonspecific binding or squelching, cellular homeostasis may be disturbed, yielding pleiotropic effects. Dominant-negative approaches described earlier are also prone to yielding pleiotropic effects.

Transient gene expression methods have also emerged as important tools for functionally testing TFs. Of note, *v*irus-*i*nduced *g*ene *s*ilencing (VIGS) offers the potential to test large numbers of candidate genes rapidly, without the need to generate stable transgenic lines [111].

8.2.4.2 ERFs

Individual *ERF* genes have been overexpressed in a variety of transgenic plants including *Arabidopsis*, tobacco, tomato, and hot pepper (Table 8.2). In most instances, these modifications resulted in enhanced levels of disease resistance to a selected range of pathogens, including *Botrytis cinerea*, *Fusarium oxysporum*, and *Plectosphaerella cucumerina*, which elicit the JA/ET-defense pathways in the infected host [14,93,112]. Enhanced resistance was not compromised by mutants blocking early steps of ET signaling, indicating that the downstream events in this pathway were constitutively activated [93].

Testing of pathogens in which resistance is not mediated primarily by the JA/ET signaling pathways yielded variable results. For example, overexpression of tomato *Pti4* in *Arabidopsis* enhanced tolerance to the biotrophic bacterial pathogen

TABLE 8.2
Effect of ERF Gene Overexpression on Disease Resistance and *PR* Gene Expression

Gene	Host	Disease resistance[a]	Marker gene expression[b]	Ref.
Tomato *Pti4*	*Arabidopsis*	*Erysiphe orontii* *Pseudomonas syringae* pv. *tomato* DC3000 (tolerance)	*PR-1, PR-2, PR-3, PR-4, PDF1.2, Thi2.1*	113
Tomato *Pti5*	*Arabidopsis*	No enhanced resistance to *E. orontii* or *Pst* DC3000	*PR-1, PR-2, PR-3, PR-4, PDF1.2* (weaker than 35S:Pti4)	113
Tomato *Pti6*	*Arabidopsis*	No enhanced resistance to *E. orontii* or *Pst* DC3000	*PR-1, PR-2, PR-3, PR-4, PDF1.2, Thi2.1* (weaker than 35S:Pti4)	113
Tomato *Pti5*	Tomato	*P. syringae* pv. *tomato*	Enhanced expression of *GluB* and *Catalase* after pathogen challenge SA-regulated *PR1a1* and *PR1b1* not expressed	114
VP16:*Pti5*	Tomato	*P. syringae* pv. *tomato* (resistance similar to *Pti5* alone)	Higher levels of *GluB* and *Catalase* than observed with *Pti5* alone	114
Arabidopsis ERF1	*Arabidopsis*	*Botrytis cinerea* and *Plectosphaerella cucumerina* Enhanced susceptibility to *Pst* DC3000	*Basic chitinase* and *PDF1.2*	93
Arabidopsis ERF1	*Arabidopsis*	*Fusarium oxysporum* sp. *conglutinans* and *F. oxysporum* f. sp. *lycopersici*	*PDF1.2*	112
Tobacco *Tsi1*	Tobacco	*P. syringae* pv. *tabaci*[c]	*PR-2, PR-3, PR-4, Osmotin, SAR8.2, PR-1*	74
Tobacco *Tsi1*	Hot pepper	Pepper mild mottled virus, Cucumber mosaic virus, *Phytophthora capsici*, and *Xanthomonas camperstris* pv. *vesicatoria*	*PR-1, PR-2, PR-4, PR-5, PR-10, PinII, LTP1, SAR8.2*	196
Pepper *CaPF1*	*Arabidopsis*	*P. syringae* pv. *tomato* DC3000[d]	*GST & PDF1.2*	44
Tobacco *ERF5*	Tobacco	Tobacco mosaic virus No enhanced resistance to *P. syringae* pv. *tabaci*, *P. syringae* pv. *pisi*	No enhanced expression of *PR1a, PR1b, PR3*	98
Tomato *TSRF1*	Tomato	*Ralstonia solanacearum* and *P. syringae* pv. *tomato*	*PR2, PR3*	197

continued

TABLE 8.2 (continued)
Effect of ERF Gene Overexpression on Disease Resistance and *PR* Gene Expression

Gene	Host	Disease resistance[a]	Marker gene expression[b]	Ref.
Tomato *TSRF1*	Tobacco	*R. solanacearum*	*PR1, PR2* & *PR3*	197
Tomato *OPBP1*	Tobacco	*Phytophthora parasitica* var. *nicotianae* and *P. syringae* pv. *tabaci*[e]	*PR-5d, PR-1a*	95
Pepper *CaERFLP1*	Tobacco	*P. syringae* pv. *tabaci*[e]	*β-glucanase, osmotin, HMG-CoA reductase, cysteine protease*	198

[a] Unless otherwise noted, overexpression of ERF gene resulted in enhanced disease resistance.

[b] Unless otherwise noted, overexpression of ERF gene resulted in constitutive expression of the genes listed.

[c] Plants also display enhanced tolerance to osmotic stress.

[d] Plant also display enhanced freezing tolerance.

[e] Plants also display enhanced salt tolerance.

Pseudomonas syringae [113], but overexpression of *Arabidopsis ERF1* increased susceptibility to this pathogen [93]. Such results may reflect examples of negative or positive cross-talk, respectively, between the JA/ET and SA signaling pathways mediated by transgenic expression of the ERF protein. It is also notable that overexpression of tomato *Pti5* in *Arabidopsis* failed to increase resistance to *P. syringae* or *Erysiphe orontii* [113], but this TF was effective against the former pathogen when overexpressed in tomato [114]. These apparently conflicting results may be attributed to distinct functionalities between proteins (in the case of ERF1 and Pti4) or of the proteins in different host plants (*Arabidopsis* vs. tomato), as well as to variations in experimental designs between studies.

Enhanced resistance to pathogens conferred by overexpression of *ERF* genes was usually correlated with enhanced levels of marker *PR* gene expression following pathogen challenge or with their constitutive expression [93,113,114]. Expression of a tomato Pti5:VP16 fusion protein was found to elicit higher constitutive levels of marker *PR* genes than the native Pti5, but did not substantially increase disease resistance compared to native Pti5 [114]. Lack of enhanced disease resistance in *Arabidopsis* plants overexpressing *Pti5* or *Pti6* was correlated with weak to no constitutive expression of marker genes [113].

Most marker genes analyzed in the preceding studies contained GCC-boxes in their promoters and were known to be responsive to ET or JA. In some cases, only a subset of GCC-box-containing marker genes tested was activated in disease-resistant transgenic plants [114]. SA-inducible genes were also tested and found to be activated in some cases [113], but not in others [98]. As discussed later (Section 8.2.6), transcript profiling of transgenic plants overexpressing *ERF* genes revealed that these TFs potentially regulate large sets of genes [100,115,116].

Overexpression of *Pti4* and *ERF1* in *Arabidopsis* was correlated with developmental abnormalities. The most common phenotype observed is reminiscent of the so-called ethylene triple response, referring to the inhibition of hypocotyl and root elongation and an exaggerated curvature of the apical hook observed in wild-type seedlings exposed to ET. Seedlings of *Arabidopsis* plants overexpressing *Pti4* or *ERF1* display the inhibition of hypocotyl elongation in the absence of ET [104,113,115]; those overexpressing *ERF1* also display inhibition of seedling root elongation [104]. In no case was the curvature of the apical hook found to be affected. Interestingly, *HOOKLESS1* (*HLS1*), a gene regulating the apical hook curvature, contains a GCC-box in its promoter. Although ERF proteins can bind *in vitro* to a promoter containing a multimer of the *HLS1* GCC-box [49], *HLS1* expression is not activated in *Pti4* or *ERF1* overexpressing plants [104,113,115].

Plants overexpressing *Pti4* and *ERF1* also display phenotypes associated with ET exposure at the adult stage, including smaller size, greener leaves, and, in the case of *ERF1* overexpressors, inhibition of cell enlargement, wilting, and death before bolting [104]. Together, these results indicate that subsets of ET responses are constitutively activated by overexpression of these *ERF* genes. Overexpression of the other *ERF* genes tested to date has not been reported to induce developmental abnormalities.

The phenotypic consequences associated with loss of *ERF* gene function have yet to be determined. Analysis of the SIGnAL T-DNA database (http://signal.salk.edu) revealed the presence of insertions in or near many members of the *Arabidopsis* AP2/ERF family [117]. However, there have been no published reports of their characterization. Similarly, antisense expression of the tomato *Pti5* was reported to have no effect on race-specific resistance [114], although no data were shown.

8.2.4.3 WKRY Factors

Similar to the ERF family, there are few reports describing the phenotypic consequences of mutations in *WRKY* genes. Mutations in more than 40 *Arabidopsis WRKY* genes have been identified by reverse genetics, but most do not appear to display visible mutant phenotypes [63]. The consequences of altering levels of WRKY factors on disease resistance have been reported for only five *Arabidopsis* genes. VIGS has also been applied to study the involvement of WRKY genes from solanaceous plants in *R*-gene mediated resistance pathways.

Overexpression of *WRKY70* enhanced resistance to virulent strains of the bacterial pathogens *Erwinia carotovora* and *P. syringae*, but antisense-mediated reduction of *WKRY70* resulted in enhanced susceptibility to these pathogens [43]. Overexpression was correlated with increased expression of SA-inducible *PR* genes and decreased expression of JA/ET-regulated defense-related genes. Conversely, antisense suppression of *WRKY70* resulted in constitutive expression of JA/ET-regulated marker genes. Together, these results indicate that WRKY70 acts as a positive regulator of disease resistance and it was proposed to represent a convergence point for the SA and JA signaling pathways [43].

Overexpression of *WRKY18* also resulted in enhanced resistance to *P. syringae* and increased expression of SA-inducible *PR* genes [118]. However, these phenotypes were only observed in older plants (e.g., 5 weeks old). Interestingly, mutations at the *NPR1* locus had little effect on *PR* gene expression in the *WKRY70* overexpressors

[43]; however, they abolished potentiation of *PR* genes and enhanced disease resistance observed in older plants overexpressing *WKRY18* [118].

Despite increasing levels of *PR* genes, overexpression of *WRKY6* had no appreciable effect on resistance to virulent or avirulent strains of *P. syringae* [80]. Transposon-induced mutation of *WRKY6* also induced changes in gene expression consistent with a possible role in defense responses, but was not associated with any visible mutant phenotype [80]. Thus, although results implicate WKRY6 and WRKY18 in mediating the expression of defense genes, they also indicate that neither gain nor loss (in the case of WRKY6) of these TFs by itself is sufficient to affect disease resistance.

In all cases reported, stable transgenic overexpression of *WRKY* genes resulted in developmental abnormalities, including stunted growth, altered leaf morphology, and changes in flowering time [43,80,118]. Overexpression of *WRKY6* was also associated with development of necrotic areas on leaves and loss of apical dominance [80]. Reduction of *WRKY70* levels resulted in larger plants and early flowering (overexpression of *WRKY70* delays flowering) [43], but reduction or loss of *WRKY6* did not result in any developmental abnormalities [80]. Expression of *WRKY6* and *WRKY18* transgenes also decreased levels of corresponding endogenous genes [80,118], suggesting that these TFs are involved in regulating expression of their genes and, possibly, other members of the family. Transcript profiling of plants with altered levels of *WRKY6* and *WRKY70* also supports the notion that these TFs can activate and suppress target gene expression [43, 80].

Transient overexpression of *AtWRKY29* in *Arabidopsis* leaves reduced disease symptoms caused by *P. syringae* and *B. cinerea* [119]. However, the effects of altering levels of this *WRKY* gene have not been tested in stable transgenic or mutant plants. VIGS has also implicated the tobacco *WRKY1*, *WRK2*, and *WRKY3* genes as being required for full *N* gene-mediated resistance to TMV [120].

8.2.4.4 Whirly Factors

AtWhy1 is the only Whirly factor functionally characterized to date [37]. Two missense mutant alleles in this gene were recovered by *t*argeting *i*nduced *l*ocal *l*esions *in* *g*enomes (TILLING). Based on the crystal structure of StWhy1, one mutation mapped to the single-stranded DNA-binding domain, and the other was located in the region implicated in tetramerization. Mutant plants exhibited reduced AtWhy1 DNA-binding activity and *PR-1* expression following SA treatment. Both mutant backgrounds were hypersusceptible to a virulent strain of the biotrophic oomycete *Peronospora parasitica*. The *atwhy1.2* mutant also showed intermediate susceptibility to an avirulent strain of *P. parasitica* and was unable to mount an effective SAR response following treatment with SA [37]. Thus, AtWhy1 constitutes a positive regulator of SA-dependent disease resistance. However, SA-induced Whirly DNA-binding activity does not require functional NPR1, suggesting that Whirly activation occurs through a distinct, NPR1-independent pathway.

8.2.4.5 TGA Factors

The most compelling evidence implicating TGA factors as mediators of defense responses comes from the analysis of the triple *tga2tga5tga6* loss-of-function

Arabidopsis mutant [36]. This mutant is compromised in SAR against virulent strains of *P. syringae* and *P. parasitica*. Interestingly, basal resistance to these pathogens was not compromised. The mutant also failed to express *PR-1* in response to SA, but displayed higher basal levels of *PR-1* in the absence of SA elicitation. Although no developmental abnormalities were reported, seedlings of the triple mutant were hypersensitive to the toxic effects of SA, a phenotype also observed in *npr1* mutants (see Section 8.3.1.1).

Attempts to study the role for TGA factors in mediating disease and *PR* gene expression using dominant-negative versions of TGA factors have yielded conflicting results. Expression of a dominant-negative *Arabidopsis TGA2* gene in *Arabidopsis* compromised basal resistance against *P. syringae* pv. *maculicola* [83]; expression of a similar dominant-negative version of the same *Arabidopsis* gene in tobacco enhanced SAR against *P. syringae* pv. *tabaci* [121]. In these studies and others [73,122], the transgenic plants were also monitored for expression of marker genes containing *as-1* elements. Two types of genes were considered, based on the timing of their expression following SA treatment: early genes and late genes.

PR-1 is considered a late gene, and was the only maker gene evaluated in one study [83]. Loss of basal resistance observed by transgenic expression of dominant-negative *TGA2* was correlated with reduced levels of *PR-1* in *Arabidopsis* [83], while enhanced SAR in tobacco was associated with increased levels of *PR-1a* but reduced levels of early gene transcripts [121]. In contrast, a dominant-negative tobacco *TGA2.2* resulted in reduced levels of early and late genes, while (over)expression of wild-type *TGA2.2* increased levels of early genes but had no affect on *PR-1a* [73]. Finally, plants expressing wild-type or dominant-negative versions of the closely related tobacco *TGA2.1* gene displayed enhanced and reduced levels of early genes, respectively, but showed no changes in *PR-1a* levels [122].

Based on the observation that dominant-negative *TGA2* has opposite effects on early and late gene expression, it was proposed that TGA factors have both positive and negative roles in mediating plant defense responses [121]. Additional evidence in support of a dual role for TGA factors comes from RNAi analysis, which revealed a negative role for TGA4 and a positive role for TGA5 in regulating a reporter gene under the control of a multimerized *cis*-element related to *as-1* [123]. The consequences of (over)expressing wild-type or dominant-negative *TGA2.1* and *TGA2.2*, or the *TGA4* and *TGA5* RNAi transgenes on disease resistance have yet to be determined.

In the only study that appears to have observed altered disease resistance as a consequence of (over)expressing TGA factor genes, Kim and Delaney [124] found that transgenic *Arabidopsis* plants overexpressing *TGA5* displayed enhanced resistance to a virulent strain of *P. parasitica*, but reduced levels of *PR* genes. In contrast, neither sense nor antisense overexpression of *TGA2* affected resistance to this oomycete [124].

Overall, most analyses to date have focused on group II TGA factors, which include TGA2, TGA5, and TG6. The only study demonstrating a role for group I TGA factors in mediating disease resistance comes from VIGS of tomato TGA1 homologs, which compromised *Pto*-mediated resistance to *P. syringae* pv. *tomato* harboring *avrPto* [125].

8.2.4.6 R2R3 MYB Proteins

The R2R3MYB gene *BOS1* was recovered in a genetic screen aimed at identifying *Arabidopsis* genes required for resistance to *B. cinerea* [32]. Loss of BOS1 function resulted in enhanced susceptibility to necrotrophic fungi, including *B. cinerea* and *Alternaria brassicicola*. The *bos1* mutant displayed more severe disease symptoms in response to infection by the biotrophic pathogens *P. parasitica* and *P. syringae*; however, no detectable increase in growth of these pathogens was observed. The mutant accumulated more reactive oxygen species (ROS) in response to *B. cinerea* infection and was also more sensitive to a number of abiotic stresses [32]. Expression of marker genes for the SA and JA signaling pathways were not altered in the *bos1* mutant; however, *BOS1* transcripts were found to accumulate following infection with *B. cinerea*. Increased *BOS1* expression was blocked in the *coi1* mutant that mediates JA signaling, implicating BOS1 as a mediator of this defense pathway.

The *Arabidopsis MYB30* gene was initially identified based on its differential expression during a type of programmed cell death called the hypersensitive response (HR) associated with incompatible interactions [102]. Constitutive overexpression of *MYB30* in *Arabidopsis* and tobacco resulted in elevated expression of HR marker genes, accelerated HR against avirulent pathogens, and development of HR-like lesions upon infection with virulent pathogens [42]. Importantly, it increased resistance to virulent and avirulent biotrophic pathogens. Conversely, antisense suppression of *MYB30* suppressed HR marker gene expression, delayed the HR against avirulent pathogens, and decreased resistance against virulent and avirulent pathogens. Constitutive expression of *MYB30* did not result in spontaneous lesion formation, indicating that factors other than MYB30 are required to initiate the HR [42].

The tobacco *MYB1* gene was also shown to be required for *N*-mediated resistance to TMV using VIGS technology [120].

8.2.4.7 Zinc Finger-Related Proteins

The *Arabidopsis LSD1* gene was isolated in a screen aimed at identifying mutants that misregulate cell death responses [31]. Loss-of-function *lsd1* mutants initially display a normal HR in response to infection with virulent pathogens, but cannot limit the extent of the cell death, leading to a phenotype known as runaway cell death. Runaway cell death in the *lsd1* mutant is also observed following treatment with SA and SA analogs and depends on production of superoxide. It requires NPR1 as well as the positive regulators of disease resistance ENHANCED DISEASE SUSCEPTIBILITY 1 (EDS1) and PHYTOALEXIN DEFICIENT 4 (PAD4), both of which are putative lipases [126]. *lsd1* mutants are more resistant virulent strains of *P. parasitica*, but neither SA nor NPR1 is required for this phenotype [38,126]. These results indicate that LSD1 is a negative regulator of basal defense responses, somehow interpreting ROS-dependent signals triggered by the HR [126].

LOL1 was identified based on its similarity to LSD1 and the two proteins appear to have opposite roles in mediating cell death and disease resistance [38]. Reduction of *LOL1* expression in the *lsd1* mutant suppresses runaway cell death, while conditional high-level expression of LOL1 is sufficient to trigger cell death in the absence of pathogen infection or SA treatment. Furthermore, stable moderate overexpression of *LOL1* enhanced resistance to virulent strains of *P. parasitica*, while reduction of *LOL1* levels enhanced susceptibility to this pathogen [38].

Overexpression of the pepper *CaZFP1* gene in transgenic *Arabidopsis* resulted in enhanced resistance to a virulent strain of *P. syringae* and enhanced tolerance of drought [70]. Overexpressing lines also displayed developmental abnormalities. Expression of SA-regulated *PR* genes or the JA-regulated *PDF1.2* gene was not altered by *CaZFP1* overexpression, suggesting that the observed enhanced resistance was mediated through the activation of other pathways.

8.2.4.8 MYC Proteins

The *Arabidopsis jin1* mutants were identified in an ET-insensitive genetic background as being insensitive to JA and represent alleles of *AtMYC2* [33]. Mutant plants display enhanced resistance to necrotrophic pathogens, including *B. cinerea* and *P. cucumerina*. All *jin1* mutant alleles are semidominant and retain the capacity to produce the N-terminal end of the protein. It has been speculated that these could interfere with interacting proteins.

8.2.5 Post-Translational Regulation

In addition to regulation at the transcriptional level (reviewed in Section 8.2.3.4), there is overwhelming evidence for the post-translational regulation of TFs in response to pathogen challenge. Most studies have focused on reversible phosphorylation/dephosphorylation, which is a prevalent means of modulating the activity of eukaryotic TFs [127]. Reversible oxidoreduction of key cysteine residues is emerging as an important mechanism for regulating mammalian and microbial TFs, and recent studies have implicated this type of control in the regulation of plant TFs mediating defense responses [128]. Physical interactions between TFs and other proteins are also likely to be of regulatory significance.

8.2.5.1 Post-Translational Modifications

In one of the few studies aimed at monitoring changes in the nuclear pools of TFs following defense response elicitation, Turck et al. [103] visualized changes in the pattern of WRKY proteins by two-dimensional gel electrophoresis. Several proteins migrated as adjacent pearl strings, suggesting different post-translationally modified forms of the same protein. The number of proteins and the complexity of modifications rapidly increased following treatment with an elicitor. However, the nature of the putative post-translational modifications has not been determined.

As detailed later, the property of TFs most frequently shown to be influenced by post-translational modification is their ability to bind DNA. In a few cases, protein–protein interactions were shown to be affected. Few studies have directly linked post-translational modification to changes in transactivation potential. However, several have shown that transactivation is regulated in response to treatment with elicitors or growth regulators implicated in mediating defense responses; this suggests possible post-translational control. In many cases, the significance of post-translational modification for disease resistance is unknown.

8.2.5.1.1 Phosphorylation

The ability of TFs present in plant extracts to bind to several *cis*-acting elements implicated in mediating defense gene expression (e.g., W-box, *as-1*, ERE; see Table 8.1) is dramatically altered by treatment with phosphatase or protein kinase inhibitors [65,129–131]. In addition, constitutive activation of a MAPK signaling cascade in tobacco resulted in substantially more binding activity to the W-box [129].

Consistent with this result, two WRKY factors (WKRY22 and WRKY29) were identified as downstream components of an MAPK signaling cascade conferring basal resistance to bacterial and fungal pathogens in *Arabidopsis* [119]. Transcription factors and components of MAPK signaling cascades were also identified as being required for *N*-mediated resistance to TMV in tobacco [120] and *Pto*-mediated resistance to *P. syringae* (*avrPto*) in tomato [125]. However, in none of these studies was it demonstrated that the TFs were directly phosphorylated by MAPKs. One study suggested that the MAPK cascade may lead to phosphorylation and inactivation of a specific inhibitor of WRKY proteins [119]; results from another were consistent with regulation occurring at the point of *WRKY* transcription [129].

The rice ERF OsEREBP1 is the only plant TF implicated in mediating defense gene expression that has been shown to be phosphorylated by a MAPK [132]. Activity of this MAPK, BWMK1, is rapidly induced by a pathogen-derived elicitor, ET, SA, and JA. Overexpression of BWMK1 in transgenic tobacco results in constitutive expression of *PR* genes and enhanced resistance to *P. syringae* and the oomycete pathogen *Phytophthora parasitica* [132]. BWMK1 phosphorylates OsEREBP1 *in vitro*, resulting in enhanced binding to the GCC-box. Furthermore, transient coexpression of *BWMK1* and *OsEREBP1* in *Arabidopsis* protoplasts induced expression of a reporter gene under the control of the GCC-box to higher levels than either gene alone [132]. Therefore, it is likely that OsEREBP1 is regulated *in vivo* by a pathogen-activated MAPK signaling cascade.

The tomato ERF Pti4 is also regulated by phosphorylation. Pti4 physically interacts with the *Pto R*-gene product, which encodes a serine-threonine protein kinase [41]. Pto kinase activity is required for interaction with Pti4 as well as for race-specific resistance against *P. syringae* pv. *tomato* harboring *avrPto* [92]. Phosphorylation of Pti4 by Pto enhances its DNA binding to the GCC-box element *in vitro*. This effect is highly specific because two protein kinases related to Pto (Fen and Pti1) do not phosphorylate Pti4, and Fen does not stimulate Pti4 DNA-binding activity [92]. Pti4 contains ten threonine residues: two each in the putative DNA-binding and transactivation domains and one near the nuclear localization signal [41]. Pto phosphorylates at least four of these residues (but no serines) [92]. At this time it is not known which residues are phosphorylated *in vitro* or *in vivo*. It is also not known whether Pti4 may be regulated by other protein kinases during compatible plant–pathogen interactions. As discussed in Section 8.2.4.2, overexpression of Pti4 confers enhanced basal resistance to pathogens independently of the Pto–avrPto interaction.

Phosphorylation has also been implicated in the regulation of bZIP proteins. SARP (*s*alicylic *a*cid *r*esponse *p*rotein) is a cellular factor immunologically related to TGA factors implicated in mediating the rapid induction of *as-1* binding activity following treatment with SA [130]. In the absence of SA, it was proposed that SARP is sequestered by an inhibitory protein called SAI (*SA-i*nhibitor) and that release of SARP from SAI is triggered by SA-induced phosphorylation of either or both proteins. Inhibitors of *c*asein *k*inase II (CK2) suppress *as-1* binding, suggesting that CK2 may be responsible for activation of SARP [131].

More recently, the *Arabidopsis* TGA2 was shown to be phosphorylated *in vivo* by a CK2-like activity induced by SA [131]. Phosphorylation of TGA2 appeared to suppress its ability to bind to the *as-1* element. This is in contrast to previously published studies showing that phosphorylation by CK2-like kinases enhanced *as-1* binding *in vitro*. Importantly, no difference in *PR-1* expression was observed in transgenic plants overexpressing TGA2 compared to those expressing site-directed mutant with putative CK2 phosphorylation sites removed [131]. Thus, the regulatory significance of TGA2 phosphorylation by CK2-like kinases remains elusive.

G/HBF-1 is a non-TGA bZIP TF that binds to an SA and elicitor-responsive element in the promoter of the *chalcone synthase15* gene from soybean [25]. Elicitor treatment does not alter levels of G/HBF-1 transcript or protein; however, it rapidly induces activation of a protein kinase capable of phosphorylating G/HBF-1. Furthermore, phosphorylation of G/HBF-1 *in vitro* increases its DNA-binding activity. The identity of the G/HBF-1 kinase is not known.

DNA-binding activity of Whirly proteins is also regulated by phosphorylation. In particular, a kinase related to mammalian protein kinase C has been implicated [65]. Even though Whirly factors are constitutively present in the nucleus, their associated DNA-binding activity is not detectable until after pathogen challenge or treatment with SA or an elicitor [65]. Chromatographic purification of StWhy1 from uninduced nuclei activates its DNA-binding activity, possibly by removing an unidentified inhibitor protein. The possible role of phosphorylation in mediating the interaction of Whirly proteins with inhibitor proteins is unknown.

8.2.5.1.2 Oxidoreduction

Binding of nuclear factors to the *as-1* element is regulated by redox conditions (reviewed by Fobert and Després [128]). A recent study has started to unravel the possible mechanism involved.

The interaction between *Arabidopsis* TGA1 and NPR1 in plant cells is positively influenced by treatment with SA [133]. Two conserved cysteines located in the C-terminal region of TGA1 (C260 and C266) play a key role in regulating this interaction. Although wild-type TGA1 and NPR1 do not interact in the yeast two-hybrid system, mutation of these residues permits interaction with NPR1 in yeast and in *Arabidopsis* cells regardless of SA induction. Using a novel labeling strategy that distinguishes between protein sulfhydryls and disulfides, it was demonstrated that the redox status of cysteines in TGA1 and/or the closely related TGA4 shifted considerably following SA treatment to become predominantly reduced [133]. Thus, strong interaction with NPR1 is correlated with the reduced state of TGA1 and/or TGA4 cysteines.

Changing redox conditions altered the mobility of *in vitro* produced TGA1 in nonreducing gel electrophoresis, consistent with the formation of an intramolecular disulfide bridge under oxidizing conditions [133]; this could impede interaction with NPR1. Changing redox conditions has been shown to influence the DNA-binding activity of several TFs (see Fobert and Després [128]); however, the ability of TGA1 to bind the *as-1* element *in vitro* was unaltered by vast molar excess of redox-regulating compounds. Instead, redox regulation of TGA1 DNA binding required the redox-regulated recruitment of NPR1, which was proposed to act as a cofactor in stimulating TGA1 DNA-binding activity [133] (see also Section 8.3.2 and Section 8.3.3).

8.2.5.1.3. Protein Turnover/Proteolysis

TGA factors were shown to have different stabilities during tobacco development; TGA1 and TGA3, but not TGA2, were rapidly degraded in mature leaves [89]. Degradation of TGA3 appeared to be mediated via the proteosome. However, the regulatory significance, if any, of TGA factor degradation in response to SA or pathogen challenge was not assessed.

The role of targeted protein degradation in the regulation of *ERF1* transcription was discussed in Section 8.2.3.4. Some preliminary evidence also suggests regulation of defense-related TFs by proteolysis. The tobacco ERF3 was shown to interact with a ubiquitin-conjugating enzyme (NtUBC2) in the yeast two-hybrid system [134]. However, NtUBC2 did not affect the transrepression capacity of ERF3 in transient assays. The *Arabidopsis* R2R3MYB protein BOS1 contains a consensus sumolyation motif, suggesting that it may be regulated by SUMO, a ubiquitin-like modifier associated with protein stabilization [32]. Finally, mutations of the *Arabidopsis* F-box protein CORONATINE INSENSITIVE1 (COI1) block JA-mediated defense responses, resulting in increased susceptibility to several necrotrophic pathogens [135]. Although the targets of COI1 are unknown, it has been proposed that it may mediate removal of TFs tagged by JA-dependent phosphorylation.

8.2.5.2 Protein–Protein Interactions

Transcription of eukaryotic genes is achieved by synergistic interactions between combinations of TFs [9]. This combinatorial control relies on precise juxtapositioning of specific TFs through DNA–protein and protein–protein interactions using functional groupings of *cis*-regulatory elements called enhancers (or silencers). Enhancers are responsible for a subset of the total gene expression pattern.

Accordingly, a typical gene may contain several enhancer elements. The stable nucleoprotein complex containing enhancer DNA with associated bound TFs and interacting proteins is sometimes referred to as an enhanceosome [136]. Little is known about the organization of *cis*-acting modules required to create pathogen-responsive enhancer elements in plants. However, individual *cis*-acting elements implicated in defense gene regulation frequently cluster, and synergistic interactions between closely spaced elements have been reported (e.g., W-boxes), suggesting cooperation between TFs [60].

Several examples of protein–protein interactions have already been discussed in this chapter, including Pti4–Pto and TGA factors–NPR1. TGA3 has also been shown to interact with calmodulin, suggesting a possible role for calcium signaling in regulating TGA factors [137]. BZI-1, a bZIP protein not part of the TGA factor family, interacts with a novel protein (ANK1) containing ankyrin repeats [138]. As revealed by analysis of dominant-negative transgenes, BZI-1 is required for resistance to TMV. ANK1 does not appear to act as a cofactor of BZI-1 and was speculated to function in the cytosol rather than the nucleus.

Although the combinatorial model of gene expression stresses the importance of interactions between TFs, few have been reported between factors implicated in mediating plant defense responses. TGA4 was shown to interact with the ERF protein AtERF [139] and with members of the Dof family of TFs, which encode proteins with a single zinc finger [140]. The Dof proteins (OBP1-3) are capable of binding to a *cis*-element in the CaMV35S promoter and enhance the DNA-binding activity of TGA4 *in vitro*. The genes encoding OBP1-3 are inducible by SA. Overexpression of OBP3 results in numerous developmental phenotypes, but its effects on defense gene expression and disease resistance were not assessed. PRHA, a homeobox protein capable of binding to an elicitor responsive element of a *PR-10* promoter, also interacted with two putative cofactors [141].

The limited number of reported protein interactions may be attributed in part to limitations of the yeast two-hybrid system for studying TFs. First, the natural autonomous transactivation or transrepression properties of many TFs complicate their analysis in systems that rely on reconstitution of an active TF to detect reporter gene expression. Second, yeast cells may not be competent to carry out post-translational modifications required for interactions. Finally, the yeast two-hybrid system monitors binary interactions between proteins. As indicated at the beginning of this section, TFs are likely to function as part of larger protein complexes requiring multiple protein–protein and protein–DNA interactions. Approaches aimed at recovery and characterization of protein complexes *in vivo*, such as *t*andem *a*ffinity *p*urification (TAP) strategies, coupled with mass spectroscopy, promise to be useful tools in identifying proteins interacting with plant TFs [142].

8.2.6 Transcriptional Targets and Networks

Even for the best characterized plant TFs, the identity of very few target genes is known. Nevertheless, such information is very important in assigning function to individual TFs. Large-scale transcript profiling studies have identified motifs related to W-box, GCC-box, and ERE as enriched in genomic DNA upstream of genes found to be differentially expressed following exposure to pathogens, elicitors, or defense-related growth regulators, the W-box being the most prevalent [7,143]. These genes can be viewed as putative targets for WRKY, ERF, and Whirly factors, respectively. There have been no reports of enrichment for MYB binding sites, and *as-1* like elements appear to be under-represented in promoters of these genes. Several novel motifs are also enriched in promoters of these genes [7,116]; however, the identity of the TFs binding to these sites, if any, is unknown.

As discussed earlier in this chapter, TFs bind to short stretches of DNA (5 to 10 bp) that can accommodate limited sequence variability. Sequences outside the consensus *cis*-element may also influence binding—in particular, the presence and spacing of neighboring TF recognition sites. Furthermore, only a fraction of TFs within a family is likely to bind to any given promoter containing *cis*-elements that conform to the consensus binding sequence for that family. For example, a number of genes containing GCC-boxes in their promoters are not differentially regulated in plants overexpressing ERF factors [104,113,115]. Accordingly, it is not possible to identify target genes accurately based uniquely on DNA sequence information.

Genome-wide transcript profiling of plants containing altered levels of a TF offers a means of identifying putative targets for that specific factor. However, due to the existence of transcriptional cascades (see later discussion), these approaches cannot distinguish direct and indirect targets. Nevertheless, they have confirmed that many genes differentially regulated in plants with altered levels of *WRKY6* contain multiple W-boxes in their promoters [80]; those ectopically expressing *Pti4* are enriched for GCC-boxes [115,116]. One study also reported an enrichment of MYB-binding sites in *Pti4* overexpressors [116].

Many of the genes found to be differentially regulated in plants expressing altered levels of *WRKY* or *ERF* TFs genes encode proteins with functions potentially relevant to disease resistance [43,80,100,115]. These include classical PR proteins, proteins implicated in oxidative stress responses (e.g., P450s, glutathione-S-transferase), calcium signaling, and cell wall modification. Numerous genes identified in these plants, including many without known functions, are also differentially regulated in wild-type plants following treatment with pathogen, elicitors, or defense-related growth regulators, suggesting a role in mediating disease resistance.

Comparison of overall gene expression profiles between plants with altered *WRKY70* levels and signaling mutants suggests that this TF controls expression of a substantial number of genes regulated by the JA- and SA-dependent pathways (~60 to 40%, respectively); a positive correlation was observed between WRKY70- and JA-dependent expression, but the inverse was true for SA-dependent expression [43]. Similarly, more than one third of the genes induced by treatment with ET and JA were constitutively expressed in *ERF1* overexpressors [100]. The overlap increased to 80% if only genes classified as "defense related" were considered.

Another class of target genes frequently recognized in plants with altered levels of *WRKY* and *ERF* genes includes those that encode TFs. Thus, transcriptional repression or, alternatively, activation and subsequent translation, of such genes could lead to modulation of additional sets of target genes, some of which may again encode TFs. The result is a transcriptional cascade in which TFs at the top of the hierarchy have the potential to regulate entire developmental or metabolic programs [144,145]. Such proteins are commonly referred to as "master switches" or "controllers." Well-known examples include homeotic genes in drosophila and plants; mutation of these genes leads to the dramatic development of organs at inappropriate positions. Closer to the bottom of the hierarchy would be TFs controlling more focused aspects of specific programs. Research in well-characterized systems, such as yeast, has revealed the existence of a number of regulatory loops, or networks, in which the activity of one or more TFs influences that of another [146].

Results obtained with WRKY70 and ERF1 suggest that these TFs could be master controllers of defense-related pathways involving JA/SA and JA/ET signaling, respectively. Neither TF controls the expression of all genes known to be regulated by any given signaling molecule; instead, they appear to control branches of these pathways implicated in defense responses [43,100]. ERF1 has been proposed to integrate JA and ET defense pathways and WRKY70 to mediate cross-talk between SA and JA pathways. Based on expression of marker genes, the *Arabidopsis* MYC2 was recently proposed to discriminate between different JA-mediated defense responses [33]. It will be interesting to see if this claim is substantiated by genome-wide transcript profiling.

Chromatin immunoprecipitation (ChIP) permits identification of direct targets for TFs in real time and space [147]. Using this approach, StWhy1 was shown to bind to the potato *PR-10* promoter in response to wounding and elicitor treatment [37], while the recruitment of *Arabidopsis* TGA2 and TGA3 to the *PR-1* promoter was shown to depend on SA and NPR1 [148]. ChIP analysis of parsley WRKY1 revealed it was transiently recruited to regions containing W-boxes in the promoter of its own gene, as well as those of the immediate-early genes *WRKY3* and *PR1-1* following elicitor treatment [103]. Occupancy by WRKY1 was associated with reduced expression of *WRKY1* but enhanced expression of *PR1-1*.

When not occupied by WRKY1, these promoters are otherwise constitutively occupied by different WRKY factors. It was proposed that activation of *WRKY1* and *PR1-1* involves elicitor-dependent post-translational modification of WRKY factors already occupying W-boxes in these promoters; the newly produced WRKY1 would be subsequently recruited to autoregulate expression of its own gene and activate late gene expression. Post-translational modification of these WRKY factors may rely on MAPK signaling cascades as described in Section 8.2.5.1.1 [103].

ChIP analysis of a subset of *Arabidopsis* genes differentially expressed in response to Pti4 overexpression revealed that about 60% (11 of 18) were direct targets of this ERF [116]. Three promoters that were not immunoprecipitated with the Pti4 antibody contained GCC-boxes, further emphasizing the notion that the presence of cognate *cis*-elements in a gene promoter does not necessarily make it a target *in vivo*. Interestingly, seven of the promoters immunoprecipitated lacked GCC-boxes within 1 kb upstream of the coding region. It was suggested that Pti4 may be capable of binding to a novel *cis*-element other than the GCC-box or be recruited to promoters indirectly through other DNA-binding proteins [116].

New strategies such as ChIP chip [147] and *s*equence *t*ag of *g*enomic *e*nrichment (STAGE) [149] now permit identification of direct targets of a TF on a genome-wide basis. Although neither method has yet to be applied for the characterization of plant TFs implicated in mediating defense responses, a whole-genome *Arabidopsis* tiling array is available that would be suitable for ChIP chip [19].

8.3 NPR1

8.3.1 Genetic Analysis of NPR1

8.3.1.1 Isolation of *npr1* Mutants and Cloning of *NPR1* Gene

The *npr1-1* mutation was found to block SA-inducible *PR* gene expression following treatment with the SA analog 2,6-dichloroisonicotinic acid (INA) [150]. Additional *npr1* alleles were subsequently recovered in different genetic screens aimed at identifying genes required for expression of *PR* genes in response to SA (*salicylic acid insensitive1*, renamed *npr1-5* [151]), basal resistance against virulent *P. syringae* (*enhanced disease susceptibility5* and *53*, renamed *npr1-2* and *npr1-3*, respectively [152]), and INA-induced SAR against *P. parasitica* (*non-inducible immunity1*; *nim1* [153]). The *nim1* mutants have not been redesignated as *npr1* alleles and continue to be referred to by their original names. All *npr1/nim* mutants, with the exception of *nim1-5*, are recessive; however, the dominant phenotype observed in *nim1-5* plants is likely attributed to a second site mutation [154]. Because the *npr1* phenotype cannot be rescued by exogenous SA, it was proposed that NPR1 functions downstream of this metabolite in the signaling pathway.

Mutations in *NPR1* compromise basal resistance against biotrophic pathogens such as *P. syringae*, *P. parasitica*, and *E. cichoracearum*. They cannot mount effective SAR against *P. syringae* or *P. parasitica* or ISR against *P. syringae* [155]. Also, different *npr1* mutants have been found to be more susceptible to some incompatible races of *P. parasitica* [153,156,157], *P. syringae* [151], or *E. cichoracearum* [158]. Loss of *NPR1* function does not appear to affect age-related resistance against *P. syringae* [159] or basal resistance against necrotrophic pathogens, including *A. brassicicola* and *B. cinerea* [160]. However, unlike the wild-type, *npr1* mutants do not show enhanced resistance to *B. cinerea* following treatment with SA [161].

Although *npr1* mutants fail to express well-accepted marker genes for the SA-signaling pathway (*PR-1*, *PR-2*, *PR-5*) in response to treatment with SA or SA analogs, transcripts for these genes continue to be expressed in response to pathogen challenge (albeit at reduced levels [152,157]) or in combination with mutations at other loci (see, for example, Dong [162]). These observations indicate that one or more SA-dependent, NPR1-independent defense pathways exist in *Arabidopsis*.

Despite being unresponsive to exogenous SA, *npr1* mutants accumulate high titers of this metabolite following pathogen challenge [153] and *npr1* seedlings grown in the presence of SA bleach and die after developing cotyledons [163]. It has been proposed that NPR1 may be involved in feedback regulation of SA accumulation.

Cloning of the *NPR1* gene revealed that it encodes a protein with two identifiable protein–protein interaction motifs: a BTB/POZ (*b*road-complex, *t*ramtrack, and *b*ric-a-brac/*po*x virus and *z*inc finger) and an *a*nkyrin *r*epeat *d*omain (ARD) [154,163]. Several *npr1* alleles affect conserved amino acids within the ARD, suggesting that this domain is important for NPR1 function. Although one mutation (*npr1-2*) maps to the BTB/POZ, it is within a nonconserved region of the domain (unpublished observation) and thus should not be used as evidence that the NPR1 POZ/BTB is required for disease resistance. NPR1 contains no well-recognized DNA-binding motifs and cannot bind to the *as-1* element *in vitro* [40]. This suggests that NPR1 does not function as a sequence-specific TF; however, its ability to bind DNA has not been rigorously assessed.

Sequences encoding NPR1 or NPR1-related proteins can be identified in many higher plants, suggesting that the NPR1 function is well conserved. In support of this notion, VIGS analysis has demonstrated that a tobacco homolog of NPR1 is required for resistance to TMV [164] and a tomato homolog is required for resistance to *P. syringae* (*avrPto*) [125].

8.3.1.2 *npr1* Suppressors

In efforts to identify additional genes implicated in mediating defense responses, several groups have screened for mutations that restore *PR* gene expression or disease resistance in *npr1* backgrounds (i.e., genetic suppressors). Several suppressors appear to have activated defense responses constitutively; even in the absence of pathogen challenge, they contain elevated levels of SA, express *PR* genes, and display enhanced basal resistance to pathogens, typically *P. syringae* and *P. parasitica* [165–168]. Many of these suppressors also display dwarfism and spontaneous disease-like lesions. Several other mutants displaying these properties were also recovered in unrelated genetic screens. The observed phenotypes were subsequently shown to be largely independent of NPR1, prompting some researchers to classify them as *npr1* suppressors (see, for example, Kim and Delaney [169]).

Conversely, others have argued that any mutant displaying elevated levels of SA should not be considered as a true suppressor of *npr1* [162]. In fact, two suppressors of this class contain mutations in putative *R*-genes, and the phenotypes likely result from the constitutive activation of *R*-gene signaling [167,168].

A recessive mutation that restores SA-inducible *PR* gene expression and disease resistance in the *npr1-1* background is *suppressor of npr1-1 inducible1* (*sni1*) [170]. Although plants are dwarfed, they contain wild-type levels of SA, express only low constitutive levels of *PR* genes, and do not produce any spontaneous disease-like lesions. Neither *sni1 npr1-1* nor *sni1 NPR1* plants are more resistant than the wild-type in the absence of INA. SNI1 encodes a novel leucine-rich nuclear protein that likely acts as a negative regulator of SAR. It was suggested that the probable role of NPR1 in SAR is to remove SNI1 repression. Currently, no evidence suggests that NPR1 and SNI1 physically interact [170].

Increased resistance to virulent *P. syringae* and *P. parasitica* in the *suppressor of nim1* (*son1*) *nim1-1* double mutant is independent of SA and not correlated with increased levels of *PR* genes. Accordingly, it has been proposed to define a novel type of *S*AR-*i*ndependent *r*esistance (SIR) [169]. Several *PR* genes are constitutively activated in the *son1 NIM1* (i.e., *NPR1*) background, indicating that *SON1* also participates in SAR, possibly upstream of or at NPR1 [169]. *SON1* encodes an F-box protein. Given that the *son1* mutation is recessive, SON1 is likely a negative regulator of SIR and SAR. It has been proposed that SON1 may target specific positive regulators of defense responses for degradation by the ubiquitin/proteosome pathway [169].

8.3.1.3 Other Genetic Interactions

Genetic interactions between *NPR1* and genes encoding TFs were described in Section 8.2. The current section summarizes findings that link NPR1 to other regulators of plant defense responses.

8.3.1.3.1 SA-Dependent Signaling

The recessive *enhanced disease resistance1* (*edr1*) mutation leads to increased resistance against *P. syringae* and *E. cichoracearum* without constitutive activation of *PR* genes. *EDR1* codes for a MAP kinase kinase kinase [171]. This enhanced resistance is lost in the *edr1 npr1* double mutant, suggesting that EDR1 is part of a MAPK cascade operating upstream of NPR1 that negatively regulates plant defense responses [171]. MAPK4 is a MAPK that also acts as a negative regulator of SA-dependent defense responses; however, the *mapk4* mutant phenotype is not attenuated in the *npr1* background, suggesting that it is part of a MAPK cascade that acts independently or downstream of NPR1 [172].

Overexpression of two soybean calmodulin isoforms, GmCaM4-5, in *Arabidopsis* results in constitutive expression of SA-inducible *PR* genes, spontaneous micro-HR, and enhanced resistance to virulent *P. syringae* [173]. Constitutive expression of *PR* genes is lost when *GmCaM4-5* are introduced into an *npr1* mutant background, suggesting that it is mediated by NPR1. NPR1 is also partly required for the enhanced resistance to *E. cichoracearum* observed in the p*owdery* m*ildew* r*esistant4* (*pmr4*) mutant [174].

Pathogen-induced expression of SA-I*NDUCTION* D*EFICIENT2* (*SID2*), which encodes ISOCHORISMATE SYNTHASE1, a key enzyme in SA synthesis, is elevated in the *npr1* mutant [175]. In contrast, SA-dependent expression of *PAD4* is compromised in the *npr1* mutant, although *P. syringae*-induced expression of this gene is not affected [176]. These provide further evidence that NPR1 acts downstream of SA and indicate that its role as a negative regulator of SA metabolism may be achieved, at least in part, at the level of *SID2* expression or through a signal amplification loop involving PAD4 [175,176].

The double mutant between *npr1* and the gain-of-function mutation a*ccelerated* c*ell* d*eath6* (*acd6*) displayed dramatic alterations in cell enlargement and division resulting in production of abnormal growths [177]. Subsequent detailed analysis of the *npr1* single mutant revealed reduced cell numbers with high ploidy levels. Thus, in addition to a role in mediating SA defense responses, NPR1 is also required for SA-dependent control of cell growth. Such effects were not apparent upon casual observation of the *npr1* mutants.

The requirement of NPR1 for runaway cell death in the *lsd1* mutant has already been discussed (Section 8.2.7.4). HR-associated cell death is also increased in the *npr1* single mutant in response to infection with avirulent races of *P. syringae* [177,178]. In contrast, no cell death or hydrogen peroxide accumulation is detectable in the n*onrace-specific* d*isease* r*esistance1* (*ndr1*) mutant following challenge with *P. syringae avrRpt2* [178]. Genetic analyses demonstrated that *ndr1* is epistatic to *npr1* with respect to both these parameters. However, *NDR1* and *NPR1* have an additive effect on SAR and *PR-1* expression [179].

8.3.1.3.2 JA/ET-Dependent Signaling

Epistasis analysis has revealed links between NPR1 and the ET signaling pathways. Resistance to *B. cinerea* is not compromised in the *npr1-1* mutant and is moderately reduced in *ein2*, a mutant affected in the ET pathway; however, the *npr1-1 ein2* double mutant is considerably more susceptible to this necrotroph [161]. Furthermore, breakdown of *R*-gene-mediated resistance to *P. syringae* (*avrRpt2*) is more severe in the double mutant than in either single mutant alone [180].

Simultaneous application of SA with JA can inhibit JA-dependent responses in wild-type plants. This inhibition is alleviated in the *npr1* mutant, suggesting that

NPR1 acts as a negative regulator of JA-dependent signaling. Interestingly, this function of NPR1, unlike its role in SA-mediated signaling, does not require localization to the nucleus [181].

8.3.2 Biochemical Function of NPR1

The protein sequence of NPR1 provided few clues as to its function, other than it likely interacts with protein partners through its POZ/BTB or ARD. Using the yeast two-hybrid system, several groups established that NPR1 interacts with various TGA factors, including the *Arabidopsis* TGA2, TGA3, TGA5, TGA6, and TGA7 [40,124,182]. As described in Section 8.2.5.1.2, TGA1 and TGA4 are capable of interacting with NPR1 only after reduction or mutation of key cysteines. The *Arabidopsis* NPR1 was also used as bait to screen a rice cDNA library and found to interact with three TGA factors related to TGA2 and a putative homolog of the maize liguleless2 [183]. Screening of a tomato library with a putative ortholog of NPR1 from this species identified a group II TGA factor named NIF1 (*N*PR1 *i*nteracting *f*actor1) [184]. The tobacco group II factors TGA2.1 and TGA2.2 were also shown to interact with *Arabidopsis* NPR1 in directed two-hybrid tests [82].

With the exception of *nim1-4* (Chern et al. [183] and our unpublished observations) all *NPR1* mutant alleles that compromise disease resistance encode proteins that fail to interact with TGA factors in the yeast two-hybrid system. Where tested, these mutants also fail to interact with TGA factors *in vitro*. This suggests that the ability to interact with TGA factors *in vivo* is important for NPR1 function and establishes TGA factors as downstream components of the NPR1 signaling pathway.

The NPR1–TGA2 interaction observed in yeast and *in vitro* has been confirmed in plant cells. Importantly, these studies have revealed that the interaction is stimulated by SA [83,185]. Visualization of the interaction in protoplasts using a protein-complementation assay revealed weak diffuse interaction throughout the cell under uninduced conditions, but intense nuclear interaction following SA treatment [185]. This pattern is consistent with reports that TGA2 and NPR1 are predominantly localized in the nucleus following SA treatment [89,186]. In fact, nuclear localization of NPR1 was shown to be required for *PR-1* expression [186].

Several lines of evidence indicate that interaction with NPR1 is important for TGA factor function *in vivo*:

NPR1 stimulates the DNA-binding properties of interacting TGA factors *in vitro*, including the reduced form of TGA1 [40,133].

Protein extracts prepared from wild-type transgenic plants expressing a chimeric TGA2:GAL4 DNA-binding domain (DB) protein bound a probe containing GAL4 binding sites substantially better than similar extracts prepared from *npr1* mutant plants expressing the chimeric protein [83].

Activation of a reporter gene under the control of a promoter containing GAL4-binding sites was only detected in wild-type transgenic plants expressing the TGA2:GAL4 DB fusion; no expression was detected in the *npr1* background [83]. The DNA-binding enhancement and transactivation properties of the TGA2:GAL4 DB chimeric protein observed in the wild-type plants depended on SA.

The ability of TGA2 and TGA3 to bind to the *PR-1* promoter, as measured by ChIP, depended on functional NPR1 and SA [148].

Although NPR1 clearly enhances the binding of interacting TGA factors by EMSA, it does not alter the migration of the protein–DNA complex [40,83,133]. Furthermore, supershift experiments using antibodies against NPR1 antibodies indicate that NPR1 is not present in these complexes. This suggests that NPR1 stimulates the DNA-binding activity of TGA factors without binding stably to the TGA–DNA complex. It is possible that the NPR1–TGA interaction is insufficiently robust to withstand the polyacrylamide gel electrophoresis or that, upon binding DNA, the TGA factors release NPR1.

Although there are no reports of direct interactions between NPR1 and WRKY factors, several lines of evidence suggest a role for this group of TFs as regulators of *NPR1* expression and potential mediators of the NPR1-signaling pathway. First, many genes differentially regulated in the *npr1* mutant contain W-boxes in their promoters [7,143]. Second, the expression of numerous *WRKY* genes depends on NPR1 function [101]. Finally, WRKY factors bind to W-boxes in the promoter of the *NPR1* gene, and mutation of these elements compromises *NPR1* expression [101].

Thus, NPR1 appears to act as a novel cofactor to regulate *PR*-gene expression by modulating the activity of TGA factors and possibly other groups of TFs in response to pathogen challenge. However, the observation that cross-talk between the SA and JA signaling pathways does not require nuclear localization of NPR1 [181] suggests that this versatile regulator may have additional modes of action.

8.3.3 Post-Translational Regulation of NPR1

Yeast two-hybrid screens also identified a novel group of four *Arabidopsis* proteins called NIMINs (*NIM*1 *IN*TERACTORs) [187]. The interaction of NPR1 with NIMIN1 has been confirmed in plants. Overall, NIMIN proteins share limited sequence identity (14 to 44%); however, they contain short stretches of high similarity. For example, NIMIN1 and NIMIN3 possess stretches of acidic amino acids that may serve as effector domains, and NIMIN1 and NIMIN2 each contain a cluster of basic amino acids that may represent a nuclear localization signal. All NIMIN proteins contain a motif similar to the EAR domain necessary for transrepression of ERF proteins (see Section 8.2.3.2) and are nuclear localized. *NIMIN* genes are transiently expressed following treatment with SA.

Overexpression of NIMIN1 in *Arabidopsis* reduced *PR* gene expression in response to SA and avirulent *P. syringae* and compromised SAR against virulent *P. syringae* [187]. These phenotypes are reminiscent of the *npr1* mutation; however, unlike that mutant, NIMIN1 overexpressing plants were also compromised in *R*-gene-mediated resistance against *P. syringae* (*avrRpt2*). Overexpression of a NIMIN1 mutant protein unable to interact with NPR1 had no effect on *PR* gene expression or disease resistance, indicating that NIMIN1 function is mediated through NPR1 [187].

Mutation and RNAi suppression of *NIMIN1* resulted in increased expression of *PR* genes but had no measurable effects on resistance to *P. syringae*.

It was proposed that NIMIN1 may regulate only a subset of NPR1-dependent defense genes and that derepression of this specific gene set in the *nimin1* mutant and RNAi lines is insufficient to confer disease resistance. It is also possible that the closest relative of NIMIN1 (NIMIN1b) may partly compensate for loss of NIMIN1 function under some conditions [187]. Furthermore, one cannot discount the possibility that the phenotypes observed in the plants overexpressing NIMIN1 may be related to squelching (see Section 8.2.4.1), with the vast excesses of NIMIN1 protein binding all available NPR1 at the expense of other NIMIN proteins or interacting partners. Overall, functional analysis of NINIM1 indicates that it is a negative regulator of NPR1, possibly involved in modulating the amplitude of certain defense responses that may be relevant to plant fitness [187].

The fact that NPR1 is constitutively expressed but changes subcellular localization following pathogen infection or SA treatment suggests that it is regulated at the post-translational level. NPR1 shares limited sequence similarity with the mammalian transcriptional regulator Iκ-B, including conserved N-terminal lysines and serines potentially involved in ubiquitination and phosphorylation events, respectively [154]. However, there is no evidence to support the notion that NPR1 is regulated by either of these mechanisms.

In contrast, some evidence does support post-translational regulation by redox changes of conserved cysteines [188]. Using nonreducing SDS-PAGE to monitor the presence of disulfide bonds, the mobility of a protein fusion between NPR1 and the *g*reen *f*luorescent *p*rotein (GFP) was found to change dramatically in response to INA. In uninduced samples, a large cytosolic complex was detected, consistent with an NPR1 oligomer held together by intermolecular disulfide bridges. Following INA treatment, a band corresponding to the monomeric size of NPR1-GFP gradually accumulated; the fusion protein was detected in the nucleus and *PR-1* expression activated. Accordingly, it was proposed that INA triggered the reduction of NPR1 disulfide bridges, yielding biologically active monomers [188].

To identify residues affected by changes in redox status, NPR1-GFP fusions containing site-directed mutations at each of ten conserved NPR1 cysteines were expressed in transgenic plants. Mutation of two cysteines (C82 and C216) resulted in the constitutive expression of monomeric, nuclear NPR1-GFP as well as *PR-1* expression in the absence of INA. Because altered residues would no longer be capable of forming disulfide bridges, they were predicted to mimic the reduced state of the cysteine residues.

Together these results suggest that reduction of C82 and C216 is critical for NPR1 activation. Consistent with this notion, monitoring the levels and ratio of reduced to oxidized glutathione following INA treatment revealed that after an initial oxidative phase, the cellular redox status became more reducing [188]. Together with those obtained with TGA1 ([133]; see Section 8.2.5.1.2), these results indicate that properties of proteins mediating SA-dependent defense responses are regulated by SA-mediated reduction of key cysteine residues. Currently, the enzymes responsible for modulating the redox changes of TGA1 and NPR1 cysteines or how these changes specifically affect protein structure and function is not known.

8.3.4 Analysis of Transgenic Plants Overexpressing NPR1

Arabidopsis plants expressing higher levels of NPR1 protein have increased basal resistance to *P. syringae*, *P. parasitica* [189,190], and *E. cichoracearum* [189]. In one study, transgenic plants also displayed enhanced SAR following treatment with low levels of a chemical activator (benzothiadiazole; BTH) that did not elicit a response in untransformed plants [189]. Most transgenic lines did not constitutively express SA-inducible *PR* genes. In one study, increased resistance was correlated with stronger, rather than faster, *PR* gene expression [190]; in the other study, resistance was correlated with the speed, rather that the level, of *PR* gene expression [189]. A direct correlation between levels of NPR1 and disease resistance was reported in only one of the two studies [190].

The *Arabidopsis NPR1* gene has also been overexpressed in two heterologous hosts. Expression in tomato resulted in substantial resistance against virulent strains of *P. syringae* and *F. oxysporum*; moderate resistance to *Xanthamonas campestris*, *Ralstonia solanacearum*, and *Stemphylium solani*; and no enhanced resistance to *P. infestans*, cucumber mosaic virus, and tomato yellow leaf curl virus [191]. Levels of resistance to *P. syringae* and *F. oxysporum* were reported to be comparable, but not as complete, as those conferred by *R*-genes. In general, levels of NPR1 were correlated with the effectiveness of disease resistance; however, several exceptions were noted, leading the authors to propose that resistance may require a threshold level of NPR1 expression. There appeared to be no correlation between the levels of six *PR* genes tested with levels of NPR1 or disease resistance [191].

Expression of *Arabidopsis* NPR1 in rice resulted in increased resistance to virulent *Xanthamonas oryzae* [183]. These results suggest that NPR1 function and signaling are conserved among dicots, monocots, and, in particular, rice, which has very high constitutive levels of endogenous SA. Resistance to *X. oryzae* conferred by NPR1 was not as effective as *R*-gene resistance; however, substantial reduction of pathogen growth was observed in the leaf central vein, which limited bacterial spread and enhanced survival.

As suggested in tomato, it appeared that a threshold level of NPR1 was required to confer resistance in rice [183]. Closer examination of the rice transgenics under different growth conditions revealed that overexpression of NPR1 can trigger the development of spontaneous disease-like lesions [192]. This phenotype was correlated with the accumulation of hydrogen peroxide and could be potentiated by exposure to SA or BTH. The transgenic plants were found to contain lower levels of SA—once again providing a link between NPR1 and regulation of SA metabolism. Neither the production of lesions nor changes in SA levels were observed in dicotyledonous plants overexpressing NPR1 [189–191].

8.3.5 *NPR1*-Related Genes

Completion of genome sequence revealed that *Arabidopsis* encodes six NPR1-related proteins that can be separated into three discrete groups [157]. All of these proteins contain the POZ/BTB and ARD domains implicated in mediating protein–protein interactions. Every amino acid affected by known *npr1* mutations that compromise

PR gene expression and disease resistance is conserved within the family. The genes encoding the three family members most closely related to NPR1 (At4g26120, At5g45110, At4g19660) also contain introns at the same positions as *NPR1*.

Three *Arabidopsis NPR1*-related genes have recently been functionally analyzed. Of these, the product of *NPR4* (A4g19660) is the most closely related to NPR1, sharing 36% identity. Similar to NPR1, NPR4 is localized in the nucleus and can interact with the same spectrum of TGA factors. *NPR4* transcript is induced by SA and pathogen challenge and is rapidly repressed by JA [157]. Mutation of *NPR4* results in a modest increase in susceptibility to virulent *P. syringae* and *E. cichoracearum* but not *P. parasitica*. However, partial breakdown in *R*-gene resistance against some races of *P. parasitica* was reported. *PR* gene expression was only marginally reduced in the *npr4* mutant. Although these results implicate NPR4 in regulation of *PR* genes and resistance against biotrophic pathogens, they also suggest that its role is not as prominent as that of NPR1. It was suggested that NPR1 and NPR4 may have distinct roles in modulating cross-talk between SA- and JA-dependent defense signaling.

B*LADE* ON P*ETIOLE* (*BOP*) *1* and *2* are the two most distantly related members of the *NPR1*-gene family. These genes have been implicated in regulation of plant development [193,194] and their contribution to defense responses is unclear.

8.4 ... TO CROP IMPROVEMENT

Thanks in large part to use of model systems, our knowledge of TFs involved in regulating plant defense responses has increased very rapidly in recent years. It is clear that numerous TFs, encoded by different gene families, are required to ensure the highly coordinated expression of defense genes in response to pathogen challenge. Many TF genes are highly regulated at the transcription level and the products they encode can act as positive or as negative regulators of gene expression. Thus, in addition to the activation of positive factors, the elimination of negative regulators—at the transcriptional or post-translational level—is required for timely activation of plant defense responses.

Highly related TFs can be recognized within gene families. The roles of these proteins appear to overlap substantially, making assignment of function difficult. Overall, differences in DNA-binding preferences, gene expression patterns, transactivation properties, and, probably, post-translational regulation indicate that members of a TF family have diversified to fulfill unique roles. Although interactions between members of a family and between different families of TFs are likely critical for controlling defense gene expression, they remain poorly understood. The same is true of the post-translational modifications that alter the functionality of TFs following pathogen challenge.

Even though our understanding of how plant TFs regulate defense responses is far from complete, several studies have revealed that changing expression levels of individual TFs or cofactors can have a profound effect on disease resistance. Given that plant defense signaling pathways are broadly conserved across plant species [7], TFs found to be effective at enhancing resistance against specific pathogens in *Arabidopsis*, or related proteins from crop plants, may also function against the same or similar microbes in the crop species. It is noteworthy that the vast majority of studies in *Arabidopsis* were performed under controlled laboratory conditions. It will be important to determine whether overexpression of TFs in crop plants will confer commercially relevant levels of disease resistance under field conditions. A recent study indicated that *Arabidopsis* mutants constitutively expressing defense responses had reduced fitness, although NPR1 overexpressors were unaffected [195].

As discussed in Section 8.2.4, many plants overexpressing TFs implicated in mediating defense responses also display developmental abnormalities. Obviously, these undesirable side effects must be minimized or eliminated before any commercialization can be considered. A better understanding of where the TF lies within a transcription cascade may allow targeting of proteins that have minimal impact on development. Manipulation of interactions with other proteins or of post-translational modifications may also alleviate negative side effects, as would use of promoters that offer better control of transgene expression.

REFERENCES

1. Katagiri, F., A global view of defense gene expression regulation—a highly interconnected signaling network, *Curr. Opin. Plant Biol.*, 7, 506, 2004.
2. Van Loon, L.C. and van Strien, E.A., The families of pathogenesis-related proteins, their activities, and comparative analysis of PR-1 type proteins, *Physiol. Mol. Plant Pathol.*, 55, 85, 1999.
3. Salmeron, J.M. and Vernooij, B., Transgenic approaches to microbial disease resistance in crop plants, *Curr. Opin. Plant Biol.*, 1, 347, 1998.
4. Wan, J., Dunning, F.M., and Bent, A.F., Probing plant–pathogen interactions and downstream defense signaling using DNA microarrays, *Funct. Integr. Genomics*, 2, 259, 2002.
5. Rushton, P.J. and Somssich, I.E., Transcriptional control of plant genes responsive to pathogens, *Curr. Opin. Plant Biol.*, 1, 311, 1998.
6. Dangl, J.L. and Jones, J.D., Plant pathogens and integrated defense responses to infection, *Nature*, 411, 826, 2001.
7. Eulgem, T., Regulation of the *Arabidopsis* defense transcriptome, *Trends Plant Sci.*, 10, 71, 2005.
8. Biggin, M.D. and Tjian, R., Transcriptional regulation in *Drosophila*: the post-genome challenge, *Funct. Integr. Genomics*, 1, 223, 2001.
9. White, R.J., *Gene Transcription: Mechanisms and Control*, Blackwell Science, Oxford, 2001, 273.
10. Dong, X., NPR1, all things considered, *Curr. Opin. Plant Biol.*, 7, 547, 2004.
11. Pieterse, C.M. and Van Loon, L., NPR1: the spider in the web of induced resistance signaling pathways, *Curr. Opin. Plant Biol.*, 7, 456, 2004.
12. Gomez-Gomez, L., Plant perception systems for pathogen recognition and defense, *Mol. Immunol.*, 41, 1055, 2004.
13. Hammond-Kosack, K.E. and Parker, J.E., Deciphering plant–pathogen communication: fresh perspectives for molecular resistance breeding, *Curr. Opin. Biotechnol.*, 14, 177, 2003.
14. Thomma, B.P. et al., The complexity of disease signaling in *Arabidopsis*, *Curr. Opin. Immunol.*, 13, 63, 2001.

15. Oliver, R.P. and Ipcho, S.V.S., *Arabidopsis* pathology breathes new life into the necrotrophs-vs.-biotrophs classification of fungal pathogens, *Mol. Plant Pathol.*, 5, 347, 2004.
16. Durrant, W.E. and Dong, X., Systemic acquired resistance, *Annu. Rev. Phytopathol.*, 42, 185, 2004.
17. Feys, B.J. and Parker, J.E., Interplay of signaling pathways in plant disease resistance, *Trends Genet.*, 16, 449, 2000.
18. Davis, K.R. and Hammerschmidt, R., *Arabidopsis thaliana as a Model for Plant–Pathogen Interactions*, APS Press, Paul, MN, 1993, 132.
19. Mockler, T.C. and Ecker, J.R., Applications of DNA tiling arrays for whole-genome analysis, *Genomics*, 85, 1, 2005.
20. Riechmann, J.L. et al., *Arabidopsis* transcription factors: genome-wide comparative analysis among eukaryotes, *Science*, 290, 2105, 2000.
21. Ostergaard, L. and Yanofsky, M.F., Establishing gene function by mutagenesis in *Arabidopsis thaliana*, *Plant J.*, 39, 682, 2004.
22. Ohme-Takagi, M. and Shinshi, H., Ethylene-inducible DNA-binding proteins that interact with an ethylene-responsive element, *Plant Cell*, 7, 173, 1995.
23. Rushton, P.J. et al., Interaction of elicitor-induced DNA-binding proteins with elicitor response elements in the promoters of parsley PR1 genes, *EMBO J.*, 15, 5690, 1996.
24. Katagiri, F., Lam, E., and Chua, N.H., Two tobacco DNA-binding proteins with homology to the nuclear factor CREB, *Nature*, 340, 727, 1989.
25. Droge-Laser, W. et al., Rapid stimulation of a soybean protein–serine kinase that phosphorylates a novel bZIP DNA-binding protein, G/HBF-1, during the induction of early transcription-dependent defenses, *EMBO J.*, 16, 726, 1997.
26. Korfhage, U. et al., Plant homeodomain protein involved in transcriptional regulation of a pathogen defense-related gene, *Plant Cell*, 6, 695, 1994.
27. Desveaux, D. et al., PBF-2 is a novel single-stranded DNA-binding factor implicated in *PR-10a* gene activation in potato, *Plant Cell*, 12, 1477, 2000.
28. Boyle, B. and Brisson, N., Repression of the defense gene *PR-10a* by the single-stranded DNA binding protein SEBF, *Plant Cell*, 13, 2525, 2001.
29. Sugimoto, K., Takeda, S., and Hirochika, H., MYB-related transcription factor NtMYB2 induced by wounding and elicitors is a regulator of the tobacco retrotransposon *Tto1* and defense-related genes, *Plant Cell*, 12, 2511, 2000.
30. Boter, M. et al., Conserved MYC transcription factors play a key role in jasmonate signaling both in tomato and *Arabidopsis*, *Genes Dev.*, 18, 1577, 2004.
31. Dietrich, R.A. et al., A novel zinc finger protein is encoded by the *Arabidopsis LSD1* gene and functions as a negative regulator of plant cell death, *Cell*, 88, 685, 1997.
32. Mengiste, T. et al., The *BOTRYTIS SUSCEPTIBLE1* gene encodes an R2R3MYB transcription factor protein that is required for biotic and abiotic stress responses in *Arabidopsis*, *Plant Cell*, 15, 2551, 2003.
33. Lorenzo, O. et al., *JASMONATE-INSENSITIVE1* encodes a MYC transcription factor essential to discriminate between different jasmonate-regulated defense responses in *Arabidopsis*, *Plant Cell*, 16, 1938, 2004.
34. Riechmann, J.L. and Ratcliffe, O.J., A genomic perspective on plant transcription factors, *Curr. Opin. Plant Biol.*, 3, 423, 2000.
35. Zhang, J.Z., Overexpression analysis of plant transcription factors, *Curr. Opin. Plant Biol.*, 6, 430, 2003.
36. Zhang, Y. et al., Knockout analysis of *Arabidopsis* transcription factors *TGA2*, *TGA5*, and *TGA6* reveals their redundant and essential roles in systemic acquired resistance, *Plant Cell*, 15, 2647, 2003.

37. Desveaux, D. et al., A "Whirly" transcription factor is required for salicylic acid-dependent disease resistance in *Arabidopsis*, *Dev. Cell*, 6, 229, 2004.
38. Epple, P. et al., Antagonistic control of oxidative stress-induced cell death in *Arabidopsis* by two related, plant-specific zinc finger proteins, *Proc. Natl. Acad. Sci. USA*, 100, 6831, 2003.
39. Tani, H. et al., Activation tagging in plants: a tool for gene discovery, *Funct. Integr. Genomics*, 4, 258, 2004.
40. Despres, C. et al., The *Arabidopsis* NPR1/NIM1 protein enhances the DNA-binding activity of a subgroup of the TGA family of bZIP transcription factors, *Plant Cell*, 12, 279, 2000.
41. Zhou, J., Tang, X., and Martin, G.B., The Pto kinase conferring resistance to tomato bacterial speck disease interacts with proteins that bind a *cis*-element of pathogenesis-related genes, *EMBO J.*, 16, 3207, 1997.
42. Vailleau, F. et al., A R2R3-MYB gene, *AtMYB30*, acts as a positive regulator of the hypersensitive cell death program in plants in response to pathogen attack, *Proc. Natl. Acad. Sci. USA*, 99, 10179, 2002.
43. Li, J., Brader, G., and Palva, E.T., The WRKY70 transcription factor: a node of convergence for jasmonate-mediated and salicylate-mediated signals in plant defense, *Plant Cell*, 16, 319, 2004.
44. Yi, S.Y. et al., The pepper transcription factor *CaPF1* confers pathogen and freezing tolerance in *Arabidopsis*, *Plant Physiol.*, 136, 2862, 2004.
45. Durrant, W.E. et al., cDNA-AFLP reveals a striking overlap in race-specific resistance and wound response gene expression profiles, *Plant Cell*, 12, 963, 2000.
46. Meyers, B.C. et al., Methods for transcriptional profiling in plants. Be fruitful and replicate, *Plant Physiol.*, 135, 637, 2004.
47. Chen, W. et al., Expression profile matrix of *Arabidopsis* transcription factor genes suggests their putative functions in response to environmental stresses, *Plant Cell*, 14, 559, 2002.
48. Zimmermann, P. et al., GENEVESTIGATOR. *Arabidopsis* microarray database and analysis toolbox, *Plant Physiol.*, 136, 2621, 2004.
49. Fujimoto, S.Y. et al., *Arabidopsis* ethylene-responsive element binding factors act as transcriptional activators or repressors of GCC box-mediated gene expression, *Plant Cell*, 12, 393, 2000.
50. Zhang, B., Foley, R.C., and Singh, K.B., Isolation and characterization of two related *Arabidopsis* ocs-element bZIP binding proteins, *Plant J.*, 4, 711, 1993.
51. Yamamoto, S., Suzuki, K., and Shinshi, H., Elicitor-responsive, ethylene-independent activation of GCC box-mediated transcription that is regulated by both protein phosphorylation and dephosphorylation in cultured tobacco cells, *Plant J.*, 20, 571, 1999.
52. Goff, S.A. et al., A draft sequence of the rice genome (*Oryza sativa* L. ssp. *japonica*), *Science*, 296, 92, 2002.
53. Meissner, R.C. et al., Function search in a large transcription factor gene family in *Arabidopsis*: assessing the potential of reverse genetics to identify insertional mutations in R2R3 *MYB* genes, *Plant Cell*, 11, 1827, 1999.
54. Dong, J., Chen, C., and Chen, Z., Expression profiles of the *Arabidopsis* WRKY gene superfamily during plant defense response, *Plant Mol. Biol.*, 51, 21, 2003.
55. Sakuma, Y. et al., DNA-binding specificity of the ERF/AP2 domain of *Arabidopsis* DREBs, transcription factors involved in dehydration- and cold-inducible gene expression, *Biochem. Biophys. Res. Commun.*, 290, 998, 2002.
56. Gutterson, N. and Reuber, T.L., Regulation of disease resistance pathways by AP2/ERF transcription factors, *Curr. Opin. Plant Biol.*, 7, 465, 2004.

57. Allen, M.D. et al., A novel mode of DNA recognition by a β-sheet revealed by the solution structure of the GCC-box binding domain in complex with DNA, *EMBO J.*, 17, 5484, 1998.
58. Magnani, E., Sjolander, K., and Hake, S., From endonucleases to transcription factors: evolution of the AP2 DNA binding domain in plants, *Plant Cell*, 16, 2265, 2004.
59. Wuitschick, J.D. et al., Homing endonucleases encoded by germ line-limited genes in *Tetrahymena thermophila* have APETELA2 DNA binding domains, *Eukaryot. Cell*, 3, 685, 2004.
60. Eulgem, T. et al., The WRKY superfamily of plant transcription factors, *Trends Plant Sci.*, 5, 199, 2000.
61. Zhang, Y. and Wang, L., The WRKY transcription factor superfamily: its origin in eukaryotes and expansion in plants, *BMC Evol. Biol.*, 5, 1, 2005.
62. Yamasaki, K. et al., Solution structure of an *Arabidopsis* WRKY DNA-binding domain, *Plant Cell*, 17, 944, 2005.
63. Ülker, B. and Somssich, I.E., WRKY transcription factors: from DNA binding towards biological function, *Curr. Opin. Plant Biol.*, 7, 491, 2004.
64. Desveaux, D. et al., A new family of plant transcription factors displays a novel ssDNA-binding surface, *Nat. Struct. Biol.*, 9, 512, 2002.
65. Desveaux, D., Marechal, A., and Brisson, N., Whirly transcription factors: defense gene regulation and beyond, *Trends Plant Sci.*, 10, 95, 2005.
66. Jakoby, M. et al., bZIP transcription factors in *Arabidopsis*, *Trends Plant Sci.*, 7, 106, 2002.
67. Katagiri, F., Seipel, K., and Chua, N.H., Identification of a novel dimer stabilization region in a plant bZIP transcription activator, *Mol. Cell Biol.*, 12, 4809, 1992.
68. Martin, C. and Paz-Ares, J., MYB transcription factors in plants, *Trends Genet.*, 13, 67, 1997.
69. Luscombe, N.M. et al., An overview of the structures of protein–DNA complexes, *Genome Biol.*, 1, REVIEWS001, 2000.
70. Kim, S.H. et al., *CAZFP1*, Cys2/His2-type zinc-finger transcription factor gene functions as a pathogen-induced early-defense gene in *Capsicum annuum*, *Plant Mol. Biol.*, 55, 883, 2004.
71. Carey, M.F. and Smale, S.T., *Transcriptional Regulation in Eukaryotes: Concepts, Strategies, and Techniques*, Cold Spring Harbor Press, Cold Spring Harbor, NY, 2000, 640.
72. Lam, E. and Lam, Y.K., Binding site requirements and differential representation of TGF factors in nuclear ASF-1 activity, *Nucleic Acids Res.*, 23, 3778, 1995.
73. Niggeweg, R. et al., Tobacco transcription factor TGA2.2 is the main component of *as-1*-binding factor ASF-1 and is involved in salicylic, *J. Biol. Chem.*, 275, 19897, 2000.
74. Park, J.M. et al., Overexpression of the tobacco *Tsi1* gene encoding an EREBP/AP2-type transcription factor enhances resistance against pathogen attack and osmotic stress in tobacco, *Plant Cell*, 13, 1035, 2001.
75. Yang, Y. and Klessig, D.F., Isolation and characterization of a tobacco mosaic virus-inducible *myb* oncogene homologue from tobacco, *Proc. Natl. Acad. Sci. USA*, 93, 14972, 1996.
76. Krawczyk, S. et al., Analysis of the spacing between the two palindromes of *activation sequence-1* with respect to binding to different TGA factors and transcriptional activation potential, *Nucleic Acids Res.*, 30, 775, 2002.
77. Taatjes, D.J., Marr, M.T., and Tjian, R., Regulatory diversity among metazoan co-activator complexes, *Nat. Rev. Mol. Cell Biol.*, 5, 403, 2004.

78. Naar, A.M., Lemon, B.D., and Tjian, R., Transcriptional coactivator complexes, *Annu. Rev. Biochem.*, 70, 475, 2001.
79. Hanna-Rose, W. and Hansen, U., Active repression mechanisms of eukaryotic transcription repressors, *Trends Genet.*, 12, 229, 1996.
80. Robatzek, S. and Somssich, I.E., Targets of *At*WRKY6 regulation during plant senescence and pathogen defense, *Genes Dev.*, 16, 1139, 2002.
81. Jin, H. and Martin, C., Multifunctionality and diversity within the plant *MYB*-gene family, *Plant Mol. Biol.*, 41, 577, 1999.
82. Niggeweg, R. et al., Tobacco TGA factors differ with respect to interaction with NPR1, activation potential and DNA-binding properties, *Plant Mol. Biol.*, 42, 775, 2000.
83. Fan, W. and Dong, X., *In vivo* interaction between NPR1 and transcription factor TGA2 leads to salicylic acid-mediated gene activation in *Arabidopsis*, *Plant Cell*, 14, 1377, 2002.
84. Eulgem, T. et al., Early nuclear events in plant defense signaling: rapid gene activation by WRKY transcription factors, *EMBO J.*, 18, 4689, 1999.
85. Ohta, M., Ohme-Takagi, M., and Shinshi, H., Three ethylene-responsive transcription factors in tobacco with distinct transactivation functions, *Plant J.*, 22, 29, 2000.
86. Ohta, M. et al., Repression domains of class II ERF transcriptional repressors share an essential motif for active repression, *Plant Cell*, 13, 1959, 2001.
87. Hiratsu, K. et al., Dominant repression of target genes by chimeric repressors that include the EAR motif, a repression domain, in *Arabidopsis*, *Plant J.*, 34, 733, 2003.
88. Robatzek, S. and Somssich, I.E., A new member of the *Arabidopsis* WRKY transcription factor family, *At*WRKY6, is associated with both senescence- and defense-related processes, *Plant J.*, 28, 123, 2001.
89. Pontier, D. et al., Differential regulation of TGA transcription factors by post-transcriptional control, *Plant J.*, 32, 641, 2002.
90. van der Krol, A.R. and Chua, N.H., The basic domain of plant B-ZIP proteins facilitates import of a reporter protein into plant nuclei, *Plant Cell*, 3, 667, 1991.
91. Vandromme, M. et al., Regulation of transcription factor localization: fine-tuning of gene expression, *Trends Biochem. Sci.*, 21, 59, 1996.
92. Gu, Y.Q. et al., *Pti4* is induced by ethylene and salicylic acid, and its product is phosphorylated by the Pto kinase, *Plant Cell*, 12, 771, 2000.
93. Berrocal-Lobo, M., Molina, A., and Solano, R., Constitutive expression of *ETHYLENE-RESPONSE-FACTOR1* in *Arabidopsis* confers resistance to several necrotrophic fungi, *Plant J.*, 29, 23, 2002.
94. Brown, R.L. et al., A role for the GCC-box in jasmonate-mediated activation of the *PDF1.2* gene of *Arabidopsis*, *Plant Physiol.*, 132, 1020, 2003.
95. Guo, Z.J. et al., Overexpression of the AP2/EREBP transcription factor OPBP1 enhances disease resistance and salt tolerance in tobacco, *Plant Mol. Biol.*, 55, 607, 2004.
96. Broun, P., Transcription factors as tools for metabolic engineering in plants, *Curr. Opin. Plant Biol.*, 7, 202, 2004.
97. Amador, V. et al., Gibberellins signal nuclear import of PHOR1, a photoperiod-responsive protein with homology to *Drosophila* armadillo, *Cell*, 106, 343, 2001.
98. Fischer, U. and Droge-Laser, W., Overexpression of *Nt*ERF5, a new member of the tobacco ethylene response transcription factor family enhances resistance to tobacco mosaic virus, *Mol. Plant Microbe Interact.*, 17, 1162, 2004.
99. Onate-Sanchez, L. and Singh, K.B., Identification of *Arabidopsis* ethylene-responsive element binding factors with distinct induction kinetics after pathogen infection, *Plant Physiol.*, 128, 1313, 2002.

100. Lorenzo, O. et al., ETHYLENE RESPONSE FACTOR1 integrates signals from ethylene and jasmonate pathways in plant defense, *Plant Cell*, 15, 165, 2003.
101. Yu, D., Chen, C., and Chen, Z., Evidence for an important role of WRKY DNA-binding proteins in the regulation of *NPR1* gene expression, *Plant Cell*, 13, 1527, 2001.
102. Daniel, X. et al., A novel *myb* oncogene homologue in *Arabidopsis thaliana* related to hypersensitive cell death, *Plant J.*, 20, 57, 1999.
103. Turck, F., Zhou, A., and Somssich, I.E., Stimulus-dependent, promoter-specific binding of transcription factor WRKY1 to its native promoter and the defense-related gene *PcPR1-1* in parsley, *Plant Cell*, 16, 2573, 2004.
104. Solano, R. et al., Nuclear events in ethylene signaling: a transcriptional cascade mediated by ETHYLENE-INSENSITIVE3 and ETHYLENE-RESPONSE-FACTOR1, *Genes Dev.*, 12, 3703, 1998.
105. Guo, H. and Ecker, J.R., Plant responses to ethylene gas are mediated by $SCF^{EBF1/EBF2}$-dependent proteolysis of EIN3 transcription factor, *Cell*, 115, 667, 2003.
106. Potuschak, T. et al., EIN3-dependent regulation of plant ethylene hormone signaling by two *Arabidopsis* F box proteins: EBF1 and EBF2, *Cell*, 115, 679, 2003.
107. Gagne, J.M. et al., *Arabidopsis* EIN3-binding F-box 1 and 2 form ubiquitin-protein ligases that repress ethylene action and promote growth by directing EIN3 degradation, *Proc. Natl. Acad. Sci. USA*, 101, 6803, 2004.
108. Ouaked, F. et al., A MAPK pathway mediates ethylene signaling in plants, *EMBO J.*, 22, 1282, 2003.
109. Chandler, J.W. and Werr, W., When negative is positive in functional genomics, *Trends Plant Sci.*, 8, 279, 2003.
110. McElver, J. et al., Insertional mutagenesis of genes required for seed development in *Arabidopsis thaliana*, *Genetics*, 159, 1751, 2001.
111. Burch-Smith, T.M. et al., Applications and advantages of virus-induced gene silencing for gene function studies in plants, *Plant J.*, 39, 734, 2004.
112. Berrocal-Lobo, M. and Molina, A., Ethylene response factor 1 mediates *Arabidopsis* resistance to the soil-borne fungus *Fusarium oxysporum*, *Mol. Plant Microbe Interact.*, 17, 763, 2004.
113. Gu, Y.Q. et al., Tomato transcription factors Pti4, Pti5, and Pti6 activate defense responses when expressed in *Arabidopsis*, *Plant Cell*, 14, 817, 2002.
114. He, P. et al., Overexpression of *Pti5* in tomato potentiates pathogen-induced defense gene expression and enhances disease resistance to *Pseudomonas syringae* pv. tomato, *Mol. Plant Microbe Interact.*, 14, 1453, 2001.
115. Wu, K. et al., Functional analysis of tomato *Pti4* in *Arabidopsis*, *Plant Physiol.*, 128, 30, 2002.
116. Chakravarthy, S. et al., The tomato transcription factor Pti4 regulates defense-related gene expression via GCC box and non-GCC box *cis* elements, *Plant Cell*, 15, 3033, 2003.
117. Alonso, J.M. et al., Genome-wide insertional mutagenesis of *Arabidopsis thaliana*, *Science*, 301, 653, 2003.
118. Chen, C. and Chen, Z., Potentiation of developmentally regulated plant defense response by AtWRKY18, a pathogen-induced *Arabidopsis* transcription factor, *Plant Physiol.*, 129, 706, 2002.
119. Asai, T. et al., MAP kinase signalling cascade in *Arabidopsis* innate immunity, *Nature*, 415, 977, 2002.
120. Liu, Y., Schiff, M., and Dinesh-Kumar, S.P., Involvement of MEK1 MAPKK, NTF6 MAPK, WRKY/MYB transcription factors, *COI1* and *CTR1* in *N*-mediated resistance to tobacco mosaic virus, *Plant J.*, 38, 800, 2004.

121. Pontier, D., Miao, Z.H., and Lam, E., Trans-dominant suppression of plant TGA factors reveals their negative and positive roles in plant defense responses, *Plant J.*, 27, 529, 2001.
122. Kegler, C. et al., Functional characterization of tobacco transcription factor TGA2.1, *Plant Mol. Biol.*, 55, 153, 2004.
123. Foley, R.C. and Singh, K.B., TGA5 acts as a positive and TGA4 acts as a negative regulator of ocs element activity in *Arabidopsis* roots in response to defense signals, *FEBS Lett.*, 563, 141, 2004.
124. Kim, H.S. and Delaney, T.P., Overexpression of *TGA5*, which encodes a bZIP transcription factor that interacts with NIM1/NPR1, confers SAR-independent resistance in *Arabidopsis thaliana* to *Peronospora parasitica*, *Plant J.*, 32, 151, 2002.
125. Ekengren, S.K. et al., Two MAPK cascades, NPR1, and TGA transcription factors play a role in Pto-mediated disease resistance in tomato, *Plant J.*, 36, 905, 2003.
126. Aviv, D.H. et al., Runaway cell death, but not basal disease resistance, in *lsd1* is SA- and *NIM1/NPR1*-dependent, *Plant J.*, 29, 381, 2002.
127. Whitmarsh, A.J. and Davis, R.J., Regulation of transcription factor function by phosphorylation, *Cell Mol. Life Sci.*, 57, 1172, 2000.
128. Fobert, P.R. and Després, C., Redox control of systemic acquired resistance, *Curr. Opin. Plant Biol.*, 8, 378, 2005.
129. Kim, C.Y. and Zhang, S., Activation of a mitogen-activated protein kinase cascade induces WRKY family of transcription factors and defense genes in tobacco, *Plant J.*, 38, 142, 2004.
130. Jupin, I. and Chua, N.H., Activation of the CaMV *as-1 cis*-element by salicylic acid: differential DNA-binding of a factor related to TGA1a, *EMBO J.*, 15, 5679, 1996.
131. Kang, H.G. and Klessig, D.F., Salicylic acid-inducible *Arabidopsis* CK2-like activity phosphorylates TGA2, *Plant Mol. Biol.*, 57, 541, 2005.
132. Cheong, Y.H. et al., BWMK1, a rice mitogen-activated protein kinase, locates in the nucleus and mediates pathogenesis-related gene expression by activation of a transcription factor, *Plant Physiol.*, 132, 1961, 2003.
133. Després, C. et al., The *Arabidopsis* NPR1 disease resistance protein is a novel cofactor that confers redox regulation of DNA-binding activity to the basic domain/leucine zipper transcription factor TGA1, *Plant Cell*, 15, 2181, 2003.
134. Koyama, T. et al., Isolation of tobacco ubiquitin-conjugating enzyme cDNA in a yeast two-hybrid system with tobacco ERF3 as bait and its characterization of specific interaction, *J. Exp. Bot.*, 54, 1175, 2003.
135. Turner, J.G., Ellis, C., and Devoto, A., The jasmonate signal pathway, *Plant Cell*, 14 Suppl., S153, 2002.
136. Merika, M. and Thanos, D., Enhanceosomes, *Curr. Opin. Genet. Dev.*, 11, 205, 2001.
137. Szymanski, D.B., Liao, B., and Zielinski, R.E., Calmodulin isoforms differentially enhance the binding of cauliflower nuclear proteins and recombinant TGA3 to a region derived from the *Arabidopsis Cam-3* promoter, *Plant Cell*, 8, 1069, 1996.
138. Kuhlmann, M. et al., The α-helical D1 domain of the tobacco bZIP transcription factor BZI-1 interacts with the ankyrin-repeat protein ANK1 and is important for BZI-1 function, both in auxin signaling and pathogen response, *J. Biol. Chem.*, 278, 8786, 2003.
139. Buttner, M. and Singh, K.B., *Arabidopsis thaliana* ethylene-responsive element binding protein (AtEBP), an ethylene-inducible, GCC box DNA-binding protein interacts with an ocs element binding protein, *Proc. Natl. Acad. Sci. USA*, 94, 5961, 1997.

140. Kang, H.G. and Singh, K.B., Characterization of salicylic acid-responsive, *Arabidopsis* Dof domain proteins: overexpression of OBP3 leads to growth defects, *Plant J.*, 21, 329, 2000.
141. Cormack, R.S., Hahlbrock, K., and Somssich, I.E., Isolation of putative plant transcriptional coactivators using a modified two-hybrid system incorporating a GFP reporter gene, *Plant J.*, 14, 685, 1998.
142. Rubio, V. et al., An alternative tandem affinity purification strategy applied to *Arabidopsis* protein complex isolation, *Plant J.*, 41, 767, 2005.
143. Pan, Y. et al., Discovery of functional genes for systemic acquired resistance in *Arabidopsis thaliana* through integrated data mining, *J. Bioinform. Comput. Biol.*, 2, 639, 2004.
144. Coen, E.S., *The Art of Genes: How Organisms Make Themselves*, Oxford University Press, Oxford, 1999, 386.
145. Gehring, W., *Master Control Genes in Development and Evolution: The Homeobox Story*, Yale University Press, New Haven, CT, 1999, 254.
146. Lee, T.I. et al., Transcriptional regulatory networks in *Saccharomyces cerevisiae*, *Science*, 298, 799, 2002.
147. Buck, M.J. and Lieb, J.D., ChIP-chip: considerations for the design, analysis, and application of genome-wide chromatin immunoprecipitation experiments, *Genomics*, 83, 349, 2004.
148. Johnson, C., Boden, E., and Arias, J., Salicylic acid and NPR1 induce the recruitment of trans-activating TGA factors to a defense gene promoter in *Arabidopsis*, *Plant Cell*, 15, 1846, 2003.
149. Kim, J. et al., Mapping DNA–protein interactions in large genomes by sequence tag analysis of genomic enrichment, *Nat. Methods*, 2, 47, 2005.
150. Cao, H. et al., Characterizaton of an *Arabidopsis* mutant that is nonresponsive to inducers of systemic acquired resistance, *Plant Cell*, 6, 1583, 1994.
151. Shah, J., Tsui, F., and Klessig, D.F., Characterization of a salicylic acid-insensitive mutant (sai1) of *Arabidopsis thaliana*, identified in a selective screen utilizing the SA-inducible expression of the tms2 gene, *Mol. Plant Microbe Interact.*, 10, 69, 1997.
152. Glazebrook, J., Rogers, E.E., and Ausubel, F.M., Isolation of *Arabidopsis* mutants with enhanced disease susceptibility by direct screening, *Genetics*, 143, 973, 1996.
153. Delaney, T.P., Friedrich, L., and Ryals, J.A., *Arabidopsis* signal transduction mutant defective in chemically and biologically induced disease resistance, *Proc. Natl. Acad. Sci. USA*, 92, 6602, 1995.
154. Ryals, J. et al., The *Arabidopsis* NIM1 protein shows homology to the mammalian transcription factor inhibitor IκB, *Plant Cell*, 9, 425, 1997.
155. Pieterse, C.M. et al., A novel signaling pathway controlling induced systemic resistance in *Arabidopsis*, *Plant Cell*, 10, 1571, 1998.
156. McDowell, J.M. et al., Downy mildew (*Peronospora parasitica*) resistance genes in *Arabidopsis* vary in functional requirements for *NDR1*, *EDS1*, *NPR1* and salicylic acid accumulation, *Plant J.*, 22, 523, 2000.
157. Liu, G. et al., An *Arabidopsis NPR1*-like gene, *NPR4*, is required for disease resistance, *Plant J.*, 41, 304, 2005.
158. Xiao, S. et al., The atypical resistance gene, *RPW8*, recruits components of basal defense for powdery mildew resistance in *Arabidopsis*, *Plant J.*, 42, 95, 2005.
159. Kus, J.V. et al., Age-related resistance in *Arabidopsis* is a developmentally regulated defense response to *Pseudomonas syringae*, *Plant Cell*, 14, 479, 2002.

160. Thomma, B.P.H.J. et al., Separate jasmonate-dependent and salicylate-dependent defense-response pathways in *Arabidopsis* are essential for resistance to distinct microbial pathogens, *Proc. Natl. Acad. Sci. USA*, 95, 15107, 1998.
161. Ferrari, S. et al., *Arabidopsis* local resistance to *Botrytis cinerea* involves salicylic acid and camalexin and requires *EDS4* and *PAD2*, but not *SID2*, *EDS5* or *PAD4*, *Plant J.*, 35, 193, 2003.
162. Dong, X., Genetic dissection of systemic acquired resistance, *Curr. Opin. Plant Biol.*, 4, 309, 2001.
163. Cao, H. et al., The *Arabidopsis NPR1* gene that controls systemic acquired resistance encodes a novel protein containing ankyrin repeats, *Cell*, 88, 57, 1997.
164. Liu, Y. et al., Tobacco *Rar1*, *EDS1* and *NPR1/NIM1* like genes are required for *N*-mediated resistance to tobacco mosaic virus, *Plant J.*, 30, 415, 2002.
165. Nandi, A., Welti, R., and Shah, J., The *Arabidopsis thaliana* dihydroxyacetone phosphate reductase gene *SUPPRESSSOR OF FATTY ACID DESATURASE DEFICIENCY1* is required for glycerolipid metabolism and for the activation of systemic acquired resistance, *Plant Cell*, 16, 465, 2004.
166. Shah, J., Kachroo, P., and Klessig, D.F., The *Arabidopsis ssi1* mutation restores pathogenesis-related gene expression in *npr1* plants and renders defensin gene expression salicylic acid dependent, *Plant Cell*, 11, 191, 1999.
167. Zhang, Y. et al., A gain-of-function mutation in a plant disease resistance gene leads to constitutive activation of downstream signal transduction pathways in *suppressor of npr1-1, constitutive 1*, *Plant Cell*, 15, 2636, 2003.
168. Shirano, Y. et al., A gain-of-function mutation in an *Arabidopsis* Toll Interleukin1 receptor-nucleotide binding site-leucine-rich repeat type R gene triggers defense responses and results in enhanced disease resistance, *Plant Cell*, 14, 3149, 2002.
169. Kim, H.S. and Delaney, T.P., *Arabidopsis* SON1 is an F-box protein that regulates a novel induced defense response independent of both salicylic acid and systemic acquired resistance, *Plant Cell*, 14, 1469, 2002.
170. Li, X. et al., Identification and cloning of a negative regulator of systemic acquired resistance, SNI1, through a screen for suppressors of *npr1-1*, *Cell*, 98, 329, 1999.
171. Frye, C.A., Tang, D., and Innes, R.W., Negative regulation of defense responses in plants by a conserved MAPKK kinase, *Proc. Natl. Acad. Sci. USA*, 98, 373, 2001.
172. Petersen, M. et al., *Arabidopsis* MAP kinase 4 negatively regulates systemic acquired resistance, *Cell*, 103, 1111, 2000.
173. Park, C.Y. et al., Pathogenesis-related gene expression by specific calmodulin isoforms is dependent on NIM1, a key regulator of systemic acquired resistance, *Mol. Cells*, 18, 207, 2004.
174. Nishimura, M.T. et al., Loss of a callose synthase results in salicylic acid-dependent disease resistance, *Science*, 301, 969, 2003.
175. Wildermuth, M.C. et al., Isochorismate synthase is required to synthesize salicylic acid for plant defence, *Nature*, 414, 562, 2001.
176. Jirage, D. et al., *Arabidopsis thaliana PAD4* encodes a lipase-like gene that is important for salicylic acid signaling, *Proc. Natl. Acad. Sci. USA*, 96, 13583, 1999.
177. Vanacker, H. et al., A role for salicylic acid and NPR1 in regulating cell growth in *Arabidopsis*, *Plant J.*, 28, 209, 2001.
178. Zhang, C., Gutsche, A.T., and Shapiro, A.D., Feedback control of the *Arabidopsis* hypersensitive response, *Mol. Plant Microbe Interact.*, 17, 357, 2004.
179. Zhang, C. and Shapiro, A.D., Two pathways act in an additive rather than obligatorily synergistic fashion to induce systemic acquired resistance and *PR* gene expression, *BMC Plant Biol.*, 2, 9, 2002.

180. Clarke, J.D. et al., Roles of salicylic acid, jasmonic acid, and ethylene in *cpr*-induced resistance in *Arabidopsis*, *Plant Cell*, 12, 2175, 2000.
181. Spoel, S.H. et al., NPR1 modulates cross-talk between salicylate- and jasmonate-dependent defense pathways through a novel function in the cytosol, *Plant Cell*, 15, 760, 2003.
182. Zhou, J.M. et al., NPR1 differentially interacts with members of the TGA/OBF family of transcription factors that bind an element of the *PR-1* gene required for induction by salicylic acid, *Mol. Plant Microbe Interact.*, 13, 191, 2000.
183. Chern, M.S. et al., Evidence for a disease-resistance pathway in rice similar to the *NPR1*-mediated signaling pathway in *Arabidopsis*, *Plant J.*, 27, 101, 2001.
184. Zhang, Y. et al., Interaction of NPR1 with basic leucine zipper protein transcription factors that bind sequences required for salicylic acid induction of the *PR-1* gene, *Proc. Natl. Acad. Sci. USA*, 96, 6523, 1999.
185. Subramaniam, R. et al., Direct visualization of protein interactions in plant cells, *Nat. Biotechnol.*, 19, 769, 2001.
186. Kinkema, M., Fan, W., and Dong, X., Nuclear localization of NPR1 is required for activation of *PR* gene expression, *Plant Cell*, 12, 2339, 2000.
187. Weigel, R.R., Pfitzner, U.M., and Gatz, C., Interaction of NIMIN1 with NPR1 modulates *PR* gene expression in *Arabidopsis*, *Plant Cell*, 17, 1279, 2005.
188. Mou, Z., Fan, W., and Dong, X., Inducers of plant systemic acquired resistance regulate NPR1 function through redox changes, *Cell*, 113, 935, 2003.
189. Friedrich, L. et al., *NIM1* overexpression in *Arabidopsis* potentiates plant disease resistance and results in enhanced effectiveness of fungicides, *Mol. Plant Microbe Interact.*, 14, 1114, 2001.
190. Cao, H., Li, X., and Dong, X., Generation of broad-spectrum disease resistance by overexpression of an essential regulatory gene in systemic acquired resistance, *Proc. Natl. Acad. Sci. USA*, 95, 6531, 1998.
191. Lin, W.C. et al., Transgenic tomato plants expressing the *Arabidopsis* NPR1 gene display enhanced resistance to a spectrum of fungal and bacterial diseases, *Transgenic Res.*, 13, 567, 2004.
192. Fitzgerald, H.A. et al., Overexpression of *(At)NPR1* in rice leads to a BTH- and environment-induced lesion-mimic/cell death phenotype, *Mol. Plant Microbe Interact.*, 17, 140, 2004.
193. Ha, C.M. et al., *BLADE-ON-PETIOLE1* encodes a BTB/POZ domain protein required for leaf morphogenesis in *Arabidopsis thaliana*, *Plant Cell Physiol.*, 45, 1361, 2004.
194. Hepworth, S.R. et al., BLADE-ON-PETIOLE-dependent signaling controls leaf and floral patterning in *Arabidopsis*, *Plant Cell*, 17, 1434, 2005.
195. Heidel, A.J. et al., Fitness costs of mutations affecting the systemic acquired resistance pathway in *Arabidopsis thaliana*, *Genetics*, 168, 2197, 2004.
196. Shin, R. et al., Ectopic expression of *Tsi1* in transgenic hot pepper plants enhances host resistance to viral, bacterial, and oomycete pathogens, *Mol. Plant Microbe Interact.*, 15, 983, 2002.
197. Zhang, H. et al., Tomato stress-responsive factor TSRF1 interacts with ethylene responsive element GCC box and regulates pathogen resistance to *Ralstonia solanacearum*, *Plant Mol. Biol.*, 55, 825, 2004.
198. Lee, J.H., et al., The ethylene-responsive factor like protein 1 (CaERFLP1) of hot pepper (*Capsicum annuum* L.) interacts *in vitro* with both GCC and DRE/CRT sequences with different binding affinities: possible biological roles of CaERFLP1 in response to pathogen infection and high salinity conditions in transgenic tobacco plants, *Plant Mol. Biol.*, 55, 61, 2004.

9 Defense Signaling and Pathway Interactions Involved in Rice Disease Resistance

Xiangjun Zhou, Tameka Bailey, and Yinong Yang

CONTENTS

9.1 INTRODUCTION

Plants are constantly faced with a variety of biotic and abiotic challenges during their lifetime. As sessile organisms, plants have evolved elaborate mechanisms to perceive environmental cues and adjust their metabolism in response to microbial infection or abiotic stresses. The host perception of pathogen infection, which is often mediated by disease resistance (*R*) genes, triggers a cascade of signal transduction that involves protein phosphorylation, ion fluxes, reactive oxygen species, and other signaling events [1]. Subsequent transcriptional and/or post-translational activation of transcription factors leads to induction of a diverse array of plant defense genes such as those encoding pathogenesis-related (PR) proteins and phytoalexin biosynthetic enzymes.

Following early defense signaling, primary signals are typically amplified in the plant through the generation of secondary signal molecules such as salicylic acid (SA), jasmonic acid (JA), ethylene (ET), abscisic acid (ABA), hydrogen peroxide (H_2O_2), and nitric oxide (NO). Increasing evidence has shown that the defense pathways mediated by these endogenous signal molecules often cross-talk (i.e., interact with each other) and form a complex network of signal transduction that eventually leads to induced resistance such as hypersensitive response (HR) and systemic acquired resistance (SAR).

Much of what is known about the molecular mechanism of plant defense response has been derived from studies on dicotyledons, mainly *Arabidopsis* and tobacco. However, studies are increasingly conducted in rice and other economically important cereal crops. In addition to its importance as a staple food, rice has arisen as a pivotal model for cereals because of its small genome size, extensive genetic mapping data, complete genome sequences, and relative ease of transformation. Furthermore, physical mapping and genome sequencing have revealed extensive synteny and collinearity among rice, maize, wheat, and other cereals [2,3]. This conservation of gene order may allow for information gained from studies of the rice defense mechanism to allow predictions based on gene orthology to be made with respect to other cereal crop species.

In this chapter, we attempt to provide an overview of current knowledge on signal perception and transduction of the rice defense response. Emphasis will be given to signaling pathways (JA, SA, ABA, ET, brassinosteriods, and the mitogen-activated protein kinase cascade) in rice and the emerging theme of complicated cross-talk among these signaling pathways.

9.2 DISEASE-RESISTANCE GENES AND EARLY SIGNAL PERCEPTION

To date, seven rice *R* genes have been isolated by map-based cloning, specifically *Xa21*, *Xa1*, *Xa26*, *Xa27*, *xa5*, *Pi-ta*, and *Pib*. These genes encode extracellular/transmembrane receptors or cytoplasmic/nuclear factors that may recognize avirulence (avr) gene products through direct or indirect interaction [4–10]. Of these, *Xa21*, *Xa1*, *Xa26*, *Xa27*, and *xa5* confer resistance to particular races of *Xanthomonas oryzae* pv. *oryzae*, the causal agent of rice bacterial blight, which affects production in irrigated and rain-fed lowland ecosystems throughout Asia, north Australia, mainland Africa, and Latin America [4,6] and can cause yield losses of up to 50% under specific conditions [11].

Xa21 is probably the best characterized rice *R* gene. It is a member of a small multigene family of about seven members, most of which are clustered in a complex locus [12]. The *Xa21* locus confers resistance to all known races of *X. oryzae* pv. *oryzae* in the Philippines and India [13], and the *Xa21* gene specifies resistance to 29 isolates from eight countries [12]. The resistance activity of *Xa21* is developmentally controlled; its contribution to resistance increases progressively from weak or no resistance at the juvenile stage to full resistance at the adult stage [9].

Because *Xa21* encodes a transmembrane protein with an intracellular kinase domain and a leucine-rich repeat (LRR) domain exposed to the extracellular environment, it has been postulated that Xa21 may recognize the pathogen secreted AvrXa21 product at the surface of the host cell. Upon interaction between AvrXa21 and the extracellular LRR domain of Xa21, it has been proposed that the receptor kinase domain becomes phosphorylated at specific serine and threonine residues, leading to a conformational change of the protein and subsequent activation of the disease resistance response [5].

The *Xa1* gene confers resistance to Japanese race 1 of *X. oryzae* pv. *oryzae*, which is one of the dominant races present in Japanese inoculum populations. It is a single-copy gene in rice that belongs to the NBS-LRR class of plant *R* genes [6]. *Xa1* is induced by wounding or infection of virulent and avirulent *X. oryzae* pv. *oryzae* strains. It is possible that induced accumulation of Xa1 may lead to highly efficient interaction with AvrXa1.

Xa26 is constitutively expressed and confers the bacterial blight resistance at both seedling and adult stages. Xa21 and Xa26 belong to a common LRR-transmembrane-kinase class of plant R proteins with 23 and 26 imperfect LRRs respectively, but mediate resistance to different spectra of pathogen races [9]. Resistant lines carrying *Xa26* displayed significant differences in lesion length after inoculation with *X. oryzae* pv. *oryzae* strains. Moreover, transgenic plants expressing *Xa26* showed enhanced resistance compared with the donor line of the gene. Therefore, expression of *Xa26*-mediated resistance is probably significantly influenced by genetic background [9].

Most recently, *Xa27* and its corresponding *avrXa27* were isolated from rice and *X. oryzae* pv. *oryzae*, respectively [10]. *Xa27* is an intronless gene and encodes a protein of 113 amino acids with no discernible sequence similarity to proteins from other plant species. Although resistant and susceptible alleles of *Xa27* encode identical proteins, induction of *Xa27* expression only occurs in resistant lines infected with the bacterial blight pathogen carrying *avrXa27*. The avrXa27 is a member of the AvrBs3/PthA family of type III effectors and contains a highly conserved carboxy-terminal region with three nuclear localization signal motifs and a transcription activation domain [14]. Following bacterial infection, avrXa27 likely enters the rice cell nucleus and specifically triggers *Xa27* expression and disease resistance.

In contrast with *Xa21, Xa1, Xa26*, and *Xa27*, which are dominant *R* genes, *xa5* is a recessive gene conferring resistance to races containing *avrxa5*, another member of the *avrBs3/PthA* gene family that encodes nuclear localization signals and transcriptional activation domains. Interestingly, like *xa5*, *Xa27* does not conform to the typical *R* gene classes and uniquely encodes the gamma subunit of general eukaryotic transcription factor IIA [8].

Rice blast disease is a major constraint for rice production worldwide. The causal agent, *M. grisea*, is a hemibiotrophic filamentous ascomycete that infects many grasses, including cereal crops such as rice, wheat, barley, and millet [15]. To date, at least two blast resistance genes (*Pib* and *Pita*) have been isolated from rice. *Pib* confers resistance to most Japanese blast races [16] and belongs to a small gene family with additional members *PibH8*, *HPibH8-1*, and *HPibH8-2*. Members of the *Pib* gene family appear to be transcriptionally regulated by environmental conditions such as high humidity and darkness that favor pathogen infection. They are also induced by defense signal molecules such as JA, SA, and ET [17].

Pita confers resistance to *M. grisea* strains containing the *avrPita* gene and is constitutively expressed at a low level in resistant as well as susceptible rice cultivars. It encodes a predicted 928-amino acid cytoplasmic protein with an NBS motif and a leucine-rich domain at the C-terminus [18]. Transient expression of *Pita* together with *avrPita* in cells of susceptible rice lines induces a resistance response. Furthermore, interaction between avrPita and Pita proteins was observed in the yeast two-hybrid system and in an *in vitro* binding assay [7]. Single amino acid mutations in the Pita leucine-rich domain or in the AvrPita protease motif disrupt physical interaction and loss of the resistance response. Therefore, the leucine-rich domain of Pita protein is required for the protein–protein interaction, which is important for initiating the *Pita*-mediated blast resistance response.

9.3 MAJOR SIGNALING PATHWAYS INVOLVED IN RICE DEFENSE RESPONSE

9.3.1 Salicylic Acid Signaling Pathway

The role of SA as a key defense signal molecule has been well established in dicots, based on the correlation of SA accumulation, *PR* gene expression, and induction of local and systemic acquired resistance [19–22]. However, relatively little is known about the potential role of SA in mediating defense signaling in monocots.

Rice plants contain a high level of free endogenous SA. The basal levels of SA for various rice cultivars range from 5 to 30 µg/g fresh weight, in comparison to less than 0.1 µg/g fresh weight for tobacco and *Arabidopsis* [23]. Despite the high basal level of SA, rice plants have been shown to respond to exogenous SA treatment in some cases. For example, SA is capable of inducing the SA glucosyl transferase, an enzyme that conjugates free SA [24]. Exogenous application of SA can promote H_2O_2 accumulation in the veins and interveinal regions of rice leaves, suggesting that SA may induce oxidative stress through production of H_2O_2 and active oxygen species [25].

The existence of the SA-mediated signaling pathway was implicated by analysis of transgenic rice expressing NPR1, a key regulator of the SA signaling pathway and SAR in *Arabidopsis*. Transgenic rice overexpressing *AtNPR1* is more resistant to *X. oryzae* pv. *oryzae* and displays a lesion-mimic phenotype [26], which correlates with the expression of rice defense genes and accumulation of H_2O_2 [27]. Furthermore, AtNPR1 can interact with the rice TGA family of transcription factors [26]. Therefore, the molecular mechanism of NPR1-dependent SA signaling may be similar in rice and *Arabidopsis*.

On the other hand, SA levels in rice do not change significantly after infection with compatible or incompatible pathogens. Our recent study suggests that SA is not an effective signal molecule for defense gene expression because depletion of endogenous SA in transgenic rice overexpressing bacterial *NahG* gene (which encodes a salicylate hydroxylase that degrades SA) does not measurably affect *PR* gene expression [23]. Interestingly, SA-deficient transgenic rice contains elevated levels of superoxide and H_2O_2 and exhibits spontaneous lesion formation in an age- and light-dependent manner. When infected with *M. grisea*, SA-deficient rice exhibits increased susceptibility to oxidative bursts elicited by avirulent blast isolates. Furthermore, the same transgenic rice is hyper-responsive to oxidative damage caused by paraquat treatment. These results strongly suggest that the high-level endogenous SA may play an important role in modulating redox balance and protecting rice plants from oxidative stress.

9.3.2 Jasmonate Signaling Pathway

Jasmonates (JAs), including jasmonic acid (JA) and methyl jasmonate (MeJA), are signal molecules important for initiating and/or maintaining developmental processes and defense responses in various plants. In rice, the level of endogenous JA differs among various tissue and cell types and increases in response to diverse environmental stimuli such as pathogen attack and wounding. Jasmonates are rapidly biosynthesized in chloroplasts from α-linolenic acid via an inducible octadecanoid pathway [28]. Key enzymes of the *Arabidopsis* octadecanoid pathway are lipase, lipoxygenase (LOX), allene oxide synthase (AOS), allene oxide cyclase (AOC), and 12-oxo-phytodienoic acid reductase (OPR). The rice homologs of genes encoding for enzymes of the octadecanoid pathway have recently been isolated, partially characterized, and reviewed [29-34].

The biosynthesis of JA starts from α-linolenic acid, a C18 unsaturated fatty acid. Phospholipases catalyze the release of linoleic acid from plant cell membranes. Five phospholipase D isoforms (RPLD1-5) were recently isolated from rice cultivar IRBB10 [35]. Of the five *RPLD* genes, *RPLD1* shares the highest sequence similarity to *AtPLDα1*, which is involved in wound-induced JA accumulation and the activation of JA-responsive genes in *Arabidopsis* [36]. In rice, *RPLD1* and *RPLD2* are induced by wounding and *X. oryzae* pv. *oryzae* infection, suggesting their potential role in rice defense responses [35].

Lipoxygenase is the second enzyme of the octadecanoid pathway, which synthesizes 13-hydroperoxy-octadecanoic acid from linolenic acid. Homolog search reveals that at least 16 LOX genes are present in the rice genome. However, little was known about their involvement and regulation in the rice octadecanoid pathway until a full-length rice, LOX cDNA (RCI-1), was cloned [37]. The transcript of *RCI-1* accumulates in response to treatment of SAR inducers such as benzo (1,2,3) thiadiazole-7-carbothioic acid, S-methyl ester (BTH), 2,6-dichloroisonicotinic acid (INA), probenazol (PBZ), and JA. However, inoculation with compatible and incompatible races of *M. grisea* and the nonhost pathogen *Pseudomonas syringae* pv. *Syringae* failed to induce *RCI-1* expression.

Recombinant RCI-1 protein exhibits lipoxygenase activity that converts linoleic or linolenic acid to 13-hydroperoxy-octadecanoic acid. The RCI-1 protein also contains a putative chloroplast transit peptide, further indicating its involvement in the octadecanoid pathway. In addition to *RCI-1*, four other *OsLOX* genes were found to share high levels of sequence similarity with *LOX* genes involved in the octadecanoid pathway of *Arabidopsis* and other plants [38–41].

AOS enzymes convert 13-hydroperoxy-9,11,15-octadecatrienoic acid (13-HPOT) to 12,13-epoxy-9,11,15-octadecatrienoic acid (12,13-EOT), which is the first specific reaction in the biosynthesis of jasmonate. In contrast to the single-copy *AOS* gene in *Arabidopsis* [42], at least four *AOS* genes are in the rice genome [29,30,34]. These chloroplast-localized enzymes are members of the cytochrome P450 CYP74A subfamily. Recombinant AOS protein expressed in *Escherichia coli* was able to convert 13-hydroperoxylinoenic acid to allene oxide [30]. Expression of *OsAOS2* (designated as *OsAOS* previously) is weak in response to wounding as well as SA, ET, H_2O_2, or ABA treatment; JA, protein phosphatase inhibitors, and *M. grisea* infection induced strong expression of *OsAOS2*. In addition, expression of *OsAOS1* and *OsAOS4* was enhanced in response to red and far red light, suggesting that phytochrome mediates the transcriptional expression of these two genes. This upregulation of JA may in turn inhibit growth of rice coleoptiles [34].

The 12,13-EOT product synthesized by AOS is unstable and is converted immediately to 12-oxo-phytodienoic acid (OPDA) by allene oxide cyclase (AOC). Two *OsAOC* genes, *OsAOC1* and *OsAOC2*, have been isolated from rice and encode for almost identical proteins differing by a single amino acid. Comparative analysis of the N-terminal amino acid sequence of OsAOC1 and OsAOC2 with their homologs from tomato and *Arabidopsis* shows that both proteins contain a chloroplast transit peptide [43]. *OsAOC1* expression is differentially regulated by wounding and JA in addition to other signaling molecules, including SA, ABA, ET, H_2O_2, and fungal elicitors. Wounding has the most profound effect on *OsAOC1* expression, suggesting that *OsAOC* may play a role in rice defense response [31].

Reduction of OPDA to 10,11-dihydro-12-oxo-phytodienoic acid, which is the last committed enzymatic step of the octadecanoid pathway leading to JA biosynthesis, is catalyzed by 12-oxo-phytodienoic acid reductase (OPR). The rice genome encodes for at least 13 OPR genes; however, only *OsOPR1* has been characterized at transcriptional and translational levels [33]. Encoding a 380-amino acid peptide with OPR activity [44], *OsOPR1* is thought to be a regulatory signal for JA biosynthesis because application or coapplication of JA, SA, ET, ABA, H_2O_2, and *de novo* protein synthesis inhibitors results in a massive transient expression of *OsOPR1*. In addition, other genes putatively involved in the rice octadecanoid pathway, including *RPLD1*, *RPLD2*, *OsAOS*, and *OsOPR*, may also regulate JA biosynthesis during rice defense responses. Pathogen attack may enhance expression of these genes and result in rapid JA accumulation.

Currently, the mechanism of JA signal perception in rice is unclear. In *Arabidopsis*, the mode of jasmonate action was elucidated with the aid of several JA signal mutants, such as *coi1*, which exhibits JA insensitivity, pollen sterility, and enhanced resistance to a necrotrophic pathogen [45]. COI1 contains 16 imperfect LRRs and a degenerate F-box, both of which may be involved in protein–protein interactions. By searching the rice genome sequences [46,47], an ortholog of *COI1* was identified and used to generate rice *coi1* suppression lines via double-stranded RNA interference (RNAi) [48].

The rice *coi1* RNAi lines exhibit reduced sensitivity to JA and increased sensitivity to gibberellic acid. As a result, they have much longer neck internodes due to cell elongation and are significantly taller (35%) than untransformed control plants. Preliminary data indicate that disruption of JA signaling in the *coi1* RNAi lines also alters defense gene expression and disease resistance in rice.

9.3.3 Ethylene Signaling Pathway

Rice growers and researchers have long observed the phenomenon that drought stress and upland culture conditions increase lesion number, lesion size, and severity of rice blast [49,50], whereas flood conditions reduce the number of blast lesions and rate of lesion expansion as well as flatten disease gradients [51]. However, until recently, the mechanism for this phenomenon remained unknown.

Biotic and abiotic stresses, including pathogen infection and anaerobic conditions caused by water submergence, induce the biosynthesis of the simple gaseous hormone, ethylene. For instance, submergence leads to activation of *Os-ACS1* and *OS-ACS5,* two rice ethylene biosynthetic genes, thereby causing rapid ethylene accumulation in deepwater rice [52,53]. In addition, application of ethephon (2-chloroethylphosphonic acid), an ethylene-releasing chemical, significantly increases blast resistance in disease-susceptible rice cultivars [54], indicating a circumstantial association between ethylene accumulation and disease resistance.

During the past decade, genes involved in ET signaling have been successfully identified and characterized in *Arabidopsis* [55] and tomato [56] using various genetics approaches. Orthologs have also been found in other plant species, including rice. Furthermore, a dominant negative mutant of the *Arabidopsis* ET receptor gene (*ETR1*) confers ethylene insensitivity in heterologous plants, including tobacco [57], tomato, and petunia [58], suggesting the universal existence of the ethylene signaling pathway throughout the plant kingdom [59].

In addition to the *ETR1* receptor gene, *EIN2* encodes an integral membrane protein that is a key component of ethylene signaling. Loss-of-function mutations in *Arabidopsis EIN2* gene block ethylene responses completely [60]. Based primarily on sequence similarity, two rice *EIN2* orthologs, which share, respectively, 57 and 31.7% sequence identity with AtEIN2, were isolated from rice [61,62]. *OsEIN2* antisense lines and *OsEIN2-2* RNAi lines exhibit ethylene insensitivity, which is a reduced shoot elongation and a decreased expression of ethylene-responsive genes, thus suggesting that both genes are important for ethylene signaling in rice. Because suppression of *OsEIN2-2* results in increased susceptibility of the RNAi lines to *M. grisea* and *Burkholderia glumae*, ethylene signaling was demonstrated to play an important role in mediating broad-spectrum resistance against rice pathogens [62].

9.3.4 Abscisic Acid Signaling Pathway

The role of abscisic acid (ABA) in seed development and dormancy, stomatal guard cell regulation, root geotropism, bud dormancy, and environmental stress responses such as high salinity, drought, and low temperature has been well documented. In addition, several studies in tomato, *Arabidopsis*, and potato indicated that ABA has a negative effect on plant defense responses. An ABA-deficient tomato mutant, *sitens*, exhibited enhanced resistance to the necrotrophic fungus *Botrytis cinerea* in comparison to wild-type tomato [63]. In *Arabidopsis*, ABA deficiency or disruption of ABA signaling leads to increased resistance to the necrotrophic fungal pathogen *Fusarium oxysporum*, upregulation of JA/ethylene responsive defense genes [64], and reduced anthocyanin accumulation during symptom development of *Arabidopsis–Verticillium dahliae* interactions [65].

In contrast, exogenous application of ABA enhances potato and tomato disease susceptibility to *Phytophthora infestans*, *Cladosporium cucumerinum*, and *B. cinerea* [63,66]. Similarly, pretreatment of rice plants with ABA leads to increased susceptibility to *M. grisea* [67,68]. RNAi suppression of *OsMAPK5*, an ABA-inducible MAP kinase in rice, results in enhancement of broad-spectrum disease resistance and constitutive *PR* gene expression [69]. Treatment of the *OsMAPK5* RNAi lines with ABA leads to reversion from blast resistance to a blast susceptible phenotype [68]. These data suggest that ABA is probably a negative regulator in rice defense responses.

9.3.5 Brassinosteroid Signaling Pathway

BRs are a class of plant steroids that, until recently, were not regarded as plant hormones. However, they have been shown to have important regulatory roles in many developmental and physiological processes, including seed germination, stem elongation, leaf expansion, xylem differentiation, stress tolerance, and disease resistance. In rice, the most well-known physiological response to BRs is the bending of the lamina joint [70].

The *OsBRI1* gene, an ortholog of *Arabidopsis BRI1*, encodes a putative BR receptor kinase. Complementary and antisense experiments in rice show that it controls internode elongation, bending of the lamina joint, and skotomorphogenesis [71]. Recently, BRs were shown to induce rice resistance to *M. grisea* and *X. oryzae* pv. *oryzae*, but failed to induce expression of *PBZ1*, suggesting that BRs may regulate rice disease resistance through an independent pathway [72].

9.4 DEFENSE-RELATED MAP KINASES AND TRANSCRIPTION FACTORS

9.4.1 Mitogen-Activated Protein Kinase Cascade in Rice

Mitogen-activated protein kinases (MAPKs) mediate signal transduction and are involved in a plethora of biological processes, including plant growth and development as well as biotic and abiotic stress responses. As a component of an integrated signaling network, MAP kinase cascades transduce the perception of environmental cues to the intracellular milieu. The MAP kinase cascade is minimally composed of three kinase modules: MAP kinase (MAPK), MAP kinase kinase (MAPKK), and MAP kinase kinase kinase (MAPKKK).

MAPK cascades are linked in various ways to upstream receptors and downstream targets [73]. To date, about ten rice MAPK cascade components have been isolated and partially characterized. These include one MAPKKK (OsEDR1) [74], two MAPKKs (OsMEK1) [75,76], and six MAPKs (OsBWMK1, OsMAPK3, OsMSRMK3, OsMAPK4, OsMAPK5, and OsWJUMK1) [69,77–81]. Most are activated by defense signal molecules and/or pathogen infection and are implicated in the rice defense response.

OsEDR1 is an ortholog of AtEDR1, which is a MAPKKK that negatively regulates SA-inducible defense responses in *Arabidopsis* [74]. Loss of function of *AtEDR1* (*e*nhanced *d*isease *r*esistance) in *Arabidopsis* confers resistance to fungal powdery mildew disease [82]. *OsEDR1* is constitutively expressed in the leaves of rice seedlings and is drastically upregulated by key defense signal molecules, including JA, SA, ethephon, ABA, and H_2O_2, as well as fungal elicitors and temperature changes [74].

OsMEK1, which encodes an MAPKK, was initially isolated from rice suspension-culture cells treated with *M. grisea*-derived elicitor in an mRNA differential display experiment. The transcript of *OsMEK1* is specifically induced by elicitors derived from an avirulent race of *M. grisea* and thus may function in an *R*-gene-dependent manner [75]. The delayed induction of *OsMEK1* in response to rice blast elicitors suggests that this gene may be involved in late defense signaling events. However, another rice MAPKK (also named *OsMEK1*) was shown to be involved primarily in abiotic stress signaling [76].

OsBWMK1 (blast-and wound-induced MAP kinase) was the first MAPK isolated in rice. *OsBWMK1* is induced by *M. grisea* infection and mechanical wounding [77]. Recently, OsBWMK1 was shown to target and phosphorylate OsEREBP1, an ethylene responsive element binding protein transcription factor. Phosphorylation of OsEREBP1 by BWMK1 enhanced *in vitro* DNA-binding activity of the factor to a GCC box *cis*-element (AGCCGCC) present in the promoters of several basic *PR* genes. Furthermore, transgenic tobacco plants overexpressing OsBWMK1 enhance *PR* gene expression and disease resistance. These findings suggest that BWMK1 mediates defense signal transduction by phosphorylating one or more transcription factors [83].

OsWJUMK1 encodes a wound and JA-uninducible MAP kinase of approximately 65 kDa. Its expression is constitutive in seedling leaves and is enhanced by cold stress, heavy metals, and H_2O_2. Its expression also increases with plant maturity, particularly in the panicles, indicating its involvement in rice stress signaling and development [80].

OsMAPK4 and OsMSRMK3 (*m*ultiple *s*tress *r*esponsive *MA*P *k*inase), whose protein sequences are 99% identical with one another [80], are developmentally regulated. *OsMAPK4* is expressed strongly in mature leaves and weakly in young leaves and panicles [78]. *OsMSRMK3* is constitutively expressed in seedling leaves, and expression levels increase with the plant's maturity. *OsMSRMK3* expression is induced by many biotic and abiotic factors, including wounding, JA, SA, ET, ABA, H_2O_2, fungal elicitors, high salinity, and heavy metals. Similarly, OsMAPK4 expression is enhanced by high salinity as well as cold and sugar starvation [78]. These observations suggest that OsMSRMK3 and OsMAPK4 are involved in rice stress signaling and developmental processes.

OsMAPK5 (variously named as *OsMSRMK2*, *OsMAPK2*, *OsMAP1*, or *OsBIMK1*) was identified by at least five laboratories and shown to be induced at the mRNA level by multiple biotic and abiotic stresses [69,76,79,81,84]. *OsMAPK5* is a single-copy gene but can generate at least two differentially spliced transcripts. The *OsMAPK5* gene, its protein, and kinase activity are inducible by pathogen infection, ABA, salt, drought, and low temperature. Suppression of *OsMAPK5* expression and its kinase activity resulted in constitutive expression of several *PR* genes, including *PR1* and *PR10*, as well as enhanced rice resistance to rice pathogens such as *M. grisea* and *B. glumae*. These results strongly suggest that *OsMAPK5* can negatively modulate *PR* gene expression and broad-spectrum disease resistance in rice [69].

9.4.2 Transcription Factors and Defense Gene Activation

In plants, many of the biological processes are regulated at the transcriptional level. Transcriptional activation and repression of defense genes during pathogen infection ultimately determine the outcome of disease during plant–pathogen interactions [85,86]. Regulation of defense genes is mediated by change in the levels and/or activities of sequence-specific DNA-binding factors that interact with their promoters [86]. Transcription factors are divided into families based upon characteristics of their respective DNA-binding domains. In rice, at least four major families of transcription factors, including AP2/EREBP, WRKY, Myb, and bZIP, are implicated in host defense response.

AP2/EREBP genes form a large gene family in rice. The rice genome is estimated to contain approximately 143 members of the AP2/ERF/RAV family [46]. AP2/EREBP transcription factors have been shown to play a variety of roles throughout the life cycle of a plant, including growth and development as well as biotic and abiotic stress responses. AP2/EREBP transcription factors are classified according to the number of conserved AP2 (APETALA2) domains.

AP2s contain two AP2 domains and EREBPs contain one AP2 domain. Members of the EREBP subgroup are commonly induced by low temperature, water deficit, salinity, pathogen infection, or other environmental stimuli [87–91]. OsEREBP1 belongs to the EREBP subgroup of the AP2/EREBP family and was isolated by differential display from rice suspension-cultured cells treated with *M. grisea*-derived elicitor [75]. It has recently been shown that phosphorylation of OsEREBP1 by BWMK1 enhances its DNA-binding activity to a GCC *cis*-element *in vitro*, indicating its involvement in regulation of the rice defense response [83].

Until recently, WRKY genes were thought to be restricted to plants. However, it is now known that they occur in slime moulds and protists. The members of the rice WRKY family contain either no zinc finger domain (group IV) or one (group II and III) or two (group I) WRKY domains in addition to the normal features of transcription factors, including nuclear localization signals and transcriptional activation domains [92–94]. Each WRKY domain is a 60-amino acid region composed of an N-terminal WRKY motif, with the most common being WRKYGQK,

WRKYGKK, or WRKYGEK in addition to a C-terminal zinc finger-like motif [92,94–96]. The sequences outside the WRKY domains are highly divergent even for closely related WRKY transcription factors. This difference may reflect the protein's ability to function in various biological processes [92,97]. The sequence of the zinc fingers (DNA-binding motif) in the rice WRKY domain(s) of groups I and II WRKY proteins are C-H_{4-5}-C-X_{22-23}-H-X-H and C-H_{4-5}-C-X_{22-24}-H-X_{1-2}-H, respectively. These motifs are novel in comparison to the C_2H_2 zinc fingers (C-H_{2-4}-C-X_{12}-H-X_{3-5}-H) found in other transcription factors. Members of rice WRKY group III have a C_2-HC motif, C-X_{6-7}-C-X_{23-33}-H-X_1-C [92,94,95,98]. In addition to the DNA-binding function, zinc fingers of WRKY transcription factors may also play a vital role in stabilization of protein–protein interactions between WRKY transcription factors and other general transcription machinery for optimal DNA binding and transcriptional activation [99].

Many studies in *Arabidopsis*, parsley, potato, and rice have shown that WRKY genes are involved in wounding, defense response, senescence, trichrome development, and regulation of secondary metabolism. WRKY proteins bind the w-box *cis*-element is commonly found in the promoter of defense-related genes, suggesting an important role of WRKY proteins in regulating defense gene expression. The rice genome contains approximately 93 members of the WRKY transcription factor family [95]. At least two WRKY genes were shown to be associated with rice defense response. *OsWRKY1* was isolated from rice (cv. Milyang 117) suspension-culture cells and shown to be induced rapidly by a rice blast elicitor [75]. Another rice WRKY gene (*OsiWRKY*) is induced by pathogen attack and mechanical wounding [100]. When challenged with *X. oryzae* pv. *oryzae*, *OsiWRKY* was induced more drastically in IR-26, a resistant rice cultivar in comparison to the susceptible cultivar Jingang 30.

The *myb* gene was originally identified as an oncogene carried by the avian myeloblastosis virus (AMV). Animal *myb* genes encode transcription factors and play an important role in animal pathogenesis and the immune response [101,102]. Similarly, plant *myb* orthologs have been implicated in plant pathogenesis and stress responses [103–107]. Myb transcription factors comprise the largest family of transcription factors in rice [46]. Among these, *JAmyb* is JA- and pathogen-inducible and associated with fungal infection and host cell death [108].

The *JAmyb* transcript can be induced within one day of fungal infection in resistant and susceptible interactions. Its induction by the blast fungus is higher in the susceptible interaction that is accompanied by large lesions and extensive tissue damage. Significant induction of *JAmyb* has also been observed during cell death and lesion formation in certain lesion mimic mutants. Interestingly, *JAmyb* can be specifically and rapidly activated by JA or wounding, but not by other endogenous signal molecules.

In contrast to the preceding transcription factors, which regulate host gene expression, RF2a and RF2b, two bZIP proteins isolated from rice, have been found to bind Box II, an essential *cis* element in the phloem-specific promoter of rice tungro bacilliform virus (RTBV). Transgenic rice plants with reduced levels of RF2b exhibit a disease-like phenotype, demonstrating that RF2a and RF2b or other host factors regulate expression of the RTBV promoter and potentially control RTBV replication and development of tungro disease [109].

9.5 INTERACTIONS AMONG VARIOUS DEFENSE SIGNAL PATHWAYS

Genetic and molecular analyses have so far identified many important components involved in different defense signaling pathways. Increasingly, it has been shown that defense signaling is not merely mediated by parallel, linear pathways. Rather, different signal pathways often cross-talk to one another through positive or negative interactions [110]. As a result, defense signaling involves a network of interacting defense pathways that eventually determines disease resistance and susceptibility.

9.5.1 Cross-Talk between SA and JA Signaling Pathways

In most cases, JA signaling regulates plant resistance to wounding and necrotrophic pathogens, whereas SA signaling affects resistance to biotrophic pathogens. Complicated and mostly antagonistic interactions between JA and SA have been observed in *Arabidopsis* and tobacco pathosystems. In *Arabidopsis*, NPR1 and WRKY70 were identified as the nodes that mediate cross-talk between two pathways [111,112]. In rice, crosstalk between JA and SA pathways was inferred from expression analysis of some defense genes and changes in JA and SA levels in response to wounding. For example, transcript levels of *salT* and *OsIRL* genes are upregulated following blast infection, JA, or fungal elicitor treatment, but exogenous SA strongly inhibits their activation by JA or fungal elicitors [113,114]. During the early response to wounding, an inverse kinetic pattern was observed in terms of accumulation of endogenous JA and SA, suggesting a potential for negative interaction between SA and JA signaling pathways in rice stress response [115].

9.5.2 Cross-Talk between ET and ABA Signaling Pathways

ABA and ET appear to function antagonistically during the rice defense response. Exogenous application of ET enhances the level of resistance against *M. grisea* [54]. Treatment of rice plants with ABA leads to decreased endogenous ethylene content and increased disease susceptibility to *M. grisea* [68]. Suppression of an ABA-inducible MAP kinase (OsMAPK5) results in an increase of endogenous ET, constitutive expression of *PR* genes, and enhanced disease resistance. In contrast, the same RNAi lines show reduction in tolerance to drought, salt, and cold, which usually is mediated by ABA [69]. On the other hand, suppression of *OsEIN2-2* results in reduced sensitivity to ET, but hypersensitivity to ABA [62]. In comparison with wild-type plants, the *OsEIN2-2* suppression lines exhibited increased susceptibility to *M. grisea* and *B. glumae*. Interestingly, the same transgenic lines were more tolerant of cold, drought, and salt stress. Thus, *OsEIN2-2* and *OsMAPK5* may mediate antagonistic interaction between the ET and ABA pathways and inversely regulate disease resistance and abiotic stress tolerance in rice.

9.5.3 Cross-Talk between JA and ABA Pathways

Recent studies suggest an antagonistic interaction between the ABA and JA/ET signaling pathways regarding defense gene expression and disease resistance in

Arabidopsis. Exogenous ABA suppresses activation of *PDF1.2*, *CHI*, *HEL*, and *LEC* by JA and ethylene, whereas ABA deficiency upregulates their expression [116]. In addition, *jin1/myc2* and *aba2-1* mutants, which have reduced ABA sensitivity and ABA biosynthesis, respectively, show increased resistance to the necrotrophic fungal pathogen *F. oxysporum* [64]. In rice, endogenous ABA and methyl jasmonate levels in roots show a differential increase with the dose and duration of salt stress [117]. ABA and JA also regulate different sets of rice genes. ABA failed to induce a cationic peroxidase, acidic PR-1, PR-10, and salT that are markedly induced by JA and salt stress; JA did not induce an ABA-responsive Os*LEA3* protein. ABA and JA inversely affected the transcript level of s*alT* and *OsLEA3*. Together, the implication is that ABA and jasmonates antagonistically regulate responses to abiotic or biotic stresses [117].

9.5.4 Cross-Talk between MAP Kinase and Other Signaling Pathways

Many signals, including phytohormones and environmental cues, activate MAP kinase cascades. MAP kinases in turn affect, via phosphorylation, the function of target proteins (particularly with respect to activity, specificity, and protein–protein interactions) and ultimately regulate cellular responses. In rice, only OsMAPK5 has been shown to mediate cross-talk between the ABA and ET signal pathways [68].

However, increasing evidence from other plant systems indicates that the MAPK cascade plays an important role in mediating cross-talk among various defense signaling pathways—for example, SIPK, an SA-inducible MAP kinase in tobacco, that, when activated by NtMEK2, results in a dramatic increase in ethylene production [118]. MPK6, an *Arabidopsis* ortholog of tobacco SIPK, phosphorylates ACC synthases (ACS2 and ACS6), and phosphorylation of ACS2 and ACS6 leads to the stable accumulation of ACS proteins, elevated ACS activity, and ethylene production [119]. Because tobacco mosaic virus (TMV) is known to activate the NtMEK–SIPK cascade and increase ethylene biosynthesis in tobacco, the NtMEK2–SIPK/WIPK cascade is probably mediating tobacco defense response via an ET-dependent pathway [120].

9.6 CONCLUSIONS AND PROSPECTS

Rice diseases, as well as drought, salt, and cold stresses, are major constraints for rice production worldwide. Although significant progress has been made in cloning rice disease resistance genes and functional genomics in general, relatively little is known about the defense signaling and pathway interactions involved in determining disease resistance and abiotic stress tolerance. Using a combination of genetic, genomic, molecular, biochemical, physiological, and pathological approaches, future studies should not only focus on analyzing individual signaling components and specific defense pathways, but also emphasize elucidation of the complicated interactions among various defense pathways. Better understanding of defense signaling in rice should facilitate development of better breeding strategies and cultural practices for protecting rice and other cereal crops from biotic and abiotic stresses.

REFERENCES

1. Yang, Y., Shah, J., and Klessig, D.F., Signal perception and transduction in plant defense responses, *Genes Dev.*, 11, 1621, 1997.
2. Ahn, S. et al., Homologous relationships of rice, wheat and maize chromosomes, *Mol. Gen. Genet.*, 241, 483, 1993.
3. Gale, M.D. and Devos, K.M., Comparative genetics in the grasses, *Proc. Natl. Acad. Sci. USA*, 95, 1971, 1998.
4. Song, W.-Y. et al., A receptor kinase-like protein encoded by the rice disease resistance gene, *Xa21*, *Science,* 270, 1804, 1995.
5. Ronald, P.C., The molecular basis of disease resistance in rice, *Plant Mol. Biol.*, 35, 179, 1997.
6. Yoshimura, S. et al., Expression of Xa1, a bacterial blight-resistance gene in rice, is induced by bacterial inoculation, *Proc. Natl. Acad. Sci. USA*, 95, 1663, 1998.
7. Jia, Y. et al., Direct interaction of resistance gene and avirulence gene products confers rice blast resistance, *EMBO J.*, 19, 4004, 2000.
8. Iyer, A.S. and McCouch, S.R., The rice bacterial blight resistance gene xa5 encodes a novel form of disease resistance, *Mol. Plant Microbe Interact.*, 17, 1348, 2004.
9. Sun, X. et al., Xa26, a gene conferring resistance to *Xanthomonas oryzae* pv. *oryzae* in rice, encodes an LRR receptor kinase-like protein, *Plant J.*, 37, 517, 2004.
10. Gu, K. et al., R gene expression induced by a type-III effector triggers disease resistance in rice, *Nature*, 435, 1122, 2005.
11. Shen, Y. and Ronald, P., Molecular determinants of disease and resistance in interactions of Xoo and rice, *Microbes Infect.*, 4, 1361, 2002.
12. Wang, G.-L. et al., The cloned gene, Xa21, confers resistance to multiple *Xanthomonas oryzae* pv. *oryzae* isolates in transgenic plants, *Mol. Plant Microbe Interact.*, 9, 850, 1996.
13. Khush, G.S., Bacalangco, E., and Ogawa, T., A new gene for resistance to bacterial blight from *O. longistaminata*, *Rice Genet. Newsl.*, 7, 121, 1990.
14. White, F.F., Yang, B., and Johnson, L.B., Prospects for understanding avirulence gene function, *Curr. Opin. Plant Biol.,* 3, 291, 2000.
15. Howard, R.J. and Valent, B., Breaking and entering: host penetration by the fungal rice blast pathogen *Magnaporthe grisea*, *Annu. Rev. Microbiol.*, 50, 491, 1996.
16. Wang, Z.X. et al., The Pib gene for rice blast resistance belongs to the nucleotide-binding and leucine-rich repeat class of plant disease resistance genes, *Plant J.*, 19, 55, 1999.
17. Wang, Z.X. et al., Expression of the Pib rice-blast-resistance gene family is upregulated by environmental conditions favoring infection and by chemical signals that trigger secondary plant defenses, *Plant Mol. Biol.*, 47, 653, 2001.
18. Bryan, G.T. et al., A single amino acid difference distinguishes resistant and susceptible alleles of the rice blast resistance gene Pi-ta, *Plant Cell*, 12, 2033, 2000.
19. Delaney, T.P. et al., A central role of salicylic acid in plant disease resistance, *Science*, 266, 1247, 1994.
20. Friedrich, L. et al., Characterization of tobacco plants expressing a bacterial salicylate hydroxylase gene, *Plant Mol. Biol.*, 29, 959, 1995.
21. Meuwly, P. et al., Local and systemic biosynthesis of salicylic acid in infected cucumber plants, *Plant Physiol.*, 109, 1107, 1995.
22. Shah, J. and Klessig D.F., Identification of a salicylic acid-responsive element in the promoter of the tobacco pathogenesis-related beta-1,3-glucanase gene, PR-2d, *Plant J.*, 10, 1089, 1996.

23. Yang, Y., Qi, M., and Mei, C., Endogenous salicylic acid protects rice plants from oxidative damage caused by aging as well as biotic and abiotic stress, *Plant J.*, 40, 909, 2004.
24. Silverman, P. et al., Salicylic acid in rice (biosynthesis, conjugation, and possible role), *Plant Physiol.*, 108, 633, 1995.
25. Ganesan, V. and Thomas, G., Salicylic acid response in rice: influence of salicylic acid on H_2O_2 accumulation and oxidative stress, *Plant Sci.*, 160, 1095, 2001.
26. Chern, M.S. et al., Evidence for a disease-resistance pathway in rice similar to the NPR1-mediated signaling pathway in *Arabidopsis*, *Plant J.*, 27, 101, 2001.
27. Fitzgerald, H.A. et al., Overexpression of (At)NPR1 in rice leads to a BTH- and environment-induced lesion-mimic/cell death phenotype, *Mol. Plant Microbe Interact.*, 17, 140, 2004.
28. Liechti, R. and Farmer, E.E., The jasmonate pathway, *Science*, 296, 1649, 2002.
29. Agrawal, G.K. et al., Molecular cloning and mRNA expression analysis of the first rice jasmonate biosynthetic pathway gene allene oxide synthase, *Plant Physiol. Biochem.*, 40, 771, 2002.
30. Ha, S.B. et al., Molecular characterization of the gene encoding rice allene oxide synthase and its expression, *Biosci. Biotechnol. Biochem.*, 66, 2719, 2002.
31. Agrawal, G.K. et al., Cloning of novel rice allene oxide cyclase (OsAOC): mRNA expression and comparative analysis with allene oxide synthase (OsAOS) gene provides insight into the transcriptional regulation of octadecanoid pathway biosynthetic genes in rice, *Plant Sci.*, 164, 979, 2003.
32. Agrawal, G.K. et al., Diverse environmental cues transiently regulate OsOPR1 of the "octadecanoid pathway" revealing its importance in rice defense/stress and development, *Bioche., Biophys. Res. Commun.*, 310, 1073, 2003.
33. Agrawal, G.K. et al., Rice octadecanoid pathway, *Biochem. Biophys. Res. Commun.*, 317, 1, 2004.
34. Haga, K. and Iino, M., Phytochrome-mediated transcriptional upregulation of *ALLENE OXIDE SYNTHASE* in rice seedlings, *Plant Cell Physiol.*, 45, 119, 2004.
35. McGee, J.D. et al., Rice phospholipase D isoforms show differential cellular location and gene induction, *Plant Cell Physiol.*, 44, 1013, 2003.
36. Wang, C. et al., Involvement of phospholipase D in wound-induced accumulation of jasmonic acid in *Arabidopsis*, *Plant Cell*, 12, 2237, 2000.
37. Schaffrath, U., Zabbai, F., and Dudler, R., Characterization of RCI-1, a chloroplastic rice lipoxygenase whose synthesis is induced by chemical plant resistance activators, *Eur. J. Biochem.*, 267, 5935, 2000.
38. Bell, E., Creelman, R.A., and Mullet, J.E., A chloroplast lipoxygenase is required for wound-induced jasmonic acid accumulation in *Arabidopsis*, *Proc. Natl. Acad. Sci. USA*, 92, 8675, 1995.
39. Royo, J. et al., Characterization of three potato lipoxygenases with distinct enzymatic activities and different organ-specific and wound-regulated expression patterns, *J. Biol. Chem.*, 271, 21012, 1996.
40. Heitz, T., Bergey, D.R., Ryan, C.A., A gene encoding a chloroplast-targeted lipoxygenase in tomato leaves is transiently induced by wounding, systemin, and methyl jasmonate, *Plant Physiol.*, 114, 1085, 1997.
41. Voros, K. et al., Characterization of a methyljasmonate-inducible lipoxygenase from barley (*Hordeum vulgare* cv. Salome) leaves, *Eur. J. Biochem.*, 251, 36, 1998.
42. Laudert, D. and Weiler, E.W., Allene oxide synthase: a major control point in *Arabidopsis thaliana* octadecanoid signaling, *Plant J.*, 15, 675, 1998.

43. Ziegler, J. et al., Molecular cloning of allene oxide cyclase. The enzyme establishing the stereochemistry of octadecanoids and jasmonates, *J. Biol. Chem.*, 275, 19132, 2000.
44. Sobajima, H. et al., Cloning and characterization of a jasmonic acid-responsive gene encoding 12-oxophytodienoic acid reductase in suspension-cultured rice cells, *Planta*, 216, 692, 2003.
45. Xie, D.X. et al., COI1: an *Arabidopsis* gene required for jasmonate-regulated defense and fertility, *Science*, 280, 1091, 1998.
46. Goff, S.A. et al., A draft sequence of the rice genome (*Oryza sativa* L. ssp. *japonica*), *Science*, 296, 92, 2002.
47. Yu, J. et al., A draft sequence of the rice genome (*Oryza sativa* L. ssp. *indica*), *Science*, 296, 79, 2002.
48. Mei, C., Zhou, X., and Yang, Y., Rice COI1 gene mediates hormone signal interactions and negatively regulates cell elongation and plant height, presented at Annual Meeting of American Society of Plant Biologists, Orlando, July 24–29, 2004.
49. Bonman, J.M., Sanchez, L.M., and Machill, A.O., Effects of water deficit on rice blast. II. Disease—development in the field, *J. Plant Prot.*, 5, 67, 1988.
50. Gill, M.A. and Bonman, J.M., Effects of water deficit on rice blast. I. Influence of water deficit on components of resistance, *J. Plant Prot.*, 5, 61, 1988.
51. Kim, C.H., Rush, M.C., and Mackenzie, D.R., Epidemiology of rice blast under different water management practices, Annual progress report—Louisiana Agricultural Experiment Station, 77, 263, 1985.
52. Zarembinski, T.I. and Theologis, A., Expression characteristics of OS-ACS1 and OS-ACS2, two members of the 1-aminocyclopropane-1-carboxylate synthase gene family in rice (*Oryza sativa* L. cv. Habiganj Aman II) during partial submergence, *Plant Mol. Biol.*, 33, 71, 1997.
53. Van Der Straeten, D. et al., A comparative molecular-physiological study of submergence response in lowland and deepwater rice, *Plant Physiol.*, 125, 955, 2001.
54. Singh, M.P. et al., Mediation of partial resistance to rice blast through anaerobic induction of ethylene, *Phytopathology*, 94, 819, 2004.
55. Wang, K.L., Li, H., and Ecker, J.R., Ethylene biosynthesis and signaling networks, *Plant Cell*, 14, S131, 2002.
56. Wilkinson, J.Q. et al., An ethylene-inducible component of signal transduction encoded by never-ripe, *Science*, 270, 1807, 1995.
57. Knoester, M. et al., Ethylene-insensitive tobacco lacks nonhost resistance against soil-borne fungi, *Proc. Natl. Acad. Sci. USA*, 95, 1933, 1998.
58. Wilkinson, J.Q. et al., A dominant mutant receptor from *Arabidopsis* confers ethylene insensitivity in heterologous plants, *Nat. Biotechnol.*, 15, 444, 1997.
59. Bleecker, A.B., Ethylene perception and signaling: an evolutionary perspective, *Trends Plant Sci.*, 4, 269, 1999.
60. Alonso, J.M. et al., EIN2, a bifunctional transducer of ethylene and stress responses in *Arabidopsis*, *Science*, 284, 2148, 1999.
61. Jun, S.H. et al., OsEIN2 is a positive component in ethylene signaling in rice, *Plant Cell Physiol.*, 45, 281, 2004.
62. Zhou, X., Mei, C., and Yang, Y., A rice *EIN2*-like gene mediates ethylene and abscisic acid crosstalk and inversely regulates disease resistance and abiotic stress tolerance, presented at Second International Symposium on Rice Functional Genomics, Tucson, AZ, November 15–17, 2004.
63. Audenaert, K., De Meyer, G.B., and Hofte, M.M., Abscisic acid determines basal susceptibility of tomato to *Botrytis cinerea* and suppresses salicylic acid-dependent signaling mechanisms, *Plant Physiol.*, 128, 491, 2002.

64. Anderson, J.P. et al., Antagonistic interaction between abscisic acid and jasmonate-ethylene signaling pathways modulates defense gene expression and disease resistance in *Arabidopsis*, *Plant Cell*, 16, 3460, 2004.
65. Veronese, P. et al., Identification of a locus controlling *Verticillium* disease symptom response in *Arabidopsis thaliana*, *Plant J.*, 35, 574, 2003.
66. Henfling, J.W.D.M., Bostock, R., and Kuc, J., Effect of abscisic acid on rishitin and lubimin accumulation and resistance to *Phytophthora infestans* and *Cladosporium cucumerinum* in potato tuber tissue slices, *Phytopathology*, 70, 1074, 1980.
67. Koga, H., Dohi, K., and Mori, M. Abscisic acid and low temperatures suppress the whole plant-specific resistance reaction of rice plants to the infection of *Magnaporthe grisea*, *Physiol. Mol. Plant Pathol.*, 65, 3, 2004.
68. Bailey, T. and Yang, Y., A stress responsive MAP kinase mediates abscisic acid and ethylene signal interaction in rice plants, presented at Second International Symposium on Rice Functional Genomics, Tucson, AZ, November 15–17, 2004.
69. Xiong, L. and Yang, Y., Disease resistance and abiotic stress tolerance in rice are inversely modulated by an abscisic acid-inducible mitogen-activated protein kinase, *Plant Cell*, 15, 745, 2003.
70. Cao, H. and Chen, S., Brassinosteroid-induced rice lamina joint inclination and its relation to indole-3 acetic acid and ethylene, *Plant Growth Regul.*, 16, 189, 1995.
71. Yamamuro, C. et al., Loss of function of a rice brassinosteroid insensitive 1 homolog prevents internode elongation and bending of the lamina joint, *Plant Cell*, 12, 1591, 2000.
72. Nakashita, H. et al., Brassinosteroid functions in a broad range of disease resistance in tobacco and rice, *Plant J.*, 33, 887, 2003.
73. Jonak, C. et al., Complexity, cross talk and integration of plant MAP kinase signaling, *Curr. Opin. Plant Biol.*, 5, 415, 2002.
74. Kim, J.A. et al., Molecular cloning and mRNA expression analysis of a novel rice (*Oryza sativa* L.) MAPK kinase kinase, OsEDR1, an ortholog of *Arabidopsis* AtEDR1, reveal its role in defense/stress signaling pathways and development, *Biochem. Biophys. Res. Commun.*, 300, 868, 2003.
75. Kim, C.Y. et al., Identification of rice blast fungal elicitor-responsive genes by differential display analysis, *Mol. Plant Microbe Interact.*, 13, 470, 2000.
76. Wen, J.Q., Oono, K., and Imai, R., Two novel mitogen-activated protein signaling components, OsMEK1 and OsMAP1, are involved in a moderate low-temperature signaling pathway in rice, *Plant Physiol.*, 129, 1880, 2002.
77. He, C. et al., BWMK1, a novel MAP kinase induced by fungal infection and mechanical wounding in rice, *Mol. Plant Microbe Interact.*, 12, 1064, 1999.
78. Fu, S.F. et al., Transcriptional regulation of a rice mitogen-activated protein kinase gene, OsMAPK4, in response to environmental stresses, *Plant Cell Physiol.*, 43, 958, 2002.
79. Song, F. and Goodman, R.M., OsBIMK1, a rice MAP kinase gene involved in disease resistance responses, *Planta*, 215, 997, 2002.
80. Agrawal, G.K. et al., Novel rice MAP kinases OsMSRMK3 and OsWJUMK1 involved in encountering diverse environmental stresses and developmental regulation, *Biochem. Biophys. Res. Commun.*, 300, 775, 2003.
81. Agrawal, G.K., Rakwal, R., and Iwahashi, H., Isolation of novel rice (*Oryza sativa* L.) multiple stress responsive MAP kinase gene, OsMSRMK2, whose mRNA accumulates rapidly in response to environmental cues, *Biochem. Biophys. Res. Commun.*, 294, 1009, 2002.

82. Frye, C.A., Tang, D., and Innes, R.W., Negative regulation of defense responses in plants by a conserved MAPKK kinase, *Proc. Natl. Acad. Sci. USA*, 98, 373, 2001.
83. Cheong, Y.H. et al., BWMK1, a rice mitogen-activated protein kinase, locates in the nucleus and mediates pathogenesis-related gene expression by activation of a transcription factor, *Plant Physiol.*, 132, 1961, 2003.
84. Huang, H.J. et al., Expression of *Oryza sativa* MAP kinase gene is developmentally regulated and stress-responsive, *Physiol. Plant.*, 114, 572, 2002.
85. Nimchuk, Z. et al., Recognition and response in the plant immune system, *Annu. Rev. Genet.*, 37, 579, 2003.
86. Kim, C.Y. and Zhang, S. Activation of a mitogen-activated protein kinase cascade induces WRKY family of transcription factors and defense genes in tobacco, *Plant J.*, 38, 142, 2004.
87. Yamaguchi–Shinozaki, K. and Shinozaki, K., A novel *cis*-acting element in an *Arabidopsis* gene is involved in responsiveness to drought, low-temperature, or high-salt stress, *Plant Cell*, 6, 251, 1994.
88. Ohme-Takagi, M. and Shinshi, H., Ethylene-inducible DNA-binding proteins that interact with an ethylene-responsive element, *Plant Cell*, 7, 173, 1995.
89. Stockinger, E.J., Gilmour, S.J., and Thomashow, M.F., *Arabidopsis thaliana* CBF1 encodes an AP2 domain-containing transcriptional activator that binds to the C-repeat/DRE, a *cis*-acting DNA regulatory element that stimulates transcription in response to low temperature and water deficit, *Proc. Natl. Acad. Sci. USA*, 94, 1035, 1997.
90. Zhou, J., Tang, X., and Martin, G.B., The Pto kinase conferring resistance to tomato bacterial speck disease interacts with proteins that bind a *cis*-element of pathogenesis-related genes, *EMBO J.*, 16, 3207, 1997.
91. Sakuma, Y. et al., DNA-binding specificity of the ERF/AP2 domain of *Arabidopsis* DREBs, transcription factors involved in dehydration- and cold-inducible gene expression, *Biochem. Biophys. Res. Commun.*, 290, 998, 2002.
92. Eulgem, T. et al., The WRKY superfamily of plant transcription factors, *Trends Plant Sci.*, 5, 199, 2000.
93. Robatzek, S. and Somssich, I.E., Targets of AtWRKY6 regulation during plant senescence and pathogen defense, *Genes Dev.*, 16, 1139, 2002.
94. Xie, Z. et al., Annotations and functional analyses of the rice WRKY gene superfamily reveal positive and negative regulators of abscisic acid signaling in aleurone cells, *Plant Physiol.*, 137, 176, 2005.
95. Maeo, K. et al., Role of conserved residues of the WRKY domain in the DNA-binding of tobacco WRKY family proteins, *Biosci. Biotechnol. Biochem.*, 65, 2428, 2001.
96. Dong, J., Chen, C., and Chen, Z., Expression profiles of the *Arabidopsis* WRKY gene superfamily during plant defense response, *Plant Mol. Biol.*, 51, 21, 2003.
97. Rushton, P.J. et al., Interaction of elicitor-induced DNA-binding proteins with elicitor response elements in the promoters of parsley *PR1* genes, *EMBO J.*, 15, 5690, 1996.
98. Kalde, M. et al., Members of the *Arabidopsis* WRKY group III transcription factors are part of different plant defense signaling pathways, *Mol. Plant Microbe Interact.*, 16, 295, 2003.
99. de Pater, S. et al., Characterization of a zinc-dependent transcriptional activator from *Arabidopsis*, *Nucleic Acids Res.*, 24, 4624, 1996.
100. Guo, Z.J. et al., Characterization of a rice WRKY gene whose expression is induced upon pathogen attack and mechanical wounding, *Acta Bot. Sin.*, 46, 955, 2004.
101. Graf, T., Myb: a transcriptional activator linking proliferation and differentiation in hematopoietic cells, *Curr. Opin. Genet. Dev.*, 2, 249, 1992.

102. Thompson, M.A. and Ramsay, R.G., Myb: an old oncoprotein with new roles, *BioEssays*, 17, 341, 1995.
103. Urao, T. et al., An *Arabidopsis* myb homolog is induced by dehydration stress and its gene product binds to the conserved MYB recognition sequence, *Plant Cell*, 5, 1529, 1993.
104. Yang, Y. and Klessig, D.F., Isolation and characterization of a tobacco mosaic virus-inducible myb oncogene homolog from tobacco, *Proc. Natl. Acad. Sci. USA*, 93, 14972, 1996.
105. Vailleau, F. et al., A R2R3-MYB gene, AtMYB30, acts as a positive regulator of the hypersensitive cell death program in plants in response to pathogen attack, *Proc. Natl. Acad. Sci. USA*, 99, 10179, 2002.
106. Mengiste, T. et al., The *BOTRYTIS SUSCEPTIBLE1* gene encodes an R2R3MYB transcription factor protein that is required for biotic and abiotic stress responses in *Arabidopsis*, *Plant Cell*, 15, 2551, 2003.
107. Liu, Y., Schiff, M., and Dinesh–Kumar, S.P., Involvement of MEK1 MAPKK, NTF6 MAPK, WRKY/MYB transcription factors, COI1 and CTR1 in N-mediated resistance to tobacco mosaic virus, *Plant J.*, 38, 800, 2004.
108. Lee, M.W., Qi, M., and Yang, Y., A novel jasmonic acid-inducible rice *myb* gene associates with fungal infection and host cell death, *Mol. Plant Microbe Interact.*, 14, 527, 2001.
109. Dai, S. et al., RF2b, a rice bZIP transcription activator, interacts with RF2a and is involved in symptom development of rice tungro disease, *Proc. Natl. Acad. Sci. USA*, 101, 687, 2004.
110. Kunkel, B.N. and Brooks, D.M., Cross-talk between signaling pathways in pathogen defense, *Curr. Opin. Plant Biol.*, 5, 325, 2002.
111. Spoel, S.H. et al., NPR1 modulates cross-talk between salicylate- and jasmonate-dependent defense pathways through a novel function in the cytosol, *Plant Cell*, 15, 760, 2003.
112. Li, J., Brader, G., and Palva, E.T., The WRKY70 transcription factor: a node of convergence for jasmonate-mediated and salicylate-mediated signals in plant defense, *Plant Cell*, 16, 319, 2004.
113. Kim, S.T. et al., A rice isoflavone reductase-like gene, *OsIRL*, is induced by rice blast fungal elicitor, *Mol. Cells.*, 16, 224, 2003.
114. Kim, S.T. et al., Expression of a salt-induced protein (SALT) in suspension-cultured cells and leaves of rice following exposure to fungal elicitor and phytohormones, *Plant Cell Rep.*, 23, 256, 2004.
115. Lee, A. et al., Inverse correlation between jasmonic acid and salicylic acid during early wound response in rice, *Biochem. Biophys. Res. Commun.*, 318, 734, 2004.
116. Abe, H. et al., *Arabidopsis* AtMYC2 (bHLH) and AtMYB2 (MYB) function as transcriptional activators in abscisic acid signaling, *Plant Cell*, 15, 63, 2003.
117. Moons, A. et al., Antagonistic effects of abscisic acid and jasmonates on salt stress-inducible transcripts in rice roots, *Plant Cell*, 9, 2243, 1997.
118. Zhang, S. and Liu, Y., Activation of salicylic acid-induced protein kinase, a mitogen-activated protein kinase, induces multiple defense responses in tobacco, *Plant Cell*, 13, 1877, 2001.
119. Liu, Y. and Zhang, S., Phosphorylation of 1-aminocyclopropane-1-carboxylic Acid synthase by MPK6, a stress-responsive mitogen-activated protein kinase, induces ethylene biosynthesis in *Arabidopsis*, *Plant Cell*, 16, 3386, 2004.
120. Kim, C.Y. et al., Activation of a stress-responsive mitogen-activated protein kinase cascade induces the biosynthesis of ethylene in plants, *Plant Cell*, 15, 2707, 2003.

10 Identification of Heat-Shock Factor Regulated Genes and Pathways

Friedrich Schöffl, Wolfgang Busch, Mukesh Kumar, and Tressa J. Panikulangara

CONTENTS

10.1 INTRODUCTION

The heat-shock response is a conserved cellular defense mechanism activated by a variety of cytotoxic stimuli, such as elevated temperature and a number of chemical stressors. It is characterized by a rapid reprogramming of gene expression pattern, leading to a transient accumulation of heat-shock proteins (HSP), most of which act as molecular chaperones [1,2]. The central regulator of HSP expression is the heat shock transcription factor (HSF). HSF specifically binds to heat-shock promoter elements (HSE) and activates the transcription of heat-shock genes [3,4]. HSFs display a modular structure with a highly conserved N-terminal DNA-binding domain (DBD), which is characterized by a central helix–turn–helix motif, and an adjacent bipartite oligomerization domain (HR-A/B) composed of hydrophobic heptad repeats.

HSF trimerization via formation of a triple-stranded α-helical coiled coil is a prerequisite for high-affinity DNA binding and subsequently for transcriptional activation of heat-shock genes [5]. Other functional domains of HSF include clusters of basic amino acid residues (NLS) essential for nuclear import, leucine-rich export sequences in the HR-C region (NES) and a less conserved C-terminal activator domain (CTAD) rich in aromatic, hydrophobic, and acidic amino acids—the so-called AHA motifs [6].

In *Arabidopsis thaliana*, 21 different HSF genes have been identified by screening cDNA libraries [7] and genomic databases [8]. Based on amino acid homology, plant HSFs are assigned to three classes: A, B, and C [6,8]. In contrast to class-B and nonplant HSFs, class-A and -C HSFs show an insertion of 21 or 7 amino acids, respectively, separating hydrophobic regions HR-A and -B. A striking peculiarity of class-B HSFs is their CTAD's lack of AHA motifs, which are crucial for activator function of class-A HSFs [9]. In contrast to class-A HSFs, transient overexpression of several class-B HSFs was not sufficient for activation of heat-shock promoters in tobacco protoplasts [10,11], and class-B HSFs failed to rescue the yeast *Hsf1* mutation [12]. These results together with the structural differences led to the hypothesis that class-B HSFs represent inert or repressor HSFs [10,11].

An exclusive feature of the plant HSF family is the finding that expression of several members is heat induced, suggesting a multistep mechanism of HSF involvement in the heat-shock response [6]. The high number of class-A HSFs (15 members in *Arabidopsis*) suggests that functional diversification and/or genetic redundancy have evolved in plants. In *Arabidopsis* certain heat-susceptible mutants [13,14] were not caused by mutation in an *Hsf* gene. Only one mutation occurs naturally in plants: the rice spotted leaf gene *Spl7* known in an *Hsf* gene; however, the mutant allele carrying a point mutation is dispensable for heat-shock gene expression [15].

The role of *Arabidopsis* class A-HSFs as positive regulators of heat-shock gene expression was implicated by transgenic overexpression of *AtHsfA1a* (previously designated *Hsf1* in Wunderlich et al. [24] and Lohmann et al. [28]) or *AtHsfA1b* (previously designated *Hsf3* in Lohmann et al. [28], Prändl et al. [17], Panchuk et al. [26], and Panikulangara et al. [40]) constructs in *Arabidopsis*, which resulted in a constitutive expression of HSP and enhanced levels of thermotolerance [16,17]).

Identification of target genes for central transcriptional regulators is crucial to understanding the stress response network in plants. *Arabidopsis* as a model system provides the best available access to genetic and molecular analysis of regulation and function of the stress response. Some evidence suggests that the knowledge gained in this model plant can be exploited, at least to some extent, for generation of stress tolerance in agronomically important plants.

10.2 TRANSGENIC HSF LOSS-OF-FUNCTION MUTANTS OF *ARABIDOPSIS*

10.2.1 Loss of Function by Overexpression of Heat-Shock Factor Constructs

The perplexing multiplicity of HSF genes in *Arabidopsis* renders it difficult to identify mutant phenotypes. One approach to generating transdominant negative mutants of HSFs was the transgenic expression of a protein fusion construct, EN-AtHSFA1a, consisting of the *Drosophila* engrailed inhibitor domain (EN) and *Arabidopsis* AtHsfA1a. *Drosophila* engrailed has been successfully used as an inhibitor of gene expression in different vertebrate cells [18–20]. The *en* gene encodes a homeodomain-type transcription factor that is required for cell fate specification throughout *Drosophila* development. The EN protein is an active repressor of transcription capable of repressing basal transcription and transcription enhanced by a variety of activators in a manner that requires EN-binding sites in target genes, but does not require competition for activator-binding sites [21,22].

In addition to its use in *Drosophila* and vertebrate systems, chimeric constructs between the EN repressor domain and plant transcription factors have been successfully used to generate phenocopies of loss-of-function mutants or to identify biological functions [23] in transgenic *Arabidopsis* plants. One possibility is that the chimeric EN-AtHsfA1a molecules act as true transcriptional repressors. Expressed from the strong *CaMV 35S* promoter, the vast excess may displace the native *AtHsfA1a* gene products from target gene promoters, thus explaining transdominance over the native HSF.

In contrast to active repression, however, competitive modes can also be envisioned. Nonfunctional chimeric EN-AtHsfA1a proteins included in multimeric protein complexes may displace functional native complexes from target genes or excessive EN-AtHSFA1a proteins may interfere with assembly or function of HSF complexes. The latter phenomenon is generally considered *squelching*. Independently of whether the dominant-negative EN-AtHsfA1a version acts via transcriptional repression of target gene promoters or is based on squelching, the underlying mechanism must explain how individual target genes are affected differently by increasing levels of EN-AtHsfA1a protein.

10.2.2 Functional Analysis of EN-Hsf Lines in *Arabidopsis*

The strategy of turning a plant transcriptional activator (AtHsfA1a) into a repressor by adding the *Drosophila* EN repressor domain was successful considering the following criteria [24]:

Transgenic plants expressing EN-AtHsfA1a at different levels were partially or severely impaired in their ability to induce high levels of transcripts of HSF-dependent heat-shock genes, indicating a transdominant negative effect on HSF activity.

The reduced level of heat-induced HSP correlated with a lower level of thermotolerance.

The reduction of heat-inducible heat-shock gene expression correlated with expression levels of the *En-AtHsfA1a* transgene.

Most strongly affected was expression of *Hsp18.2*; less pronounced was expression of other heat-shock genes. The least reduction was observed for *Hsp101*. The differences in target gene response indicate that heat-shock promoters responded individually to excess of EN-AtHsfA1a. EN-AtHsfA1a overexpression also had a negative effect on the heat-inducible transcript levels of *AtHsfB1* (previously designated HSF4 in Prändl et al. [17], Wunderlich et al. [24], and Lohmann et al. [28]), which were largely reduced in different lines. This was the first evidence for HSF-dependent regulation of *AtHsfB1* expression.

AtHsfB1 is a member of the class-B HSFs, which all lack certain structural features of the class-A activator HSFs [8]; for example, the heat-dependent expression of *AtHsfB1* suggests that its function is not required for induction, but possibly for delayed effects or shut-off of the heat-shock response. If AtHsfB1 is a negative regulator of the heat-shock response, its activity may require heat-dependent activation because transgenic overexpression of AtHsfB1 and AtHsfB1-GUS fusion proteins had no detectable effect on expression of heat-shock genes [17].

Interestingly, it was not possible to shut down expression of HSP completely in EN-HSF transgenic plants. Two possible reasons need to be considered:

A complete knockout of HSF-dependent target gene expression may be lethal; some evidence indicates that HSF is essential in certain stages of development in *Drosophila* [25] and yeast [4]. HSF may also be required for embryo development and seed maturation in plants (for an overview, see Schöffl et al. [2]).

The EN-AtHsfA1a fusion protein may not be completely devoid of residual activator function. However, owing to strong effects in compromising heat-shock gene expression, it was concluded that the negative effect dominates the molecular function of EN: AtHsfA1a.

Although a dominant mutant phenotype of HSF has been achieved as shown by the effects on heat-shock gene expression, there was relatively little impact on thermotolerance. Basal and acquired thermotolerance was impaired, but the lethal temperature was only 2°C lower than that of WT plants. This may be explained by the fact that, despite a large reduction of mRNA levels, expression of HSP is much less impaired; this suggests that increased translational efficiency may partially compensate the reduction of HSP mRNA during heat stress in mutant lines.

Another example of dominant negative effects on the heat-shock response was reported for *LpHSF-A1* cosuppression in tomato [27]. The effects are reminiscent of EN-AtHsfA1a repression in *Arabidopsis*, but are referred to as gene silencing of *LpHSF-A1*, which was caused by inverse tandem duplication at the insertion site in the tomato genome. In this study, it was suggested that *LpHSF-A1* is the major regulator of the heat-shock response in tomato.

However, it seems possible that cosuppression was not only specific for *LpHSF-A1* because it has not been tested whether other constitutive HSFs are still expressed and active in the cosuppression line. At present no evidence suggests any major dominant HSF in *Arabidopis*. Transgenic overexpression of AtHsfA1b [17] or AtHsfA1a fusion proteins [16] resulted in cases in gain of function, a derepression of the heat-shock response, which indicates functional redundancy. The analysis of individual gene knockout mutations is required to identify and discriminate clearly functional roles of different HSFs in plants.

10.3 *ARABIDOPSIS* HEAT-SHOCK FACTOR KNOCKOUT MUTANTS

10.3.1 Gene Knockout Mutants in Early Active Heat-Shock Factors

Considering the high complexity of HSF genes exemplified by the 21 HSF genes in *Arabidopsis* and a relatively large number of HSF genes in other plant species, isolation of HSF knockout mutants is crucial for determining their functional roles and biological importance in plants. At present, genetic analysis is focused on loss-of-function mutants in *Arabidopsis*. T-DNA insertions in individual class-A HSF genes *AtHsfA1a* or *AtHsfA1b* had no detectable effects on the heat-shock response or on any physiological or morphological characteristics of mutant plants. However, in both mutant lines there is a complete but selective loss of DNA-binding capacity of the respective HSF [28].

Evidence suggests that AtHsfA1a and AtHsfA1b are the major heat-activated DNA-binding factors that recognize conserved HSE sequences during early stages of the heat-shock response. The deficiency of AtHsfA1a/AtHsfA1b resulted in inability to form the major heat-induced HSE–HSF-binding complex in the *hsfA1a/1b* double knockout plants. The role of AtHsfA1a as a heat stress-activated promoter-binding factor has been confirmed *in vivo* by UV-laser cross-linking to *Hsp18.2* and *Hsp70* promoters in *Arabidopsis* cells [29].

The HSF mutant analysis suggests that genetic redundancy of class-A HSF in *Arabidopsis* compensates for deficiency of a single HSF. Although the double knockout line *AthsfA1a/1b* was little affected in the thermotolerance phenotype, it exerts clear negative effects in the immediate induction of expression of HSF targets, HSP, and heat-inducible HSF genes. There is also a strong negative effect on induction of *Apx2* gene expression that has been identified as a novel HSF-controlled heat-shock gene [26], suggesting that expression of all potential HSF-target genes may be downregulated in mutant plants.

The effect on HSF–DNA-binding complex formation was only detectable in the very early phase (10 to 60 min) of heat shock, in which mRNA accumulates to maximum levels in WT but not in *AthsfA1a/1b* knockout plants. After 2 h heat shock, there is no difference in mRNA levels of HSF target genes between WT and *AthsfA1a/1b* double-mutant plants. The kinetics of mRNA accumulations indicate that loss of AtHSFA1a/1b is compensated in mutant plants by a slow but steady increase that, after 2 h heat shock, reaches the same levels as WT, where mRNA levels have already declined considerably at this time.

This indicates that AtHSFA1a or AtHSFA1b is necessary and sufficient for early onset of heat-shock gene expression at the transcriptional level, but they are not the sole regulators of the heat-shock response. In the HSF double-mutant plants, other HSFs must still be active or, alternatively, the function of AtHSFA1a/AtHSFA1b is replaced by activity of other "late-acting HSFs; this may cause slow induction of heat-shock gene transcription. In addition to the low level of transcription in *AthsfA1a/1b* plants, translational compensation seems to take place, as indicated by a large discrepancy between strongly impaired mRNA accumulation and substantial amounts of sHSP accumulating after 2 h heat shock. This may explain why acquisition of thermotolerance is only marginally affected in mutant plants. Impaired thermotolerance could be detected only by electrolyte leakage assay after a severe heat pulse.

At present, there is no evidence that a single HSF dominates regulation of the heat-shock response in *Arabidopsis*. At least two genes, *AtHsfA1a* and *AtHsfA1b*, previously identified as positive transcriptional regulators of heat-shock gene expression (e.g., by a gain of function upon transgenic overexpression [16,17]) seem to be necessary for immediate early regulation of the heat-shock response. In contrast to vertebrate cells, which clearly show deficiencies in the heat-shock response when *Hsf1* or *Hsf3* genes have been disrupted [30–32], plants may have a stronger backup and more complex control of the stress response system.

In *AthsfA1a/1b* plants, differences in heat-induced mRNA levels evident for different heat-shock genes further suggest that not all heat-shock genes have the same requirements of HSF and/or synergistic effects between them. Except for *Hsp18.2*, transient accumulation of mRNA indicates that during sustained stress the transcription of heat-shock genes is negatively regulated in WT, probably via negative regulation of HSF activity. It is not known whether this negative regulation is still implemented in *AthsfA1a/1b* double mutants, but there is no indication for a transient expression. Potential negative regulators of HSF activity are HSP70 and HSP90, which are associated in a complex with HSF in its inactive form in vertebrate cells [33–36]. The interactions between HSP70 and AtHSFA1a and HSF–DNA-binding complexes suggest that HSP70 proteins may be involved in feedback regulation in plants [37].

The late DNA-binding-complexes appearing after 1 h heat shock [28] cannot result from a plain association of a negative regulator to the larger early activator complex. The lower molecular weight and strong HSE-binding activity of late complexes suggest an altered composition of participating HSF and/or other proteins. At present, no direct evidence suggests involvement of class-B HSF in formation of these complexes.

Heat-dependent expression of class-B HSFs (*AtHsfB1* and *AtHsfB2b*), which is controlled by class-A HSFs, suggests that their function is required for delayed effects of the heat-shock response. Members of class-B HSF, which lack certain structural features of the class-A activator HSF [8], are implicated as transcriptional repressors or attenuators of the heat-shock response [10,11]. However, as indicated for tomato, certain class-B HSFs may serve as coactivators of gene expression during prolonged heat stress or during recovery [38]. Isolation and analysis of further HSF knockout mutants is necessary to investigate functional roles of class-B HSFs in plants.

In conclusion, loss-of-function mutant analysis provided evidence that two transcription factors, AtHsfA1a and AtHsfA1b, are fast response regulators, which are important for coordination of stress gene expression and generation of stress tolerance under rapidly changing environmental conditions in nature. HSF functions seem to be required for fast induction of stress genes (e.g., HSP, HSF) and consequently also for timing of negative feedback regulation causing transient expression in wild-type.

10.3.2 Heat Shock and Heat-Shock Factor-Dependent Transcriptome of *Arabidopsis*

AtHsfA1a and AtHsfA1b are considered to be key regulators for the fast and strong but also transient heat-shock-induced transcription. Loss-of-function mutants of individual HSF genes—for example, in *AtHsfA1a* or *AtHsfA1b* genes—had no detectable effects on the heat-shock response. Only *AthsfA1a/AthsfA1b* double mutants (previously designated *hsf1/3*) were significantly impaired in early (during the first hour of heat shock) transient mRNA accumulation of a number of *Hsp* genes [28].

Identification of the complete set of AtHsfA1a/1b target genes was crucial for determining the functional roles and biological importance of HSFs and for understanding the molecular mechanism involved in generation of stress tolerance in *Arabidopsis*. Using the Affymetrix ATH1 microarrays to conduct whole transcriptome expression profiling, it was possible to discriminate between directly recognized AtHsfA1a/1b target genes and heat-shock-induced genes that may be regulated by other factors or secondary effects of heat shock at a very early time point (1 h heat stress) in the heat-shock response.

Analysis of the transcriptomes from heat-stressed leaves of *Arabidopsis* WT and *AthsfA1a/1b* double-knockout plants revealed differential expression of a large number of genes for the immediate early heat-shock response, but few of them were direct targets for the transcription factors AtHsfA1a/1b [39]. From the total numbers of differentially expressed genes (control vs. heat shock) in WT (2567 genes) and *AthsfA1a/1b* knockout plants (3056 genes), 112 genes were identified as potential AtHsfA1a/1b targets. Comparison of observed expression levels between microarray and qRT-PCR quantifications confirmed that all heat-shock genes tested showed clearly heat-inducible transcript levels by both methods.

In addition to several of the known HSP genes, a number of novel genes have functions in different pathways. The highest ranking unconventional heat-shock gene was *GolS1*; its mRNA is 122-fold induced by heat shock in WT and represents the highest score as an unconventional AtHsfA1a/1b-dependent heat-shock gene. The *GolS1* mRNA levels in WT and *AthsfA1a/1b* knockout plants were confirmed by Northern hybridizations and qRT-PCR [40]. General comparison of microarray and qRT-PCR quantification data indicates a higher probability for genes to score as a differentially expressed gene in qRT-PCR analysis than in chip hybridizations. The numbers of differentially expressed genes detected by chip hybridizations represent a minimal but reliable set of genes regulated by heat shock and HSF, respectively, due to the stringent screening parameters applied.

Microarray expression analysis of long-term heat-shock treatment (6 h, 37°C) of *Arabidopsis* WT plants resulted in approximately tenfold lower numbers (262 upregulated, 279 downregulated) of differentially expressed genes [41]. When the same cut-off (1.5-fold $\log^2$) was applied in experiments after 1 h heat stress [39], the number of differentially expressed genes was still two- to fourfold higher (576 upregulated, 1116 downregulated). These differences reflect the fact that expression of many heat-shock-upregulated genes is transient, with a maximum after 1 to 2 h heat shock followed by strong decline [28]. The larger number of downregulated genes is probably the result of transient transcriptional repression of many non-heat-shock genes [42]. The long-term heat shock may lead to adjustment of the steady-state levels of mRNAs to control levels for a large number of transiently up- or downregulated genes. It should be noted that, in addition to the duration of heat shock, several other experimental parameters may have profound effects on gene expression.

10.4 IDENTIFICATION OF HEAT-SHOCK FACTOR TARGET GENES

10.4.1 AtHsfA1a/1b-Dependent Genes

The criterion for AtHsfA1a/1b-dependent expression of genes was a significant quantitative difference of heat-induced mRNAs between WT and *AthsfA1a/1b* mutant plants. The 112 AtHsfA1a/1b-dependent genes represent only a small fraction of heat-shock-regulated WT genes [39]. The majority (105 genes) is downregulated in the mutant and thus activated by AtHsfA1a/1b. This confirmed classification of these HSFs as transcriptional activators [16,17,28]. Both belong to the class-A HSF, which contains members for initiation of the heat-shock response.

The majority of AtHsfA1a/1b-upregulated target genes belong to the group of the most strongly induced heat-shock genes in WT. However, there is no correlation with the presence of HSE sequences in the promoter regions of those genes [39]. Whereas 47% of the AtHsfA1a/1b-regulated genes contain HSE sequences, no preferential representation is found among highly ranked heat-shock- or HSF-regulated genes. The other HSF-regulated genes contain variant HSEs, which are known to bind HSF *in vitro* [8] and may also function as AtHSFA1a/1b binding sites *in vivo*.

The presence of HSE is not a sufficient criterion for predicting heat-shock- or HSF-regulated expression in plants. Perfect and slightly altered HSE sequences have been identified in the promoter regions of all 21 HSF genes of *Arabidopsis* [8], but only the expression of 6 of them is regulated by heat shock and/or HSF [39]. A genome-wide analysis of the mammalian heat-shock response has also shown no strict correlation among the presence of HSE, HSF1-binding, and heat-induced transcription. However, they are independent post-transcriptional mechanisms that regulate accumulation of a significant number of heat-shock-elevated transcripts [43].

In the analysis of Busch et al. [39], many of the highly ranking heat-shock-induced genes have been annotated as heat-stress-related genes. A total of 111 genes represented on the chip is related to heat-shock/HSP functions; expression of 44 of those genes was induced by heat shock and 10 of them appeared to depend on AtHsfA1a/1b. Within the family of sHSP, which comprises 18 members, induction levels of six of the heat-shock-inducible sHSP genes (37.5%) clearly depended on AtHsfA1a/1b. Expression of nine other sHSP genes was also lower, but not significantly impaired in the *AthsfA1a/1b* mutant. Two sHSP genes, *Hsp15.4-CI(r)* and *Hsp14.2-P(r),* which are only distantly related to the other sHSP genes [44], were not activated by heat shock; *Hsp15.4-CI(r)* was downregulated. This may indicate that these two genes/proteins are related sHSP isoforms, which are not required under heat-shock conditions in leaves.

Other HSF-dependent genes include genes involved in protein biosynthesis/degradation, membrane transport, oxidative stress response, and signaling [39]. Involvement of ABA, SA, ethylene, and oxidative burst signaling in acquisition of thermotolerance has been implicated by studies using the respective signaling mutants [45,46]. The biological function of most of these genes during heat shock is not known; however, some have been connected with environmental stress responses. Another striking connection exists to N-myristoylation, which has been described as required for SOS3 function in plant salt tolerance [47]. The involvement of these processes in plant protection is not unexpected; however, these are not only pathways and mechanisms dealing with protein synthesis and stability but also direct and early targets of HSF in the heat-shock response.

The strong but not strict correlation between high induction levels and AtHsfA1a/1b-dependent expression suggests that other factors and mechanisms that cause strong induction of gene expression must be activated upon heat shock. This can also be predicted for expression of other heat-shock-activated, but not AtHsfA1a/1b-dependent, genes in WT.

Different HSFs may cooperate in immediate early expression of different HSF target genes. Expressions of AtHsfA1a/1b-dependent genes rely exclusively on these HSFs during early stages of the heat-shock response. Genes that scored as AtHsfA1a/1b independent in expression profiling use other as yet unknown HSFs for induction of transcription. AtHsfA1a and/or AtHsfA1b may also participate in expression of some of these genes. The importance of AtHsfA1a/1b for initiation of HSP expression is supported by the gain of function effects of HSF overexpression in transgenic plants, which resulted in constitutive expression of a number of heat-shock genes tested [16,17].

Only 11 genes showed HSF-dependent expression at normal temperature [39]. Surprisingly, basal levels but not heat-induced expression of these genes was negatively affected in the *AthsfA1a/1b* plants. This indicates that HSFs are involved in basal expression for a small number of heat-shock genes but may become replaced by other HSFs upon heat stress.

10.4.2 AtHsfA1a/1b-Regulated Expression of Other Transcription Factors during Heat Stress

AtHsfA1a/1b seems to regulate expression of a small number of transcription factor genes and hence seems also to be responsible for secondary changes in gene expression after heat shock. Most striking is the AtHsfA1a/1b-dependent expression of other HSFs, in particular *AtHsfB1*, *-B2a*, and *-A7a*, which represent three out of six heat-inducible HSF genes. qRT-PCR experiments offer evidence for AtHsfA1a/1b-dependent expression of *AtHsfB1* and *AtHsfB2b (HSF7),* which are heat shock induced in WT but to lower levels in *AthsfA1a/1b* plants [28]. Both HSFs are members of the class-B factors, which lack HSF-typical transcription activator domains.

It is interesting to note that *AtHsfB1* mRNA induction has been identified by expression profiling analyzing the wounding response in *Arabidopsis* [48]. Identification of *AtHsfA7a* (class A) as an HSF-dependent gene indicates that not only class-B factors are involved in controlling delayed functions in the heat-shock response. After long-term heat shock, the steady-state levels of HSF mRNAs [41] differ significantly from the profile after short-term treatment [39]. Of the early heat-shock-induced HSFs, only the mRNAs of AtHsfB1, AtHsfA2, and AtHsfC are present at enhanced levels after long-term heat shock. Alterations in the expression profiles of HSF indicate that AtHsfA1a/1b-dependent regulation of HSF expression is transient and that other HSFs and/or regulatory mechanisms are involved in the sustained expression of certain HSFs during continuous heat shock.

In addition to HSF genes, other transcription factors previously connected to stress responses are controlled by AtHsfA1a/1b. WRKY7 is connected to disease resistance [49]; scarecrow factors are involved in developmental processes but are also affected in expression by salt stress in *Arabidopsis* [50] and osmotic stress in white spruce [51]. Dof zinc finger proteins have been associated with stress response [52,53] and with control of seed germination [54]. Members of the AP2/EREBP factor family play a variety of roles throughout the plant life cycle; they are implicated in regulation of developmental processes and environmental stress responses [55]. All DRE-binding proteins identified to date belong to the AP2/EREBP family; the factors DREB1 and DREB2 are involved in regulation of drought-, salt-, and cold-inducible genes [56].

Other examples are IAA2, a transcription factor involved in auxin signaling [57], and AtERF4, an active repressor of transcription shown to be induced by wounding, cold, high salinity, and drought stress [58]. Interestingly, both genes and two other signaling related transcription factors are upregulated in *AthsfA1a/1b* knockout plants.

It is striking that expression of relatively few environmental stress-related transcription factor genes of large multigene families appears to be affected by heat shock and/or AtHsfA1a/1b. The majority of the heat-shock-regulated genes of these families are downregulated upon heat shock and only one gene of each family appears to be regulated by HSF. This demonstrates that alterations in gene expression in the heat-shock response are dominated by HSF. Secondary effects, based on HSF-dependent expression of other transcription factors, seem to play a minor role. The 21 AtHsf family members represent a very small fraction of the more than 1500 potential transcription factor genes of *Arabidopsis* [59].

10.4.3 AtHsfA1a/1b-Controlled Pathway for Raffinose Family Oligosaccharides

A major category of AtHsfA1a/1b-regulated genes concerns enzymes involved in carbohydrate metabolism. All enzymes required for synthesis of RFO precursors leading to galactinol and RFO synthesis are members of gene families. Expression of at least one member of each family was found to be positively affected by heat shock, several of them in an AtHsfA1a/1b-dependent fashion.

Two enzymes of this pathway were previously linked to stress responses. Myo-inositol-1-phosphate-synthase was shown to be induced by cold, drought, and salt stress in *Arabidopsis* [60] and involved in salt stress response and osmoprotection in the halophyte *Mesembryanthemum crystallinum* [61]. Two of seven galactinol synthase genes, *GolS1 and GolS2*, were heat inducible, but only *GolS1* was activated via AtHsfA1a/1b. Taji et al. [62] had shown that both genes are induced by drought and high salinity stress, but the expression was obviously not regulated by the transcription factor DREB.

It will be interesting to see whether HSFs are involved in induction by drought and salt stress, which would suggest that HSFs integrate different signaling pathways that lead to expression of *GolS1*. HSF-dependent expression of *GolS1* is correlated with an increase in the level of RFO and knockout mutations of *GolS1* are unable to accumulate stress-induced galactinol and raffinose levels in leaves [40]. It has been shown that enhanced levels of galactinol and raffinose generated in leaves of transgenic plants by overexpression of GolS2 caused improved drought tolerance of *Arabidopsis* [62].

This is the first example of a highly adjustable pathway in carbohydrate metabolism that can be alternatively driven by a set of stress-responsive genes. The expression/function of these genes overlaps between different environmental responses that require protection by osmoprotective solutes. In contrast to the pathway leading to RFO, in many cases, heat-shock-/HSF-dependent expression affects only single genes of complex biochemical pathways.

10.5 GENERAL ASPECTS OF EXPRESSION PROFILING FOR IDENTIFICATION OF TARGET GENES

10.5.1 Experimental Design Affecting Detected Target Gene Expression

Microarray expression profiling is frequently used as a tool in identifying stress-responsive genes. A number of analyses have identified sets of heat-shock-activated or -downregulated genes in *Arabidopsis*. In all analyses, heat-shock-upregulated genes include HSP and HSF genes, but representation of these genes is inconsistent. These differences are demonstrated by the analysis of available data sets. Figure 10.1 and Figure 10.2 show comparisons of expression profiles of sHSP (Figure 10.1) and HSF (Figure 10.2) genes of *Arabidopsis* after one 1 h heat stress in two different sets of experiments. The expression data were analyzed by comparing experiments conducted by Busch et al. [39] on cut leaves of wild-type *Arabidopsis thaliana* (Wassilewskija) after applying heat shock in buffer and data available about heat shock applied to the whole shoot of *Arabidopsis thaliana* (Columbia-0) in *AtGen-Express* (http://www.arabidopsis.org/servlets/Tair Object).

The profiles determined for sHSP and HSF are completely different. The heat-shock induction of sHSP is clearly obvious for 15 out of 18 genes in Wassilewskija

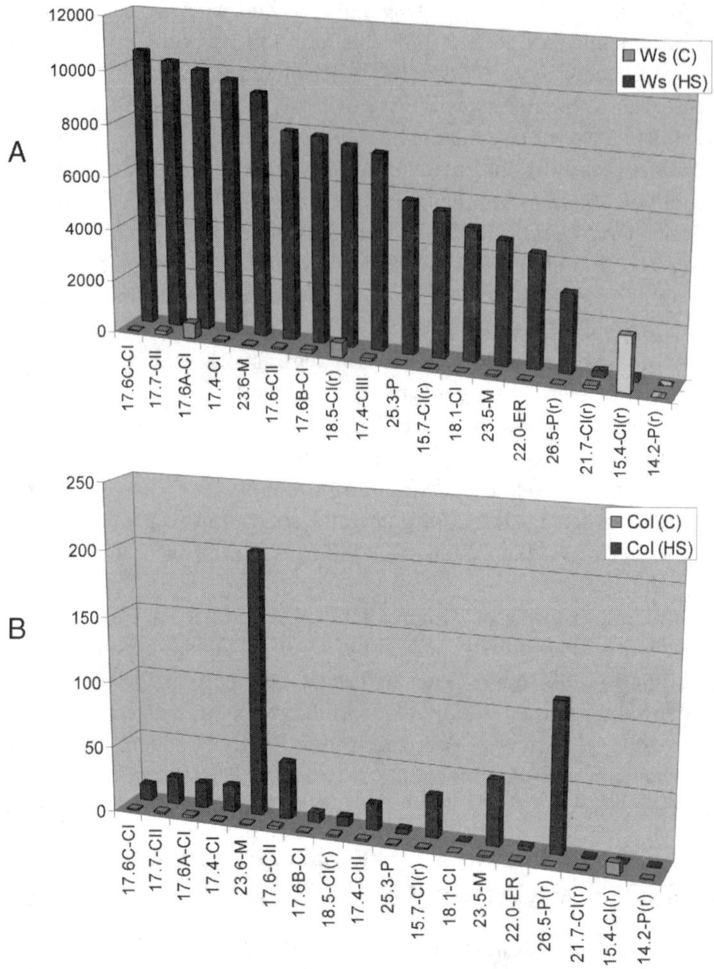

FIGURE 10.1 Expression estimates of *Arabidopsis* small heat-shock proteins (sHSP). The expression levels are measured in relative units representing the normalized signals of chip hybridization analysis. The data were retrieved from chip hybridizations obtained with mRNA probes from *Arabdiopsis* ecotypes. (A) leaves of Wassilewskija (reprinted from Busch, W. et al., *Plant J.*, 41, 1, 2005, with permission from Blackwell Publishing); shoot tissue Columbia-0 (*AtGenExpress,* http://www.arabidopsis.org/servlets/Tair Object), which had been (B) incubated under control conditions (C) or subjected to 1 h heat stress (HS).

leaves, but very few of these genes reach high expression levels in Columbia-0 shoots. However, heat-induced expression of 18 sHSP genes and heat-stress-dependent down-regulation of *sHsp15.4-C* are common to both analyses. Comparison of HSF profiles shows that 6 out of 21 HSF genes are heat shock induced in Wassilewskija, whereas in Columbia-0 shoots, almost no induction of *AtHsfB1* and *AtHsfA4* and a strong induction of *AtHsfA7B* are observed. Despite these quantitative differences, the heat

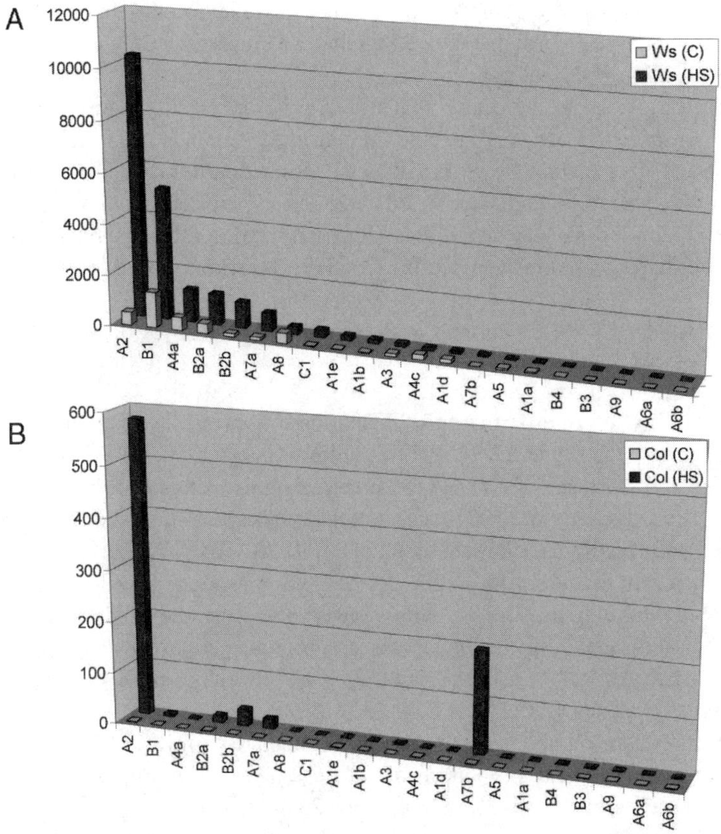

FIGURE 10.2 Expression estimates of *Arabidopsis* heat-shock factors (HSF). The expression levels are measured in relative units representing the normalized signals of chip hybridization analysis. The data were retrieved from chip hybridizations obtained with mRNA probes from *Arabdiopsis* ecotypes. (A) leaves of Wassilewskija (reprinted from Busch, W. et al., *Plant J.*, 41, 1, 2005, with permission from Blackwell Publishing) and shoot tissue of Columbia-0 (*AtGenExpress,* http://www.arabidopsis.org/servlets/Tair Object) (B) incubated under control conditions (C) or subjected to 1 h heat stress (HS).

stress induction of at least four (*AthsfA2, -B2a, -B2b, -A7a*) out of a total of seven upregulated genes is evident in both analyses. In Wassilewskija, only three out of six heat-shock-induced HSF genes are controlled by AtHsfA1a/1b [39]. Thus far, no data are available for target genes of these HSFs in Columbia-0.

These discrepancies, exemplified for two multigene families, demonstrate that differences in experimental conditions, developmental stage, plant tissue, and, possibly, genotype have an enormous impact on expression of genes during heat shock. Discrepancies in expression profiles of heat-shock-regulated genes are also evident for other genes and gene families. It should be noted that differences in processing and filtering of expression data sets may also influence the significance of detection, especially, if data sets are compared on a fold-change basis only. In our comparison, we used two data sets normalized by RMA and we emphasize only major differences in the expression levels of genes.

The conclusion can be drawn that the expression profiles determined for heat shock in *Arabidopsis* are very specific and strongly dependent on the experimental design. Such data cannot be extrapolated to other conditions and ecotypes. Even within a given ecotype and with strictly defined experimental conditions, it is not possible to identify primary target genes for transcription factors such as HSF if data are not validated by expression profiling of HSF knockout mutants.

10.5.2 AtHsfA1a/1b Target Genes Exhibit Two Different Expression Profiles

The availability of microarray mRNA expression profiling data of *Arabidopsis thaliana*, ecotype Col 0 (*AtGeneExpress*), allows analysis of expression profiles for heat stress and recovery of AtHsfA1a/1b target genes in this ecotype. The analysis (Figure 10.3) is based on 105 genes upregulated upon heat shock in wild-type compared to *AthsfA1a/1b* knockout plants [39]. It was possible to identify two different types of profiles within this group. The subgroup A profile shows a strong induction up to 3 h heat stress with slowly declining expression levels during subsequent recovery. Only a few genes are represented in this group, including *AtHsp22.0-ER, AtHsp18.1-Cl, and AtHsp25.3-P.* The subgroup B profile comprises the majority of genes that exhibit a biphasic profile.

Expression during heat stress is rapidly induced, too, followed by a fast decline during recovery but with a second peak after 12 h recovery. Representatives of the profile-B genes are: *AtHsp 26.5-P(r), AtHsp23.6-M, AtHsp15.7-CL(r), AtHsfA7A*, and *AtHsfB1*. Further analysis revealed a number of other genes with similar profiles A and B that are not AtHsfA1a/1b-dependent genes in *Arabidopsis thaliana*, ecotype Wassilewskija. Among such genes are other sHSP genes (profile A) and *AtHsfA2* and *Apx2* (profile B).

Provided that AtHsfA1a/1b–dependent target genes are the same in both ecotypes, the two different kinetics (A, B) of HSF-dependent genes and the existence of similar profiles of genes that appear to be independent AtHsfA1a/1b in heat shock-induced expression have two major implications. First, similarity in expression profiles cannot be taken as a criterion for HSF control of target gene expression. Second, differences in expression profiles are not a strict criterion for excluding a given gene from the group of HSF-regulated genes.

This analysis underscores the strict requirement of analyzing expression profiles of gene knockout mutations in respective transcription factors (e.g., HSF) for reliable identification of stress-related target genes and pathways.

10.6 GENERAL CONCLUSIONS

The analysis of HSF gain- and loss-of-function mutants has shown that the heat-shock response is a complex regulatory system, with primary and secondary target genes. Overexpression of activator HSF results in a general increase of heat-shock

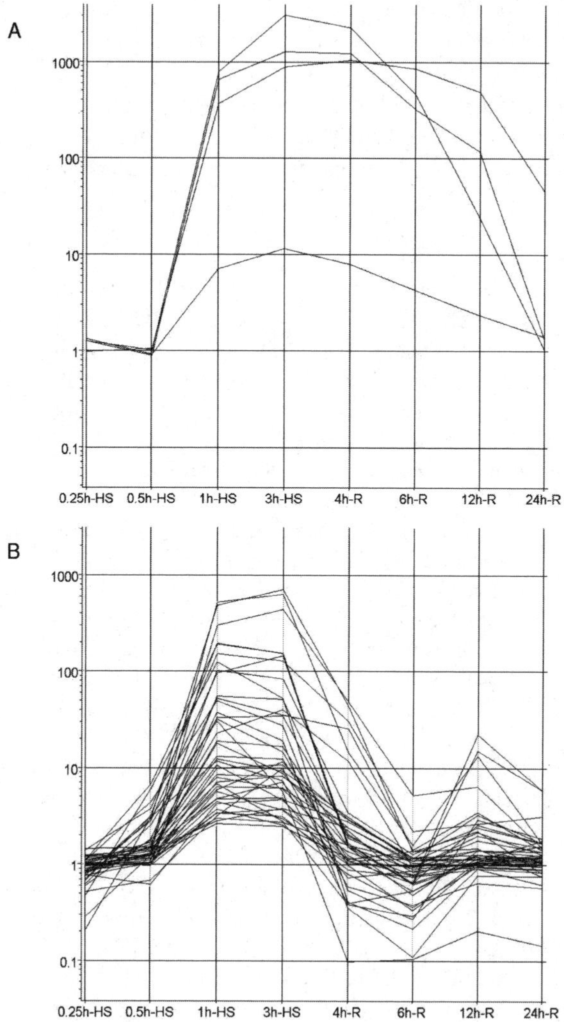

FIGURE 10.3 Expression estimates of *Arabidopsis* HsfA1a/1b target genes. Subgroup A (A) and subgroup B (B) profiles at different time points of heat stress (HS) and subsequent recovery (R). The expression levels are measured in relative units representing the normalized signals of chip hybridization analysis. The data were retrieved from chip hybridizations obtained with mRNA probes from shoot tissue of *Arabdiopsis* ecotype Columbia-0 (*AtGen-Express,* http://www.arabidopsis.org/servlets/Tair Object).

gene expression at normal temperatures and enhanced basal thermotolerance in plants. Thus, HSF manipulation appears to be a suitable target for creation of thermotolerant plants. At present, it is not known whether there is a penalty on growth and yield of plants that constitutively express a large number of stress proteins. Expression analysis has shown that thousands of genes are up- and downregulated by heat stress and more than 100 genes are putative targets of two early active HSF. Two scenarios can be envisioned for future breeding of stress tolerant plants:

HSF-manipulation that causes more stable performance and predictable yield of agricultural plants under adverse environmental conditions such as increased global climate warming

Selective manipulation of primary or secondary HSF target genes for special stress-dependent pathways and functions, e.g. certain chaperones, osmolytes, carbohydrate transport, ROS scavenging, or pathogen resistance genes.

REFERENCES

1. Morimoto, R.I., Tissieres, A., and Georgopoulos, C. Progress and perspectives on the biology of heat shock proteins and molecular chaperones, in *The Biology of Heat Shock Proteins and Molecular Chaperones*. Morimoto, R.I., Tissieres, A., and Georgopoulos, C. (Eds.). Cold Spring Harbor Laboratory Press, Cold Spring Harbor, New York, 1, 1994.
2. Schöffl, F., Prändl, R., and Reindl, A. Regulation of the heat shock response. *Plant Physiol.,* 117, 1135, 1998.
3. Wiederrecht, G. et al. The *Saccharomyces* and *Drosophila* heat shock transcription factors are identical in size and DNA-binding properties. *Cell*, 48, 507, 1987.
4. Sorger, P.K. and Pelham, H.R.B. Yeast heat shock factor is an essential DNA-binding protein that exhibits temperature-dependent phosphorylation. *Cell,* 54, 855, 1988.
5. Wu, C. Heat shock transcription factors: structure and regulation. *Annu. Rev. Cell Dev. Biol.*, 11, 441, 1995.
6. Nover, L. et al. The HSF world: classification and properties of plant heat stress transcription factors. *Cell Stress Chaperones*, 1, 215, 1996.
7. Schöffl, F. and Prändl, R. Derepression of the heat shock protein synthesis in transgenic plants, in *Plant Responses to Environmental Stress.* Smallwood, M., Calvert, C., and Bowles, D. (Eds.). BIOS Scientific Publishers Limited, Oxford, 65, 1999.
8. Nover, L. et al. *Arabidopsis* and the heat stress transcription factor world: how many heat stress transcription factors do we need? *Cell Stress Chaperones*, 6, 177, 2001.
9. Doring, P. et al. The role of AHA motifs in the activator function of tomato heat stress transcription factors HsfA1 and HsfA2. *Plant Cell*, 12, 265, 2000.
10. Czarnecka-Verner, E. et al. Plant heat shock transcription factors: positive and negative aspects of regulation. *Acta Physiol. Plant*, 19, 529, 1997.
11. Czarnecka-Verner, E. et al. Plants contain a novel multimember class of heat shock factors without transcriptional activator potential. *Plant Mol. Biol.*, 43, 459, 2000.
12. Boscheinen, O. et al. Heat stress transcription factors from tomato can functionally replace HSF1 in the yeast *Saccharomyces cerevisiae*. *Mol. Gen. Genet.*, 255, 322, 1997.
13. Burke, J.J., O'Mahony, P.J., and Oliver, M.J. Isolation of *Arabidopsis* mutants lacking components of acquired thermotolerance. *Plant Physiol.*, 123, 575, 2000.
14. Hong, S.W. and Vierling, E. Mutants of *Arabidopsis thaliana* defective in the acquisition of tolerance to high temperature stress. *Proc. Natl. Acad. Sci. USA*, 97, 4392, 2000.
15. Yamanouchi, U. et al. A rice spotted leaf gene, Spl7, encodes a heat stress transcription factor protein. *Proc. Natl. Acad. Sci. USA*, 99, 7530, 2002.
16. Lee, J.H., Hübel, A., and Schöffl, F. Derepression of the activity of genetically engineered heat shock factor causes constitutive synthesis of heat shock proteins and increased thermotolerance in transgenic *Arabidopsis*. *Plant J.*, 8, 603, 1995.

17. Prändl, R. et al. HSF3, a new heat shock factor from *Arabidopsis thaliana*, derepresses the heat shock response and confers thermotolerance when overexpressed in transgenic plants. *Mol. Gen. Genet.*, 258, 269, 1998.
18. John, A., Smith, S.T., and Jaynes, J.B. Inserting the Ftz homeodomain into the engrailed creates a dominant transcriptional repressor that specifically turns off Ftz target genes *in vivo*. *Development*, 121, 1801, 1995.
19. Conlon, S.G. et al. Inhibition of Xbra transcription activation causes defects in mesodermal patterning and reveals autoregulation of Xbra in dorsal mesoderm. *Development*, 122, 2427, 1996.
20. Lyon, J.J. and Watson, R.J. Interference of Myb transactivation activity by a conditional dominant negative protein: functional interference in a cytotoxic T-cell line results in G_1 arrest. *Gene*, 182, 123, 1996.
21. Jaynes, J.B. and O'Farrel, P.H. Active repression of transcription by the engrailed homeodomain protein. *EMBO J.*, 10, 1427, 1991.
22. Han, K. and Manley, J.L. Functional domains of the *Drosophila* engrailed protein. *EMBO J.,* 12, 2723, 1993.
23. Markel, H., Chandler, J., and Werr, W. Translational fusions with the engrailed repressor domain efficiently convert plant transcription factors into dominant negative functions. *Nucleic Acids Res.*, 30, 4709, 2002.
24. Wunderlich, M., Werr, W., and Schöffl, F. Generation of dominant negative effects on the heat shock response in *Arabidopsis thaliana* by transgenic expression of a chimeric HSF1 protein fusion construct. *Plant J.*, 35, 442, 2003.
25. Jedlicka, P., Mortin, M.A., and Wu, C. Multiple functions of *Drosophila* heat shock factor *in vivo*. *EMBO J.*, 16, 2452, 1997.
26. Panchuck, I., Volkov, R., and Schöffl, F. Heat stress and HSF-dependent expression of ascorbate peroxidase in *Arabidopsis*. *Plant Physiol.*, 129, 838, 2002.
27. Mishra, S.K. et al. In the complex family of heat stress transcription factors, HsfA1 has a unique role as master regulator of thermotolerance in tomato. *Genes Dev.*, 16, 1555, 2002.
28. Lohmann, C. et al. Two different heat shock transcription factors regulate immediate early expression of stress genes in *Arabidopsis*. *Mol. Gen. Genom.*, 271, 11, 2004.
29. Zhang, L. et al. Heat stress-dependent DNA binding of *Arabidopsis* heat shock transcription factor HSF1 to heat shock gene promoters in *Arabidopsis* suspension culture cells *in vivo*. *Biol. Chem.*, 384, 959, 2003.
30. McMillan, D.R. et al. Targeted disruption of heat shock transcription factor 1 abolishes thermotolerance and protection against heat-inducible apoptosis. *J. Biol. Chem.*, 273, 7523, 1998.
31. Tanabe, M. et al. Disruption of the HSF3 gene results in the severe reduction of heat shock gene expression and loss of thermotolerance. *EMBO J.*, 17, 1750, 1998.
32. Xiao, X. et al. HSF1 is required for extra-embryonic development, postnatal growth and protection during inflammatory responses in mice. *EMBO J.*, 18, 5943, 1999.
33. Baler, R., Zou, J., and Voellmy, R. Evidence for a role of Hsp70 in the regulation of the heat shock response in mammalian cells. *Cell Stress Chaperones*, 1, 33, 1996.
34. Mosser, D.D., Duchaine, J., and Massie, B. The DNA-binding activity of the human heat shock transcription factor is regulated *in vivo* by hsp70. *Mol. Cell Biol.*, 13, 5427, 1993.
35. Satyal, S.H. et al. Negative regulation of the heat shock transcriptional response by HSBP1. *Genes Dev.*, 12, 1962, 1998.
36. Zou, J. et al. Repression of heat shock transcription factor HSF1 activation by HSP90 (HSP90 complex) that forms a stress-sensitive complex with HSF1. *Cell*, 94, 471, 1998.

37. Kim, B.H. and Schöffl, F. Interaction between *Arabidopsis* heat shock transcription factor 1- and 70 k-Da heat shock proteins. *J. Exp. Bot.*, 53371, 2002.
38. Bharti, K. et al. Tomato heat stress transcription factor HsfB1 represents a novel type of general transcription co-activator with a histone-like motif interacting with the plant CREB binding protein ortholog HAC1. *Plant Cell*, 16, 1521, 2004.
39. Busch, W., Wunderlich, M., and Schöffl, F. Identification of novel heat shock factor dependent genes and biochemical pathways in *Arabidopsis thaliana*. *Plant J.*, 41, 1, 2005.
40. Panikulangara, T.J. et al. *Galactinol synthase 1*, a novel heat-inducible and HSF-target gene responsible for heat-induced synthesis of raffinose family oligosaccharides in *Arabidopsis*. *Plant Physiol.*, 136, 3148, 2004.
41. Rizhsky, L. et al. When defense pathways collide. The response of *Arabidopsis* to a combination of drought and heat stress. *Plant Physiol.*, 134, 1683, 2004.
42. Schöffl, F., Rossol, I., and Angermüller, S. Regulation of the transcription of heat shock genes in nuclei from soybean (Glycine max) seedlings. *Plant Cell Environ.*, 10, 113, 1987.
43. Trinklein, N.D. et al. The role of heat shock transcription factor 1 in the genome-wide regulation of the mammalian heat shock response. *Mol. Biol. Cell.*, 15, 1254, 2004.
44. Scharf, K.D., Siddique, M., and Vierling, E. The expanding family of *Arabidopsis thaliana* small heat stress proteins and a new family of proteins containing a-crystallin domains (Acd proteins). *Cell Stress Chaperones*, 6, 225, 2001.
45. Clarke, S.M. et al. Salicylic acid dependent signaling promotes basal thermotolerance but is not essential for acquired thermotolerance in *Arabidopsis thaliana*. *Plant J.*, 38, 432, 2004.
46. Larkindale, J. et al. Heat stress phenotypes of *Arabidopsis* mutants implicate multiple signaling pathways in the acquisition of thermotolerance. *Plant Physiol.*, 138, 882, 2005.
47. Ishitani, M. et al. SOS3 function in plant salt tolerance requires N-myristoylation and calcium binding. *Plant Cell*, 12, 1667, 2000.
48. Cheong, Y.H. et al. Transcriptional profiling reveals novel interactions between wounding, pathogen, abiotic stress, and hormonal responses in *Arabidopsis*. *Plant Physiol.*, 129, 661, 2002.
49. Maleck, K. et al. The transcriptome of *Arabidopsis thaliana* during systemic acquired resistance. *Nat. Genetic.*, 26, 403, 2000.
50. Ueda, A. et al. Analysis of salt-inducible genes in barley roots by differential display. *J. Plant Res.*, 115, 119, 2002.
51. Stasolla, C. et al. The effects of polyethylene glycol on gene expression of developing white spruce somatic embryos. *Plant Physiol.*, 131, 49, 2003.
52. Chen, W., Chao, G., and Singh, K. The promoter of a H_2O_2-inducible, *Arabidopsis* glutathione S-transferase gene contains closely linked OBF- and OBP-binding sites. *Plant J.*, 10, 955, 1996.
53. Kang, H.G.W. and Singh, K. Characterization of salicylic acid-responsive, *Arabidopsis* Dof domain proteins: overexpression of OBP3 leads to growth defects. *Plant J.*, 21, 329, 2000.
54. Papi, M. et al. Identification and disruption of an *Arabidopsis* zinc finger gene controlling seed germination. *Genes Dev.*, 14, 28, 2000.
55. Riechmann, J.L. and Meyerowitz, E.M. The AP2/EREBP family of plant transcription factors. *Biol. Chem.*, 379, 633, 1998.
56. Kizis, D., Lumbreras, V., and Pagès, M. Role of AP2/EREBP transcription factors in gene regulation during abiotic stress. *FEBS Lett.*, 498, 187, 2001.

57. Reed, J.W. Roles and activities of Aux/IAA proteins in *Arabidopsis. Trends Plant. Sci.*, 6, 420, 2001.
58. Fujimoto, S.Y. et al. *Arabidopsis* ethylene-responsive element binding factors act as transcriptional activators or repressors of GCC box-mediated gene expression. *Plant Cell*, 12, 393, 2000.
59. Riechmann, J.L. et al. *Arabidopsis* transcription factors: genome-wide comparative analysis among eukaryotes. *Science*, 290, 2105, 2000.
60. Kreps, J.A. et al. Transcriptome changes for *Arabidopsis* in response to salt, osmotic, and cold stress. *Plant Physiol.*, 130, 2129, 2002.
61. Ishitani, M. et al. Coordinate transcriptional induction of myo-inositol metabolism during environmental stress. *Plant J.*, 9, 537, 1996.
62. Taji, T. et al. Important roles of drought- and cold-inducible genes for galactinol synthase in stress tolerance in *Arabidopsis thaliana. Plant J.*, 29, 417, 2002.

11 Improving Low-Temperature Tolerance in Plants

Markku K. Aalto, Pekka Heino, and E. Tapio Palva

CONTENTS

11.1 INTRODUCTION

Throughout their growth period, plants are exposed to abiotic and biotic stresses. This has led to plants' development of adaptive strategies to survive these stresses. Low temperature is one of the most important abiotic factors limiting growth, productivity, and distribution of plants [1,2]. Low temperature decreases biosynthetic activity of plants, inhibits normal function of physiological processes, and may cause permanent injuries leading to death.

Plants vary widely in their ability to tolerate low temperature [3]. Chilling-sensitive tropical species can be severely damaged even at temperatures significantly higher than freezing [4]. Chilling-tolerant but freezing-sensitive plants are able to survive temperatures somewhat below 0°, but are damaged upon ice formation in the tissues. On the other hand, freezing-tolerant plants are able to survive variable levels of freezing temperatures; the degree of tolerance depends on species, developmental stage, and duration of stress [3,4].

Chilling and freezing are stresses that affect the germination and seedling stages of the plants during early summer, as well as flowering, fruiting, and maturing of the crop during the growth period. Furthermore, because the successful minimization of damage during overwintering is the prerequisite for efficient growth the next season, low-temperature stress tolerance during winter is important for perennials.

Increasing low-temperature tolerance is not only important for securing crops against sudden periods of colder climate but also provides a possibility of extending the present geographical limits of cultivated varieties, and even species, to colder climates. Another very important application is the increased cold tolerance extending the storage period of fruits and vegetables [5].

Increased cold tolerance appears to be antagonistic to optimal growth, so it is important to analyze cost and benefits in crop plant breeding and find the perfect combination of preferred traits [6]. This can only be achieved by directly testing candidate genes. It also underlines the fact that molecular biology and especially transformation offer powerful tools for breeding programs [7].

This chapter focuses on current knowledge obtained largely from transgenic experiments studying adaptive strategies used by plants to tolerate freezing stress and the genetic engineering applications that can be used to increase the tolerance.

11.1.1 Freezing Stress

Plants from temperate regions are commonly exposed to freezing temperatures seasonally and during their growth season. Plants encountering freezing temperatures have two general strategies to survive freezing stress: avoidance or tolerance of freezing [2]. The former is mainly achieved by supercooling tissue water. However, this mechanism has limited value because it mainly occurs in special organs such as seeds, overwintering buds, or parenchymal cells of temperate trees [8,9]. Therefore, tolerance of freezing is the dominant mechanism by which plants survive freezing stress.

Exposure of plants to subzero temperatures leads to freezing of tissue water. Due to the higher freezing point and presence of more active ice nucleators in apoplastic solution compared to the cytoplasm, freezing invariably occurs extracellularly. Ice formation outside the cells reduces water potential of the apoplastic solution, which leads to withdrawal of water from the cells and subsequent cellular dehydration. Therefore, on a cellular level, freezing stress also leads to dehydration stress; consequently, tolerance of freezing is correlated to tolerance of dehydration [2].

Cellular membranes appear to be the main target for freezing injuries [10]. Freeze-induced dehydration can cause different types of structural perturbations on membranes. These include membrane fusions and lamellar to hexagonal II phase transitions [11]. In addition to dehydration, other factors also contribute to the damage. Growing ice crystals can cause mechanical damage and low temperature per se can have direct effects on cellular processes due to denaturation of proteins and disruption of macromolecular complexes. In addition, low temperature, especially in combination with high light intensities, generally leads to increased production of reactive oxygen species (ROS), which can damage different macromolecules in the cells [12,13].

Reactive oxygen species (ROS) such as O_2, H_2O_2, and OH– are mainly formed in plastids, microsomes, and mitochondria and they damage macromolecules and membranes [14,15]. Plants have developed an elaborate repertoire of defense mechanisms to scavenge these and other radical compounds. These include specific enzymes such as catalase, superoxide dismutase (SOD), ascorbate peroxidase (APX), and glutathione reductase, as well as nonenzyme molecules such as ascorbate, glutathione, carotenoids, and anthocyanins [16,17].

Osmolytes [18,19], proteins like dehydrin [20], phenolic metabolites, and isoprenoids also function as ROS scavengers [21–24]. Membranes are protected by α-tocopherol and specific glutathione peroxidases [25–27]. Also, many sugars are active ROS scavengers; for example, mannitol is active specifically against hydroxyl radicals, the most dangerous form of activated oxygen species [28].

However, a complex pattern of antioxidant signaling and redox homeostasis involved in cell death and acclimation responses is only beginning to be understood [29]. For example, the nature of ROS formation may be different under low temperatures compared to other stress types because the gene network (regulon) contains different genes or they are differentially expressed [16].

11.1.2 Cold Acclimation

Plants species native to temperate regions need to adjust to daily and seasonal fluctuations in temperature; the seasonal acclimation is important for overwintering herbaceous and woody plants. Several plant species have the ability to increase their level of freezing tolerance in response to low nonfreezing temperatures by a process known as cold acclimation [2,3]. These species utilize environmental signals to trigger processes that lead to increased freezing tolerance.

In overwintering woody species, acclimation is a two-step process in which the first phase is triggered by shortening of the photoperiod below a critical length. This leads to growth cessation and dormancy development and is accompanied by a moderate increase in freezing tolerance. The second stage of acclimation, which is also essential for obtaining full winter hardiness, is induced by subsequent exposure to low temperatures.

In annual and in most overwintering herbaceous plants, freezing tolerance is not affected by the photoperiod but acclimation is triggered solely by low nonfreezing temperatures. In these species, cold acclimation is a rather rapid process; for example, in *Arabidopsis* a substantial increase in freezing tolerance can be achieved after 1-day exposure to low temperature, although full acclimation requires more than a week. In addition to low nonfreezing temperatures, cold acclimation can be triggered by exposing plants to moderate water stress or by exogenously applied abscisic acid (ABA) [30–32].

The level of freezing tolerance obtained through cold acclimation is not static, but is rapidly lost upon return to warm nonacclimating temperature. Cold acclimation is associated with several physiological and biochemical alterations in plants. The best characterized changes include changes in hormone levels; increases in soluble sugars, amino acids, and organic acids; and accumulation of osmoprotectants and protective proteins, as well as modification of membrane lipid composition [33–35] (see also other reviews listed later). Most of these changes are derived from altered gene expression and, although the causal relationship of many of these changes to increased freezing tolerance is still unclear, some of the genes and regulons involved in cold acclimation are being unraveled [35–38].

11.2 COLD SIGNALING IN PLANTS

11.2.1 Signal Transduction at Low Temperature

To respond to low temperature, plants must perceive the stress, transduce the signal to the nucleus, and activate expression of specific genes required for cold acclimation and metabolic adjustments needed for growth at low temperatures. The exact mechanisms by which this occurs are not clear, even though an increasing number of components involved in the signal processes have been characterized. However, the nature of the cold sensor still needs to be established.

Calcium is a ubiquitous second messenger in plants and also mediates abiotic stress signaling [39]. The specificity of signaling is derived from specific Ca^{2+}-signatures evoked by different types of stresses and generated by a combined action of Ca^{2+}-channels, -pumps, and -transporters as reviewed [40]. Transient increase in cytosolic Ca^{2+}-concentration in the early stages of cold acclimation has been shown to be necessary for development of freezing tolerance [41,42].

The source of Ca^{2+}- and the specific channels for Ca^{2+}-influx are currently unknown; however, by using pharmacological approaches, it has been demonstrated that rigidification of membranes and cytoskeletal rearrangements are needed for the Ca^{2+}-influx in alfalfa and *Brassica napus*, indicating involvement of mechanosensitive channels in this process [43,44].

Downstream signaling from the Ca^{2+}-signature is generally mediated by diverse types of Ca^{2+}-binding proteins, like calmodulin (CaM), calcineurin B-like proteins (CBLs), and Ca^{2+}-dependent protein kinases (CDPKs). The proteins are activated by Ca^{2+}-binding and, in activated forms, they regulate activity of proteins involved in generation of signal-specific responses [40,45]. The involvement of CaM and CDPKs in cold acclimation was initially demonstrated by Monroy et al. [41], who showed that treatment of alfalfa suspension cultures with CaM or CDPK antagonists resulted in inhibition of low-temperature responsive gene expression and development of freezing tolerance.

Genes encoding CaM and CaM-related proteins as well as genes encoding CDPKs have also been shown to be low temperature responsive, indicating a positive role in regulation of low-temperature tolerance [42,46,47]. However, Townley and Knight have shown that overexpression of the CaM gene in *Arabidopsis* leads to inhibition of low-temperature responsive gene expression, indicating that CaM can act as a negative regulator of cold acclimation [48]. CDPKs have also been shown to be involved in acquisition of chilling tolerance. Saijo et al. have demonstrated that overexpression of the *OsCDPK* gene leads to increased low-temperature tolerance in rice [47].

CBLs are a family of Ca^{2+}-binding proteins that mediate the activation of protein kinases called CIPKs (CBL-interacting protein kinases). A gene encoding a member of the CBL family, *CBL1*, has been shown to be responsive to abiotic stresses, including low temperature in *Arabidopsis* [49]. However, *cbl1* null mutant exhibits enhanced freezing tolerance and overexpression of *CBL1* leads to reduced level of tolerance, indicating that CBL1 acts as a negative regulator of cold responses [50]. A CIPK, CIPK3, has been demonstrated to be involved in activation of low-temperature responsive genes in *Arabidopsis* [51]. *cipk3* mutant plants exhibited delayed induction of gene expression but no differences in the maximum level of induction, indicating that CIPK3 acts at the early stages of cold acclimation [51].

Mitogen-activated protein kinases (MAPKs) are a family of proteins involved in transduction of diverse signals in eukaryotic cells. Activation of MAPKs is mediated through a phosphorylation cascade where activated MAPKKKs phosphorylate MAPKKs, which then phosphorylate MAPKs. Activated MAPKs typically phosphorylate and modulate the activity of specific transcription factors, participating in generation of signal-specific transcription patterns.

Arabidopsis harbors about 60 putative MAPKKKs, 10 MAPKKs, and 20 MAPKs, which suggests that MAPKKs form a cross-talk node between different signals mediated through the cascade. However, very little information exists concerning the components involved in transmission of specific signals [52].

11.2.2 Cold-Regulated Gene Expression

Cold acclimation is associated with marked alterations in gene expression; several of the alterations are likely to be significant for development of freezing tolerance. The cold-responsive genes are often also induced by other environmental stimuli, like drought and salt, indicating common protective mechanisms in these stresses. Several, but not all, of the low-temperature responsive genes are also induced by exogenous ABA, suggesting ABA involvement in generation of stress responses [35,53,54].

The effects of low temperature on *Arabidopsis* transcriptome have been analyzed in several different studies and the results indicate that up to 25% of the genes are responsive to cold [55–57]. The temporal patterns of low-temperature responsive gene expression are complex and the genes involved clearly belong to different regulons.

11.2.3 The CBF/DREB1 Regulon

The emerging insight in cold acclimation is that this mechanism is complex, consisting of several distinct signal pathways that appear to interact with each other [35,36,53,58–61]. Moreover, it is becoming evident that the low-temperature signal pathways are converging and interacting with other stress-induced pathways, such as those mediating dehydration and abscisic acid responses [54,62,63] (Figure 11.1).

Our earlier work demonstrated the presence of ABA-independent and ABA-mediated pathways in low-temperature signaling [64,65]. Interestingly, recent data indicate that there is cross-talk even between these two pathways of cold signaling [66].

Work done mainly by the Thomashow and Shinozaki laboratories has established the central role of the CBF or DREB1 family of transcription factors (TF) in controlling ABA-independent responses to low temperature [58,61,67,68]. Identification of these AP2/EREBP-types of DNA-binding proteins was initiated by recognizing a common binding site in many promoters of the low-temperature responsive genes.

We originally suggested the identity of such a low-temperature responsive element (LTRE) in the *LTI78/RD29A* promoter in *Arabidopsis* [65]. This 9-bp element, TACCGACAT, with a core sequence of CCGAC, was subsequently demonstrated by deletion analysis to confer responsiveness to low temperature, drought, and high salinity, but not to ABA [69]. This dehydration-responsive element designated as DRE also occurs in several other cold-responsive promoters and has been referred to as the C-repeat (CRT) [70]. The TF binding to the DRE/CRT/LTRE element and

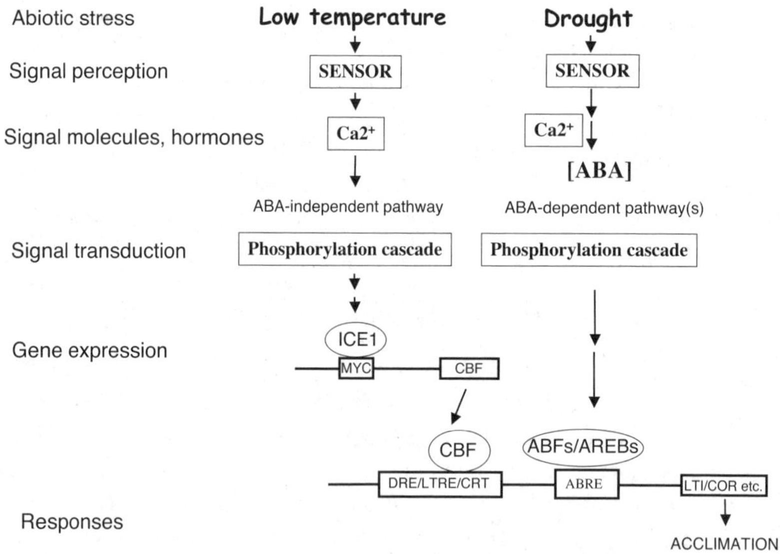

FIGURE 11.1 Simplified scheme of *Arabidopsis thaliana* signal transduction cascades involved in cold acclimation. Regulation and cross-talk are omitted for simplicity. (Picture based on Heino, P. and Palva, E.T., in *Plant Responses to Abiotic Stress*, vol. 4, Hirt, H. and Shinozaki, K., Eds., Springer-Verlag, Berlin, 2003, 151–186; and Shinozaki, K. et al., *Curr. Opin. Plant Biol.*, 6, 410–417, 2003.)

activating cold-induced gene expression was first identified by Stockinger et al. [71] and designated CBF1 (C-repeat binding factor 1).

Additional genes were subsequently isolated and shown to encode a small family of related TFs called CBF1, CBF2, and CBF3 or DREB1B, DREB1C, and DREB1A (DRE-binding protein) in *Arabidopsis* [67,68,71,72]. An additional member of this TF family, CBF4, was isolated [73] but may be more related to drought response.

Interestingly, the *CBF1-3/DREB1A-C* genes are transiently regulated by low temperature [67,68,74]. Rapid activation of *CBF* expression (within 15 to 30 min) is followed by cold-regulated target genes that define the CBF/DREB1 regulon. An efficient mutant screen by the J.-K. Zhu laboratory led to isolation of the first *ICE* (inducer of CBF) gene encoding an MYC-type bHLH transcription factor [75]. As expected, expression of *ICE1* appears constitutive.

Overexpression of *CBF/DREB1* leads to the constitutive expression of a number of target genes with promoters containing the DRE/CRT/LTRE element and to improved freezing, drought, and salt tolerance of nonacclimated plants [72,76]. In addition, overexpression of CBF3 leads to elevated levels of proline and sugars that are normally associated with cold acclimation [77]. In several recent studies, transcriptome analysis has been used to identify genes of the CBF/DREB1 regulon in *Arabidopsis* [38,55,56,78,79].

Of the hundreds of cold-regulated genes identified, 85 were upregulated by CBF2 and, of the 25 most highly cold-induced genes, the majority were under CBF2 control [38]. These studies demonstrate that the CBF/DREB1-controlled genes constitute a low-temperature responsive regulon central to the plant cold acclimation response and development of plant freezing tolerance and could be employed for engineering plant cold tolerance [80].

The conserved nature of the CBF/DREB1 regulon in cold-tolerant plants further supports the notion that this regulon has an important role in the cold acclimation process. For example, in wheat, the differences in Cbf expression were associated with variation in frost tolerance [81]. Orthologs of CBF genes have been found in a number of herbaceous species, including *Brassica napus* [82], barley [83], wheat [84], rice [85], tomato [86], sour cherry and strawberry [87], sweet cherry [88], and even woody species such as silver birch [89; Welling et. al., in preparation].

11.2.4 ABA-Mediated Gene Expression

ABA is a phytohormone that regulates diverse aspects of plant development and growth, including stress responses. Genes' responsiveness to abiotic stresses is partly mediated by pathways that are ABA dependent [53,63,90,91]. ABA-deficient mutants of *Arabidopsis* are not able to increase their freezing tolerance to wild-type levels during cold acclimation [92,93] and ABA deficiency also decreases cold-responsive gene expression [93,94]. However, the growth rate at low temperatures is not regulated by ABA but by gibberellic acid (GA) and salicylic acid (SA) [95,96].

Recently, Zhu et al. [97] identified a gene encoding HOS10, a MYB-type transcription factor needed for development of freezing tolerance in *Arabidopsis*. HOS10 appears to be required for stress induction of *NCED*, which encodes the rate-limiting enzyme in ABA biosynthesis. This indicates a role for ABA in generating protection against freeze-induced dehydration [97].

ABA-responsive gene expression has been shown to be mediated by ABFs/AREBPs, basic leucine zipper transcription factors binding to ABA response elements (ABREs) present in the promoter regions of ABA responsive genes [98,99]. Characteristic of ABREs is that more than one copy of the element is necessary for gene activation. Alternatively, a coupling element replaces one of the ABREs [100].

A low-temperature and ABA-responsive gene encoding a C_2H_2-type zinc finger protein in soybean has been characterized [101]. The encoded protein, SCOF1, regulates low-temperature responsive gene expression and development of freezing tolerance. SCOF1 was found to enhance binding of the bZIP transcription factor SGBF1 to ABREs in stress-responsive genes, suggesting that SCOF1 regulates ABA-mediated gene expression in low temperatures [101].

11.3 ENGINEERING TOLERANCE

There are many reports about genes that, when overproduced, give plants enhanced protection against low-temperature stress. However, definition of their roles as part of temperature stress tolerance mechanisms and possible overlapping functions needs to be elucidated. Indeed, an obvious explanation for the wide range of different proteins in low-temperature stress is manifested in the many different forms of stress, each with its own specific protection mechanisms evolved. This may also be the most important motivation for studying low-temperature stress for plant crop improvement purposes. In yeast, for example, the best selection strategy to obtain highly improved multiple-stress-resistant strains was found to be batch selection for freezing–thawing stress [102].

11.3.1 Target Genes

By 1985 it was established that altered gene expression at low, nonfreezing temperatures correlated with cold acclimation [103]. Subsequently, a great number of cold-induced genes has been isolated and characterized from a variety of plant species [33,34,104]. These genes code for a number of different proteins, including enzymes involved in metabolic pathways, proteins with a protective role, and proteins affecting signaling pathways. Among these are fatty acid desaturases, chaperones, lipid transfer proteins, enzymes in osmoprotectant biosynthesis, antifreeze proteins, transcription factors, kinases, and phosphatases [35,54]. Furthermore, the functions of many cold-induced proteins are not clearly defined. However, in many cases, corresponding genes have been found in a number of different plant species, indicating that many proteins induced by low temperature are conserved.

In addition to specifically induced target genes, a number of regulation mechanisms are only beginning to be unraveled. These include signaling cascades leading to specific transcription factors, as well as more global regulation of whole genomes. Thus, in addition to well-characterized regulatory cascades, including CBF/DREB genes that may have a function at the level of histone acetylation [105,106], chromosome structure is regulated following low-temperature stress [107–110].

TABLE 11.1
Expressed LEA Proteins Increasing Freezing Tolerance

LEA group	Origin	Model organism	Ref.
Group 2	CoCOR19—citrus	In tobacco, leads to a slight degrease in ion leakage during chilling and freezing stress	126
Group 2	CAP85—spinach	Small improvement in freezing tolerance of tobacco in time-course experiment	296
Group 2	Wcor410—wheat	Acidic dehydrin improves strawberry freezing tolerance	297
Group 2	COR15a—*A. thaliana*	Increases freezing tolerance in *Arabidopsis*	139
Group 3	HiC6—*C. vulgaris*	Significant suppression of chilling injury in tobacco	298
Group 3	WCS19—wheat	Group 3 chloroplast targeted protein increased freezing tolerance in *Arabidopsis*	299

11.3.2 LEA Proteins

One of the most common categories of target genes includes a number of loosely related groups of genes called LEA (*l*ate *e*mbryogenesis *a*bundant) proteins [111,112]. Many of these genes are induced in response to any environmental influence that has a dehydration component (such as drought, low temperature, or salinity) by ABA or during the late stages of embryogenesis. These proteins are thought to be involved in protecting cellular structures from effects of water loss—for example, by retaining water together with sugars [113] or directly protecting other proteins or membranes including renaturation of unfolded proteins and sequestration of ions in a wide range of higher plants [111,114–116]. Table 11.1 shows the transgenic LEA proteins expressed in plants.

LEA proteins have been identified primarily from plants and comprise at least six different subgroups [112]. These proteins do not seem to have a common origin; rather, they have adopted the same strategy for protection of cells from a common problem [117].

Group 1 family members are unstructured in solution. Their sequence is conserved especially within the hydrophobic internal 20-amino acid motif, which may be repeated. Group 1 proteins are induced by osmotic stress or ABA and they have been postulated to function as general water stress protectants in nonembryonic tissues [118].

Group 2 (also called dehydrins) is the largest group [112]. Purified maize DHN1 proteins from this group have been shown to bind with phospholipids *in vitro* [119]. The purified recombinant GmDHN1 exhibits a highly extended conformation at low temperatures, which could constitute the basis of the functional role in prevention of freezing, desiccation, ionic, or osmotic stress-related damage to macromolecular structures ([120]; see also a review by Allagulova et al. [121]).

A member of group 3 has been shown to have *in vitro* antiaggregation activity due to water stress, which was synergistic with trehalose. This group also contains members outside plants, as referred in Goyal et al. [122].

Group 4 has also been proposed to form amphiphilic α-helices that may interact with ions or membrane [111,123].

Group 5 proteins are embryogenesis specific.

Group 6 contains atypical hydrophopic LEA proteins [124].

Although LEA proteins have been known for quite some time (as reviewed in Cuming [115]), the functions of these proteins have remained somewhat unanswered. Despite some obvious enhancement in prediction tools [125], overexpression of genes in plants has not been conducted systematically. Among the published reports, Hara et al. [126] have recently shown that overproduction of a citrus dehydrin (CoCOR19) in tobacco leads to a slight decrease in ion leakage during chilling and freezing stress. Interestingly, the hydroxyl and peroxyl radical scavenging activity of CoCOR19 was found *in vitro* to be equal to serum albumin, a known antioxidant protein in mammals [20]. However, a hot-pepper protein, CaLEA6, tested in tobacco showed enhanced tolerance to dehydration and NaCl but not to chilling [124]. Furthermore, a much-studied Group 3 barley HVA1 has not been reported to be tested for cold tolerance [127–129], although it is naturally cold inducible [130].

11.3.3 LEA Proteins and Yeast

The baker's yeast *Saccharomyces cerevisiae* has been a popular model organism in studying functions of many LEA group proteins, although yeast also has LEA-like proteins [117,131]. Table 11.2 depicts the LEA proteins expressed in yeast. Barley HVA1 (group 3) displayed a shortened lag period after transfer to high-NaCl or KCl media, whereas tomato le4 (group 2) increased tolerance only to KCl, maybe by stabilizing KCl-sensitive structures. However, both increased freezing tolerance, supporting the idea of specialized functions of different LEA proteins during low-temperature stress [128]. Thus, different LEA proteins appear to have separate functions and protect cells against different stresses, even in yeast.

11.3.4 Functional Divergence of LEA Proteins

Many dehydrins are hydrophilic, containing random coil or α-helices [120], and remain soluble after boiling [132,133]. However, some atypical LEA proteins are

TABLE 11.2
Plant LEA Genes Expressed in Yeast

Group 1	Em—wheat	Enhanced osmotic tolerance	118
Group 2	le4—tomato	Enhanced freezing and KCl tolerance	128
Group 3	HVA1—barley	Enhanced freezing and salt tolerance	128
Group 3	hiC6—*Chlorella vulgaris*	Enhanced freezing tolerance	300
Group 4	le25—tomato	Increased salt and freezing tolerance	123

hydrophobic [112,124], suggesting a different function despite involvement in dehydration response. Even though the exact function of these proteins has not been established, their biochemical properties and accumulation patterns suggest that dehydrins could work in stabilizing cellular structures during dehydration stress [120], molecular chaperones [122], sequesters of iron [116,134] and calcium [135], and scavenging oxygen radicals [20].

The genome of *Arabidopsis* [136] has at least ten dehydrin genes, suggesting that each of the family members has specified functions. This specialization might be one reason why overexpression of an LEA protein does not increase tolerance to a particular type of stress—for example, chilling tolerance [124] or when expression of three different LEA proteins did not increase tobacco's drought tolerance [137].

Only a few comparative studies have been made between different plant protective assets. One study compared dehydration, salt, and heat tolerance of bacterial otsA, a trehalose-6-phosphate synthase, the first protein in making of a nonreducing disaccharide of two glucose units called trehalose, and an atypical LEA protein, CaLEA [138]. In this case, the LEA protein was found to be slightly better.

Several transgenic plant lines overexpressing different low-temperature responsive genes have been created, but only a marginal effect on freezing tolerance has been demonstrated [139,140]. Similar results have been obtained in studies of water stress tolerance. Therefore, this type of target gene engineering has thus far met with limited success. The approach is most likely hampered by redundancy of the system, where an individual gene product makes a minor contribution to overall tolerance. Thus, this approach would require pyramiding a number of target genes providing additive effects on tolerance.

This is further emphasized by the fact that several LEA proteins are activated together by their common transcription factors and correlate with stress tolerance in transgenic plants [72,76,82] Recent results from our group in expressing pairwise combinations of four different endogenous dehydrin genes in *Arabidopsis* [136] was an attempt to test this. Thus, RAB18, which accumulates in response to ABA, drought, and low temperatures [133,141], was put together with COR47, which accumulates primarily in response to low temperature, but also to ABA and salt stress [133].

On the other hand, LTI29, responding primarily to low temperature but also to ABA and salt stress, was put together with LTI30, which accumulates mainly under cold stress [133,142]. Using both constructs, we were able to show enhanced freezing tolerance that we did not see with RAB18 expressed alone [143] (data unpublished) and without any detectable deleterious effect to the plants [136]. These proteins have differing functions inside the cell, as witnessed by partitioning of LTI29 from being mainly cytoplasmic to mainly membrane localized following cold acclimation; LTI30 was not detected at the cytoplasm. Thus, pyramiding protective proteins with differing functions is clearly one way of proceeding with research.

11.3.5 Membrane-Binding Proteins

In general, plant survival during freezing may be more limited by stability of cellular membranes rather than by soluble proteins [144–146]. Indirect evidence for interactions with membranes or partially denatured proteins has been shown for the chloroplast-located hydrophilic COR15a protein, which is not a dehydrin but is predicted to be largely composed of amphipathic α-helices.

Interaction between the protein and the plastidic inner envelope, which is thought to be mediated through amphipathic α-helices, is correlated with decrease in membrane leakage [147]. Formation of the inverted hexagonal phase membrane structure can be prevented by the cold acclimation process and, in part, by overexpressing Cor15a [147,148].

Moreover, two other reports about membrane binding proteins have been reported. Osmotin-like protein from bittersweet nightshade (*Solanum dulcamara*) stabilizes kale (*Brassica oleracea*) protoplasts during freeze–thaw cycles [149]. Cryoprotectin, a lipid transfer protein homolog, shows cryoprotective activity stabilizing isolated thylakoids from freeze–thaw cycles [150]. The protective function of the protein lies in its capability to bind and immobilize thylakoid lipids [151]. Also, maize DHN1 binds to lipid vesicles, pointing to the possibility that the mode of action of some dehydrins is to support membranes [119]

11.3.6 Antifreeze Proteins

Antifreeze proteins (AFPs) were first identified from fish and found later from other species, including plants, as reviewed in Griffith and Yaish [152]. Their function is to bind to the growing surface of ice crystals and prevent them from growing. Roughly, animal and insect AFPs exhibit a substantial thermal hysteresis activity that is a noncolligative depression of the freezing point of an ice-containing solution below its melting point—the freeze avoidance strategy. AFP-producing plants and bacteria that cannot avoid freezing rely on another strategy by showing ice recrystallization inhibition.

Thus, controlling growth of an ice crystal during freeze–thaw cycles instead of preventing their formation is the freeze-tolerant strategy [153]. Transgenic attempts have been made with limited success in plant, synthetic, insect, and fish genes to express AFPs in frost-susceptible crops to increase their freezing tolerance [152]. In plants, the additional effect of these proteins is to function as pathogenesis-related proteins. Thus, plants are acquired with systemic, nonspecific, pre-emptive defense against psychrophilic pathogens that would otherwise prosper under snow cover at subzero temperatures [152,154].

11.3.7 Antioxidants and Detoxification

The importance of reactive oxygen species (ROS) scavenging proteins is that they have been shown to increase success rates not only in controlled experiments but also in overwintering plants in field trials, like the expression of superoxide dismutase within chloroplast, which did not affect freezing tolerance but rather enhanced recovery from stress [155,156]. Reactive oxygen species control many different processes in plants [16], so transgenic approaches may not always produce clear improvement, although it might just reflect a problem in expression or localization [157,158]. For example, using SODs creates enhanced hydrogen peroxide production, which may be alleviated in plants with upregulation of H_2O_2 scavenging enzymes [159,160].

ROS scavenging pathways include the water–water cycle, ascorbate–glutathione cycle, glutathione peroxidase cycle, and catalase, as reviewed in Mittler et al. [16] and Mittler [161]. The ascorbate–glutathione cycle plays an important role in regulating cellular ROS levels [162]. In this cycle, ascorbate peroxidase (APX) reduce H_2O_2 using ascorbic acid (AsA) as an electron donor generating monodehydroascorbate (MDA), which can be reduced back to AsA by MDA reductase using NAD(P)H. MDA can also spontaneously produce dehydroascorbate (DHA), which is reduced back to AsA by DHA reductase with the help of glutathione, GSH, oxidized to GSSG (the oxidized form of glutathione). The cycle closes with glutathione reductase converting GSSG back to GSH with NAD(P)H.

Plants contain a family of ascorbate peroxidases and overexpression of thylakoidal APX (tAPX) in tobacco gives chilling tolerance. Antisense lines were not obtained, suggesting an essential function [163]. However, tAPX overexpression in *Arabidopsis* did not increase resistance to low temperatures but rather to paraquat-induced oxidative stress and nitric oxide [164]. Furthermore, antisense constructs with 50% tAPX activity left showed no symptoms under normal conditions.

Expression of cytosolic cAPX has been reported to confer tolerance to chilling in tomato [165]. However, in soybean, the cAPX deficiency is associated with chilling tolerance in cultivated species [166]. Also, downregulating cAPX activity in tobacco, BY-2 cell lines showed increased tolerance to heat and salt but cells grew more slowly. The APX activity in BY-2 cells is much higher than in *Arabidopsis*, the residual activity resulting as marginal increase in cellular ROS levels triggering tolerance mechanisms was discussed as an explanation for increased tolerance [167]. It has been suggested that excessive levels of cAPX scavenge ROS too actively, thus hindering expression of defense genes, which need a certain level of stress for signaling [166].

11.3.8 Membranes

Membrane fluidity depends on composition of lipid molecular species, degree of lipid saturation where increase in the level of unsaturation increases fluidity, and temperature environments [168–170]. Thus, temperature-induced changes in membrane fluidity represent a potential site for cold perception [43,171], as also discussed in Sung et al. [172]. The proportion of unsaturated fatty acids in the lipid acyl chains is particularly high in chloroplast membranes [173]. Several genetic loci involved in fatty acid desaturation of lipids in chloroplast or microsomal membranes in *Arabidopsis* have been identified [174] and the genetic engineering of plant membrane lipids has been recently reviewed in Iba [175].

It seems that increase of the content of unsaturated lipids gives better chilling tolerance in tobacco [176,177]. A decrease in synthesis of unsaturated trienoic fatty acid increases high-temperature tolerance [178]. Glycerol-3-phosphate acyltransferase (GPAT) from *Arabidopsis* does not change proportions of individual lipid classes but increases the level of unsaturation, leading to increased chilling tolerance in tobacco and rice [179,180], whereas overexpression of squash GPAT in tobacco and rice leads to increase of saturated species of phosphatidylglycerol, which results in plants more sensitive to chilling [179,181]. Thus, it is possible to engineer tolerance to low or high temperature, but not both at the same time [175]. In addition to membrane unsaturation, it appears that lipid asymmetry in the membrane also contributes to low-temperature tolerance. Internalization of phosphatidylserine from the outer leaflet of plasma membrane by an *Arabidopsis* ALA1, a putative aminophospholipid translocase, in a yeast *dsr1* mutant was tightly linked to rescue in cold [182].

11.3.9 Enzyme Engineering

A completely different approach for achieving low-temperature tolerance is to engineer protein structure instead of engineering expression pattern, the main topic of this chapter. Pyruvate orthophosphate dikinase, a key enzyme in C4 pathway in maize, loses its activity under 12°C due to dissociation. Expressing engineered genes from *F. brownii* in maize that does not dissociate as readily gives a small improvement, which may be enhanced if the endogenous gene can be downregulated or engineered [183]. Nevertheless, it represents a potentially very interesting approach in trying to enhance the photosynthetic rate at low temperatures.

11.4 METABOLIC ENGINEERING

Compared to engineering expression of protective proteins, metabolic engineering of osmoprotectants is another way of improving plant stress tolerance. A group of target genes that have been successfully modified for increasing low-temperature tolerance in plants are those participating in producing certain small organic molecules called compatible and counteracting solutes, compensatory solutes, chemical chaperones, or osmoprotectants [184,185].

Induction of osmoprotectant biosynthesis is part of the plant response to drought, salinity, and low temperatures [186]. Compatible solutes occur in all organisms from archaebacteria to higher plants and animals [187]. These are highly soluble compounds that carry no net charge at physiological pH and do not perturb macromolecules such as proteins [188]. Some of these molecules are already present in unstressed tissues. Synthesis is often only increased or degradation decreased during stress [189,190]. The primary function of these compatible solutes is understood to maintain cell turgor and thus the driving gradient for water uptake [191,192]. For this purpose, by definition they should be able to exist at high concentrations in cells without any deleterious effects.

It seems that compatible solutes also function as free-radical scavengers or chemical chaperones and directly stabilize membranes and/or proteins [28,193,194]. Furthermore, the biosynthetic flux of the compatible solute may help to maintain redox balance and affected sugar levels may provide a signal for adaptive regulation throughout the plant [186]. Compatible solutes fall into several major groups: amino acids (e.g. proline), quaternary amines (e.g., betaines), tertiary sulfoniums (e.g., dimethylsulfoniopropionate), and polyol/sugars (e.g., mannitol, trehalose, sucrose, etc.) [195]. However, out of many potentially interesting compounds [28,187,196], relatively few have ever been tested by genetic engineering.

Of the few compounds that have been tested, in a number of publications the increase in cold tolerance was obtained by introducing simple metabolic traits from other organisms into plants that are not natural accumulators. However, some of the compounds may not be compatible any more in species that do not naturally accumulate them, as discussed later. The other way is to try to increase and maintain a higher concentration level of the metabolite in species where it has earlier been shown to elevate naturally following stress response [184,197]. The way to achieve this in the latter case is to establish a new pathway or to increase the synthesis rate of the metabolite by increasing the gene dosage (and gene expression) of the rate-limiting enzyme [198–200] or by decreasing the degradation rate of the metabolite by downregulating the degrading enzyme-encoding gene [201]. Table 11.3 lists transgenic experiments giving enhanced low-temperature tolerance.

11.4.1 Betaines

Betaines are amino acid derivatives in which the nitrogen atom is fully methylated. In plants, a representative member of this group is glycine betaine (GB) [189], which is a well studied subject in salt stress; more than 15 articles on transgenic plants have been published [202]. In GB-accumulating plants, it is synthesized in the chloroplast through an oxidation reaction from choline to glycine betaine via betaine aldehyde intermediate. In higher plants, two enzymes—choline monoxygenase (CMO; in mammals and some bacteria, CDH) and betaine aldehyde dehydrogenase (BADH)—are responsible for these reactions. In certain bacteria, both reactions are catalyzed by one enzyme, choline oxidase (COD), as reviewed in Sakamoto and Murata [203].

Transgenic plants expressing betaines are more resistant to high salt concentrations, osmotic stress, and cold and warm temperatures [204,205]. Improved low-temperature tolerance has been reported when expressing the *Arthrobacter globiformis codA* (coding for choline oxidase, which synthesizes glycine betaine in one step) gene in *Brassica juncea* [206], tobacco [207], and tomato [208], as well as rice and *Arabidopsis* [203]. However, in the case of codA, one must consider the role of the by-product of the GB synthesis, H_2O_2, which is a signaling component inducing the ROS-response and thus gives chilling tolerance [208].

Expressing CDH alone or together with BADH in tobacco leaf discs also improved tolerance to photoinhibition under low temperature [209]. However, the main theme in all reported cases has been that the levels of GB produced are very modest compared to plants that accumulate it naturally. It is thought that this occurs because of limited supplies of precursors in nonaccumulators [199,210,211]. However, this problem seems to be possible to circumvent by using extensive engineering for precursor production.

Moreover, a novel pathway from extreme halophile bacteria for producing GB by direct methylation of glycine also gives improved osmotolerance in *E. coli* [212,213]. Using closely related enzymes in transgenic *Arabidopsis* gave much higher (10×) concentration of GB in *Arabidopsis* and *Synechococcus* than with codA. In this case, the precursor is not choline but glycine, the levels of which are only somewhat limiting [214].

Thus, it seems that the protective effect of GB is independent of species or mode of synthesis in natural accumulators. However, there are also reports about GB in non-natural accumulators giving only modest [215] or even deleterious

TABLE 11.3
Transgenic Plants Showing Enhanced Low-Temperature Tolerance

Protein and origin	Model plant	Remarks	Ref.
		Chilling stress	
desC—*S. vulcanus*	Tobacco	Thermophilic bacterium acyl-lipid Δ9-desaturase giving chilling tolerance	177
acyl-lipid Δ9-desaturase—*A. nidulans*	Tobacco	Broad specificity Δ9 desaturase "fluidizing" membrane bound lipids	305
CvFAD2 and 3—*C. vulgaris*	Tobacco	Two microsomal desaturases increasing and decreasing freezing tolerance	306
FAD7—*A. thaliana*	Tobacco	Chloroplast ω3 desaturase increases trienoic fatty acid (TA) amount and confers chilling tolerance	176
GPAT—*A. thaliana*, spinach	Tobacco, rice	Increases proportion but not amount of unsaturated phosphatidylglycerol	179, 180
Glycerol-3-P-acyl transferase (GPAT)—squash and *A. thaliana*	Tobacco, rice	Causes changes in the unsaturation of fatty acids and chilling tolerance levels	181, 307
ALA1—*A. thaliana*	*A. thaliana*	Aminophospholipid translocase involved in generating membrane lipid asymmetry	182
Glutamine synthetase—rice	Rice	Faster recovery from chilling stress due to increase in glutamine levels; increase in photorespiration capacity	224
Glutathione peroxidase—*Chlamydomonas*	Tobacco	Cytosolic and chloroplast versions increased chilling tolerance	308
MnSOD and Fe-SOD—tobacco and *A. thaliana*	Tobacco	Fe-SOD binds to membranes and protects PSII; stromal Mn-SOD less effective	157
Cu/Zn-SOD—pea	Tobacco	Chloroplast targeted SOD	309
Nt107—tobacco	Tobacco	Glutathione S-transferase/glutathione peroxidase combination	310
DHAR—human	Tobacco	Dehydroascorbate dehydrogenase reduces DHA to ascorbate	311
APX—*P. sativum*	Tomato	Cytosolic expression of ascorbate oxidase enhancing chilling tolerance	165
Ipt—*A. tumefaciens*	*F. arundinacea*	Isopentenyl transferase resulted in increased cold tolerance	312
ProDH—*A. thaliana*	*A. thaliana*	Antisense suppression of proline degradation improves tolerance to freezing	201
Invertase—*S. cerevisiae*	Potato	Apoplast-localized invertase inhibits export of sugars and thus leaves retain higher sugar content	313

otsA-otsB—*E. coli*	Rice	Fusion of two *E. coli* trehalose biosynthetic genes allowing enhanced protection and recovery without stunting growth	252, 314
TpS1 and TpS2—*S. cerevisiae*	*A. thaliana*	Trehalose synthases *TPS1* alone or together with *TPS2* gives pleiotrophic effects and enhanced stress tolerance	253
mtlD—*E. coli*	Eggplant	Mannitol production enhancing chilling tolerance in a crop plant	315
SacB—*B. subtilis*	Tobacco	Bacterial levansucrase for fructan biosynthesis gives low temperature tolerance	207
OsCDPK 7 and 13—rice	Rice	Signaling; calcium-dependent protein kinase confers tolerance; number 7 has separate pathways for cold and salt	47, 267
OSISAP1—rice	Tobacco	Signaling; zinc finger protein confers tolerance at germination stage	316
Bcl-xL—human and Ced-9—*C. elegans*	Tobacco	Plants transgenic to animal cell-death suppressors germinate at lower temperatures than wild-type	317
Ppdk—*F. bidentis*/*F. brownii*	Maize	Chimeric enzyme dissociates at lower temperature than native	183
		Freezing stress	
Fe-SOD—*A. thaliana*	Alfalfa	Enhanced recovery from stress after resuming growth	156
Mn-SOD—tobacco	Alfalfa	Increased winter survival	318
AAPT1—canola	*A. thaliana*	Phospholipid synthesis; increased resistance to damage at low growth temperatures	319
Phospholipase D—*Arabidopsis*	*A. thaliana*	PLDδ overexpression and PLDα downregulation increases freezing tolerance; involved in hydrolyzation of phospholipids	320, 321
Spermidine synthase (SPDS)—*C. ficifolia*	*A. thaliana*	Increase in polyamine content gives abiotic stress tolerance and rise in DREB levels	322
Galactosidase—*L. esculentum*	Petunia	Downregulation of hydrolytic enzyme increases raffinose concentration and freezing tolerance	249
SEX1—*A. thaliana*	*A. thaliana*	Involved in regulation of starch hydrolysis in early cold response	257
Wft1, Wft2—*T. aestivum*	*L. perenne*	Fructan (sugar polyol) biosynthesis genes increase fructan levels likely in the vacuole	247
P5C—*Arabidopsis*, *Vigna*	Tobacco	Proline, fructan, and betaine biosynthesis give freezing tolerance to tobacco	207
codA—*A. globiformis*	Tomato	Betaine production protects flowers and seeds (reported)	208
codA—*A. globiformis*	*A. thaliana*, rice, *Brassica*, tomato	Biosynthesis of glycine betaine enhances germination and cold tolerance of plants	206, 208, 323, 324

continued

TABLE 11.3 (continued)
Transgenic Plants Showing Enhanced Low-Temperature Tolerance

Protein and origin	Model plant	Remarks	Ref.
Cox—*A. pascens*	*A. thaliana*	Choline oxidase giving modest freezing tolerance, but not to canola or tobacco	215
betA, betB—*E. coli*	Tobacco	Improved tolerance to photoinhibition under low temperature in leaf discs	209
GSMT, DMT—*A. halophytica*	Cyanobacterium, *A. thaliana*	Germination yield retained high despite cold imbibition-treatment of seeds	214
Antifreeze protein—carrot	Tobacco	Increased freezing tolerance	325
MKK2—*A. thaliana*	*A. thaliana*	Signaling; MAPKK especially activated by cold	266
Osmyb4—rice	*A. thaliana*	Rice transcription factor showing significantly increased freezing tolerance in *Arabidopsis*	326
OsMAPK5—rice	Rice	Signaling; affects adversely between abiotic and pathogen tolerance	327
SCOF-1—soybean	*A. thaliana*, tobacco	Zinc finger protein improves freezing and chilling tolerance	101
Cryophyte/Los4—*A. thaliana*	*A. thaliana*	RNA helicase involved in mRNA export affects CBF and ABA signaling	272
Fiery1/Hos2—*A. thaliana*	*A. thaliana*	IP_3 turnover; mutation gives cold tolerance due to increased inositol concentration	328
NPK1—tobacco	Maize	Signaling; MAPKKK enhances freezing resistance in maize	263
ABF3—*A. thaliana*	*A. thaliana*, rice	ABA-signaling component gives increased low temperature tolerance	329, 330
ABI3—*A. thaliana*	*A. thaliana*	Signaling; transcription factor in ABA mediated processes	331
CaPF1— *C. annuum*	*A .thaliana*	Overexpression of ERF/AP2-type transcription factor allowed slightly better protection	332
ICE1—*A. thaliana*	*A. thaliana*	CBF master regulator can be overexpressed without pleiotrophic effects, giving better freezing tolerance	75
CBF/DREB—*A. thaliana*, *P. avium*	*A. thaliana*, *B. napus*, maize, tomato	Constitutive overexpression of AP2-type confers low temperature resistance, according to several papers, but causes pleiotropic side effects	68, 72, 76, 77, 82, 88, 333, 334
CBF/DREB—*A. thaliana*	Tomato, tobacco	Use of regulatable promoter confers stress tolerance without affecting yield	277, 335
CBF/DREB—*A. thaliana*	Rice	Increased cold tolerance without detrimental side effects	330

Note: Lists of transgenic plants against other abiotic stresses can be found in References 28, 80, 172, 195, 198, 200, and 301 through 304.

[216–218] effects. Furthermore, even in natural accumulators, glycine betaine is not degraded once the stress is over [211,219]. Thus, GB cannot be utilized once growth is resumed [192].

BADH and S-adenosyl-L-methionine-dependent N-methyltransferase (NMTase) are enzymes able to synthesize another betaine, β-ala-betaine, which reportedly accumulates to high levels in transgenic plants; however, low-temperature tolerance was not tested [205,220]. The levels of several other betaines (proline betaine, hydroxyproline betaine, sulfonium betaine DMSP, and choline-o-sulfate) are known to respond to cold [221,222] but have not been assayed for engineering cold tolerance.

11.4.2 Amino Acids

Two amino acids are strongly upregulated following cold in *Arabidopsis*: proline and glutamine [37,223]. Glutamine is an amide form of glutamate synthesized by the glutamine synthetase GS1. Overexpression in rice allows faster recovery from chilling stress, which may be due to increase in photorespiration capacity [224].

Proline is synthesized from glutamate via glutamic-γ-semialdehyde (GSA) and Δ^1-pyrroline-5-carboxylate (P5C). P5C synthase (P5CS) catalyzes the conversion of glutamate to P5C, followed by P5C reductase (P5CR), which reduces P5C to proline. In the reverse reaction, proline is metabolized to glutamate in a feedback manner via P5C and GSA with the aid of proline dehydrogenase (ProDH) followed by P5C dehydrogenase (P5CDH) [225].

It has been proposed that proline would not function as a compatible solute per se, but rather by regulating NADP+/NADH ratios by cycling between proline and its precursors. The flux would be an important homeostatic mechanism and potentiate activity of the oxidative pentose phosphate pathway. Following relief from the stress, oxidation of proline would also provide an important energy source for ADP phosphorylation [193]. Concentration of proline in *Arabidopsis* is actively regulated by osmotic stress and drought. Cold regulates proline levels timewise much more slowly [226,227]. In rice, the proline amount is cold inducible, and the proline synthesis knockout mutant is sensitive to cold [228].

Proline application to low-temperature-stressed chickpea improved floral retention and pod set and the concentration of proline was higher in retained vs. aborted flowers [229]. Moreover, when proline catabolism was depressed in the *Arabidopsis* antisense line, it improved tolerance to freezing [201]. However, further addition of external proline caused hypersensitivity [230,231].

11.4.3 Sugars

Intracellular concentration of soluble sugars has been shown to increase in several frost-hardy plant species upon exposure to low temperatures. These include fructans, fructose, galactinol, glucose, raffinose, stachyose, sucrose, and "sugar alcohols" (mannitol, trehalose, *myo*-inositol, and sorbitol) [77,223,227,232–237]. Many sugars like mannitol, sorbitol, D-ononitol, and D-pinitol are also active ROS scavengers [28,198,238]. Many soluble sugars need to be synthesized upon exposure to cold.

However, the simplest way to achieve soluble sugar (mannitol, sucrose, raffinose, and stachyose) accumulation inside the cell following cold sensing is to degrade starch. This was noted a long time ago [3] and also shown with populus [239] and recently with moss *Physcomitrella patens* [240] and joint *Arabidopsis* and pea experiments [241].

Sugar metabolism seems to be involved in a delicate balance affecting many different signaling pathways in plants; pleiotropic effects when engineering sugar synthesis have been reported [186,242]. Undesirable phenotypes have been associated with trehalose, as discussed in Bae [243] and Avonce [244], and accumulation of sorbitol in transgenic tobacco transformed with *stpd1* (a cDNA encoding sorbitol-6-phosphate dehydrogenase) from apple caused necrotic lesions assumed to have resulted from disturbance in carbohydrate transport and allocation [245]. Also, bacterial SacB encoding for fructan synthesis gene exhibited symptoms like stunting, necrosis, reduction in starch accumulation, and chloroplast agglutination in many plant species, although no *Arabidopsis* experiment is reported [246,247]. Lists of examples can be found in Cairns [246] and Nuccio et al. [248].

Furthermore, there may be species-specific differences in the importance of different sugars. The transgenic approach and mutant analyses in *Arabidopsis* clearly proved that raffinose is not essential for basic freezing tolerance [237]. However, in petunia, there seems to be a correlation with freezing tolerance and raffinose [249]. It was discussed that α-galactosidase engineered in petunia might have some other role that might explain some of the discrepancy [237], but it may also argue for the possible explanation that the amount of a single sugar is not important if the overall concentration of sugars remains the same.

However, there are also examples of improved low-temperature tolerance using genetic engineering: transgenic fructan production showed increased tolerance to freezing in ryegrass using wheat fructosyltransferase directed to vacuole. Localization was suggested to explain why there were no negative effects noticed, unlike with fructan synthesis using bacterial SacB [246,247].

Trehalose, disaccharide of glucose, can protect membranes and transgenic rice, tobacco, and *Arabidopsis* plants from cold and dehydration [250–253]; small amounts of trehalose accumulate in *Arabidopsis* following cold signaling [37]. The phosphorylated precursor, trehalose-6-phosphate, is considered to have an important role in sugar signaling and the photosynthesis rate in *Arabidopsis* [254]. However, pathogens also secrete trehalose in trying to redirect plant carbohydrate metabolism. Thus, it seems that trehalose induces detoxification and stress response proteins involved in pathogen attack and oxidative stress in *Arabidopsis* and further induces degradation of trehalose [243,255].

Mutations in the sugar metabolism genes have also provided fruitful information. *Arabidopsis gly1* (coding for glycerol kinase) mutant, which is unable to utilize glycerol and thus transiently accumulates glycerol, exhibits enhanced freezing tolerance [256].

Furthermore, a mutation at a starch-related α-glucan/water dikinase encoded by *STARCH EXCESS 1* (*SEX1*) in *Arabidopsis* (hypothesized to regulate starch degradation in plastids by phosphorylating starch to ensure better accessibility for the degrading enzymes) is unable to accumulate malto-oligosaccharides, glucose, and fructose during the first day of cold acclimation and shows impaired freezing tolerance. However, when low-temperature treatment is continued, the situation is normalized. Even though the cold-induced starch degradation activity was not linked to this enzyme, it enabled faster mobilization [257]. Thus, starch is important during the very early phase of cold acclimation [240,257]. Sugars seem to be important during the first week of acclimation according to an *Arabidopsis* growth room experiment, after which other metabolites, among them proline, accumulate and sugar concentration starts to decline [258].

11.4.4 Other Metabolites

Analyses of low-temperature induced metabolites have only begun [259]. Thus, analysis of polar metabolites from *Arabidopsis* identified 114 metabolites, out of 434 monitored, where the pool sizes were substantially (>5×) increased in response to low temperature. Identity of roughly 30 was determined; among them, trehalose, putrescine, and ascorbate were reported from *Arabidopsis* for the first time [37] (see also Lange and Ghassemian [260]). However, the structure of most of the cold-induced metabolites remains to be determined. This is in accordance with results obtained in birch, where we have been able to identify only a few of the cold-induced metabolites of the sugar fraction (G. Brader, A. Welling, I. Tsitko, and T. Palva, personal communication). Thus, there is a good possibility for obtaining new and even better suited metabolites for freezing protection than the few tested thus far. Many of them were originally identified when a stress type other than low temperature [28,187,196] was studied (sucrose is a notable exception) [261].

11.5 REGULON ENGINEERING

11.5.1 Stress Signaling

Engineering signaling cascades opens a possibility for recruiting many more target genes under the control of one signaling pathway. However, a delicate adjustment for proper expression pattern is needed to balance minimizing the negative effects. *Nicotiana* protein kinase (NPK1) is the uppermost signaling component in a MAP-kinase cascade [262]. Low level of expression was found to be enough for lowering freezing tolerance points by 2°C in maize without deleterious effects [263]; higher production caused detrimental effects [264]. NPK1 induced a heat-shock protein (HSP) and a GST in *Arabidopsis* and maize [262]. HSPs are known to be involved in chilling resistance of fruits [265] and GSTs are involved in oxidative stress protection.

An MAPK cascade consisting of MEKK (MAPKKK), MKK2 (MAPKK), and MPK4/MPK6 (MAPKs) has been recently shown to mediate cold responses in *Arabidopsis* [266].Overexpression of MKK2, the second kinase in the MAP kinase cascade, increases cold and salt tolerance. Overexpression resulted in increase in the transcript levels of an MAP kinase cascade, up- or downregulating 152 genes at least threefold [266].

Increased cold tolerance is achieved with overexpressing calcium-dependent protein kinases 7 and 13 of rice [47,267]. The potential functions of these proteins are to enable cross-talk between multiple signaling pathways [45]. The first gene was shown to cause induction of genes encoding salt/drought specific LEA proteins [47].

OsWRKY72 and 77 are ABA-inducible rice transcription factors; their transient expression leads to induction of HVA22 reporter construct in rice and, together with ABA, they synergistically induce HVA22 promoter. However, no data on low temperature tolerance are yet reported [268].

Early response to dehydration 15 (ERD15) [269] is a signaling component encoded by a gene induced within 30 min from onset of biotic or abiotic stress. ERD15 presumably affects the cross-talk between ABA signaling and pathogen signaling. Overexpression of ERD15 gives insensitivity to ABA signaling and RNAi silencing leads to improved freezing tolerance of transgenic *Arabidopsis* (Kariola et al., manuscript in preparation).

11.5.2 Cold-Tolerant Mutations

In some cases, silencing a gene or a mutation in a gene leads to increased low-temperature tolerance instead of overexpression. Thus, T-DNA insertion mutation of a vacuolar Ca^{2+}/H^{+} antiporter, *CAX1*, leads to improved cold acclimation. However, long-term freezing tolerance is normal. Genes encoding CBFs have been shown to be upregulated in a *CAX1* mutant that normally may play a role in reducing cytosolic Ca^{2+} concentration to resting levels. CBF transcription was also found to be dependent on Ca^{2+} [270]. CRYOPHYTE/LOS4 is a DEAD box RNA helicase regulating mRNA export from the nucleus and affecting the CBF pathway. Mutations in the gene render plants more sensitive (*los4-1*) or resistant (*los4-2*) to freezing and chilling [271,272].

Constitutive overaccumulation of proline amino acid is caused by a mutation in ESKIMO1, *esk1*, which conferred freezing tolerance to nonacclimated *Arabidopsis*. Because soluble sugars and genes tested did not show a clear difference with wild-type plants, it was suggested that proline accumulation alone could be responsible for a large proportion of *Arabidopsis* freezing tolerance. However, only a small increase in total freezing tolerance of the acclimated plants was reported [273]. Antisense downregulation of phosphatase PP2CA of *A. thaliana*, also involved in signaling, showed accelerated development of freezing tolerance but no increase in basal tolerance [274].

11.6 CONCLUSIONS

Resistance to abiotic stress such as low temperature is a very complex phenomenon in plants. In addition to the actual protective mechanisms, there is a distinct role in the repair mechanisms after the stress. Furthermore, maintenance of energy metabolism and pH stability promoting these two mechanisms is a factor contributing to both types of resistance ability [5,275]. However, increasing knowledge of those pathways is offering insights as to how and where to direct research. These, for example, include targeting detoxification pathways for obtaining plants that can be sustained under true field conditions [276] or engineering CBF/DREB and ZAT12 regulons [38]. Modification of a single gene, like CBF/DREB, resulted in significant improvements in stress tolerance in several cases [277,278]. However, changes more upstream in the pathway often lead to activation of a larger set of genes, including other than stress related [79,279].

Because resistance to abiotic stress is a polygenic trait, engineering signaling pathways offers a potent way to hit the relevant genetic target by engineering a single gene or a small number of genes. It is important to remember that, during stress, protection mechanisms are induced and growth-related genes are repressed. Thus, growth and stress resistance are inversely correlated. Possibly, stress mechanisms are expensive to build and maintain and growth-related metabolism may be sensitive to stress [6,280].

The solution for the problem would be careful regulation of the engineered trait. However, the constitutive overexpression of the CBF master regulator, ICE1, in transgenic plants did not exhibit obvious growth or developmental abnormalities [75].

Useful for all interested in genetic engineering is the article about fructan biosynthesis and the most usual problems encountered in expression of foreign genes in plants [246] and a recent list of chloroplast targeting articles [281]. Also, pyramiding several protective genes into the same organism is the next step in research now that several interesting candidates have emerged [28,282].

As important as transforming individual genes is to exposing their functions, it is also at the same time very labor intensive and time consuming (as well as expensive). New methods have recently emerged that offer tools for speeding up the research. Thus, a method called tilling consists of making up point mutations in the genome with chemicals, combined with PCR-based screening of the region of interest. Amplified region harboring mutation is hybridized with respective wild-type regions and heteroduplex is cleaved by CEL I, after which the products are resolved using electrophoresis. The method enables one to form allelic series that provide a range of phenotypic severities, which is important if the gene is essential, for example [283]. Another very interesting method is transforming large chromosomal fragments instead of individual genes. These can be based on Agrobacterium [284] or particle bombardment [285]. Importantly, these methods can be used even if the whole genome is not sequenced.

11.7 PERSPECTIVE

It is clear that some of the acclimation mechanisms needed for acquired low-temperature tolerance are different between annuals and perennials, and active growth phase and overwintering phase. For example, shortening day length under a critical photoperiod (so-called short-day [SD] signal) in birch tree *Betula pendula* potentiates low temperature-induced dehydrine expression up to five times higher that that of cold alone; in *Arabidopsis* only cold induces the same birch gene [89]. However, in many cases the molecular basis for cold acclimation is remarkably conserved, allowing us to appreciate different variations of the theme and to learn more about how the acclimation process is done and how the seasonal low temperature causes characteristic changes in the induction pattern of transcripts and metabolites in many plant species. The recurring theme in seasonal acclimation is the constant change at levels of metabolites and "division of labor" among them [286–290].

Differences in low-temperature-resistance mechanisms lie within the plant also. For example, leaves accumulate sugars different from those of roots following acclimation in *Arabidopsis*, thus signifying that, when a decision about the transgenic approach must be made, choosing the correct promoter may have a bigger effect in the phenotype than traditionally has been thought [291,292]. Choosing the right compartment for targeting the gene, whether it is to be cytoplasmic or organellar, is another important factor to consider in an attempt to achieve a useful addition in the resistance mechanisms [28,247].

One key problem to address in crop improvement is lack of well-defined biochemical indicators for tolerance that would enhance breeding programs, with or without a transgenic component. An attempt to solve this for salinity stress has been reported [293]; this would also be applicable to low temperatures. However, as pointed out by the authors, the plant physiology is so complex that, in addition to variation among species, in many cases physiological responses to stress vary from cultivar to cultivar within a single species. This fact stresses that biochemical indicators would need to be specified for individual species rather than generalized for all.

Finally, it is worth remembering that many of the genetically modified stress-tolerant plants generated to date are nonagronomic plants. The stress resistance, however useful for the plant, is important only if it results in higher crop yield. This is exemplified in the case of glycine betaine accumulators, which are more prone to fungal diseases in field trials, thus negating the effect of drought tolerance [28]. The effects of engineered traits on overwintering, flowering, or seed production are only sometimes reported [208,294,295], although flowers are the parts most susceptible to abiotic stress [208,229]. Securing the crop is, after all, one of the most used arguments for motivating plant stress research in general.

ACKNOWLEDGMENTS

The authors wish to thank all the past and present members of the group. Claire Duhazé is gratefully acknowledged for comments and suggestions about this manuscript. This work was supported by the Academy of Finland (Finnish Center of Excellence Program 2000–2005).

REFERENCES

1. Boyer, J.S., Plant productivity and environment, *Science*, 218, 443–448, 1982.
2. Sakai, A. and Larher, W., *Frost Survival of Plants: Responses and Adaptation to Freezing Stress*, Springer-Verlag, Berlin, 1987.
3. Levitt, J., *Responses of Plants to Environmental Stresses: Chilling, Freezing and High Temperature Stresses*, Academic Press, New York, 1980.
4. Graham, D. and Patterson, B.D., Responses of plants to low, non-freezing temperatures—proteins, metabolism, and acclimation, *Annu. Rev. Plant Physiol. Plant Mol. Biol.*, 33, 347–372, 1982.

5. Maldonado, R., Sanchez-Ballesta, M.T., Alique, R., Escribano, M.I., and Merodio, C., Malate metabolism and adaptation to chilling temperature storage by pretreatment with high CO_2 levels in *Annona cherimola* fruit, *J. Agric. Food Chem.*, 52, 4758–4763, 2004.
6. Jackson, M.W., Stinchcombe, J.R., Korves, T.M., and Schmitt, J., Costs and benefits of cold tolerance in transgenic *Arabidopsis thaliana*, *Mol. Ecol.*, 13, 3609–3615, 2004.
7. Edmeades, G.O., McMaster, G.S., White, J.W., and Campos, H., Genomics and the physiologist: bridging the gap between genes and crop response, *Field Crops Res.*, 90, 5–18, 2004.
8. Burke, M.J., Gusta, L.V., Quamme, H.A., Weiser, C.J., and Li, P.H., Freezing and injury in plants, *Annu. Rev. Plant Phys.*, 27, 507–528, 1976.
9. George, M.F. and Burke, M.J., Supercooling of tissue water to extreme low-temperature in overwintering plants, *Trends Biochem. Sci.*, 9, 211–214, 1984.
10. Steponkus, P.L., Uemura, M., and Webb, M.S., Membrane destabilization during freeze-induced dehydration, *Curr. Topics Plant Physiol.*, 10, 37–47, 1993.
11. Gordon-Kamm, W.J. and Steponkus, P.L., Lamellar-to-hexagonal II phase transitions in the plasma membrane of isolated protoplasts after freeze-induced dehydration, *Proc. Natl. Acad. Sci. USA*, 81, 6373–6377, 1984.
12. McKersie, B.D. and Bowley, S.R., Active oxygen and freezing tolerance in transgenic plants, in *Plant Cold Hardiness*, Li, P.H. and Chen, T.H.H., Eds., Plenum Press, New York, 1998, 203–212.
13. Pastori, G.M. and Foyer, C.H., Common components, networks, and pathways of cross-tolerance to stress. The central role of "redox" and abscisic acid-mediated controls, *Plant Physiol.*, 129, 460–468, 2002.
14. Halliwell, B. and Gutteridge, J.M.C., *Free Radicals in Biology and Medicine*, Clarendon Press, Oxford, 1989.
15. Edreva, A., Generation and scavenging of reactive oxygen species in chloroplasts: a submolecular approach, *Agric. Ecosyst. Environ.*, 106, 119–133, 2005.
16. Mittler, R., Vanderauwera, S., Gollery, M., and Van Breusegem, F., Reactive oxygen gene network of plants, *Trends Plant Sci.*, 9, 490–498, 2004.
17. Apel, K. and Hirt, H., Reactive oxygen species: metabolism, oxidative stress, and signal transduction, *Annu. Rev. Plant. Biol.*, 55, 373–399, 2004.
18. Chen, W.P., Li, P.H., and Chen, T.H.H., Glycinebetaine increases chilling tolerance and reduces chilling-induced lipid peroxidation in *Zea mays* L., *Plant Cell Environ.*, 23, 609–618, 2000.
19. Parvanova, D., Ivanov, S., Konstantinova, T., Karanov, E., Atanassov, A., Tsvetkov, T., Alexieva, V., and Djilianov, D., Transgenic tobacco plants accumulating osmolytes show reduced oxidative damage under freezing stress, *Plant Physiol. Biochem.: PPB/Soc. Francaise Physiol. Vegetale*, 42, 57–63, 2004.
20. Hara, M., Fujinaga, M., and Kuboi, T., Radical scavenging activity and oxidative modification of citrus dehydrin, *Plant Physiol. Biochem.*, 42, 657–662, 2004.
21. Grace, S.C. and Logan, B.A., Energy dissipation and radical scavenging by the plant phenylpropanoid pathway, *Philosophical Trans. R. Soc. London Series B-Biol. Sci.*, 355, 1499–1510, 2000.
22. Sakihama, Y., Cohen, M.F., Grace, S.C., and Yamasaki, H., Plant phenolic antioxidant and prooxidant activities: phenolics-induced oxidative damage mediated by metals in plants, *Toxicology*, 177, 67–80, 2002.
23. Munne-Bosch, S., Linking tocopherols with cellular signaling in plants, *New Phytol.*, 166, 363–366, 2005.

24. Penuelas, J. and Munne-Bosch, S., Isoprenoids: an evolutionary pool for photoprotection, *Trends Plant Sci.*, 10, 166–169, 2005.
25. Porfirova, S., Bergmuller, E., Tropf, S., Lemke, R., and Dormann, P., Isolation of an *Arabidopsis* mutant lacking vitamin E and identification of a cyclase essential for all tocopherol biosynthesis, *Proc. Natl. Acad. Sci. USA*, 99, 12495–12500, 2002.
26. Muller-Moule, P., Havaux, M., and Niyogi, K.K., Zeaxanthin deficiency enhances the high light sensitivity of an ascorbate-deficient mutant of *Arabidopsis*, *Plant Physiol.*, 133, 748–760, 2003.
27. Rodriguez Milla, M.A., Maurer, A., Rodriguez Huete, A., and Gustafson, J.P., Glutathione peroxidase genes in *Arabidopsis* are ubiquitous and regulated by abiotic stresses through diverse signaling pathways, *Plant J.*, 36, 602–615, 2003.
28. Bohnert, H.J. and Shen, B., Transformation and compatible solutes, *Sci. Horticul.*, 78, 237–260, 1999.
29. Foyer, C.H. and Noctor, G., Redox homeostasis and antioxidant signaling: a metabolic interface between stress perception and physiological responses, *Plant Cell*, 17, 1866–1875, 2005.
30. Lang, V., Heino, P., and Palva, E.T., Low-temperature acclimation and treatment with exogenous abscisic acid induce common polypeptides in *Arabidopsis thaliana* (L.) Heynh, *Theor. Appl. Genet.*, 77, 729–734, 1989.
31. Mantyla, E., Lang, V., and Palva, E.T., Role of abscisic acid in drought-induced freezing tolerance, cold acclimation, and accumulation of LT178 and RAB18 proteins in *Arabidopsis thaliana*, *Plant Physiol.*, 107, 141–148, 1995.
32. Guy, C.L., Freezing tolerance of plants: current understanding and selected emerging concepts, *Can. J. Bot. Rev. Can. Bot.*, 81, 1216–1223, 2003.
33. Palva, E.T., Welin, B., Vahala, T., Olson, A., Nordin–Henriksson, K., Mantyla, E., and Lang, V., Regulation of low temperature-induced genes during cold acclimation of *Arabidopsis thaliana*, *NATO ASI Ser. Ser. H, Cell Biol.*, 86, 527–542, 1994.
34. Hughes, M.A. and Dunn, M.A., The molecular biology of plant acclimation to low temperature, *J. Exp. Bot.*, 47, 291–305, 1996.
35. Thomashow, M.F., Plant cold acclimation: freezing tolerance genes and regulatory mechanisms, *Annu. Rev. Plant Physiol. Plant Mol. Biol.*, 50, 571–599, 1999.
36. Nuotio, S., Heino, P., and Palva, E.T., Signal transduction under low-temperature stress, in *Crop Responses and Adaptations to Temperature Stress*, Basra, A.B., Ed., Food Products Press, Binghamton, NY, 2001, 151–176.
37. Cook, D., Fowler, S., Fiehn, O., and Thomashow, M.F., A prominent role for the CBF cold response pathway in configuring the low-temperature metabolome of *Arabidopsis*, *Proc. Natl. Acad. Sci. USA*, 101, 15243–15248, 2004.
38. Vogel, J.T., Zarka, D.G., Van Buskirk, H.A., Fowler, S.G., and Thomashow, M.F., Roles of the CBF2 and ZAT12 transcription factors in configuring the low temperature transcriptome of *Arabidopsis*, *Plant J.*, 41, 195–211, 2005.
39. Knight, H., Calcium signaling during abiotic stress in plants, *Int. Rev. Cytol.—Survey Cell Biol.*, 195, 269–324, 2000.
40. Reddy, V.S. and Reddy, A.S.N., Proteomics of calcium-signaling components in plants, *Phytochemistry*, 65, 1745–1776, 2004.
41. Monroy, A.F., Sarhan, F., and Dhindsa, R.S., Cold-induced changes in freezing tolerance, protein phosphorylation, and gene expression (evidence for a role of calcium), *Plant Physiol.*, 102, 1227–1235, 1993.
42. Monroy, A.F. and Dhindsa, R.S., Low-temperature signal transduction: induction of cold acclimation-specific genes of alfalfa by calcium at 25°C, *Plant Cell*, 7, 321–331, 1995.

43. Orvar, B.L., Sangwan, V., Omann, F., and Dhindsa, R.S., Early steps in cold sensing by plant cells: the role of actin cytoskeleton and membrane fluidity, *Plant J.*, 23, 785–794, 2000.
44. Sangwan, V., Foulds, I., Singh, J., and Dhindsa, R.S., Cold-activation of *Brassica napus* BN115 promoter is mediated by structural changes in membranes and cytoskeleton, and requires Ca2+ influx, *Plant J.*, 27, 1–12, 2001.
45. Harper, J.E., Breton, G., and Harmon, A., Decoding Ca^{2+} signals through plant protein kinases, *Annu. Rev. Plant Biol.*, 55, 263–288, 2004.
46. Polisensky, D.H. and Braam, J., Cold-shock regulation of the *Arabidopsis* TCH genes and the effects of modulating intracellular calcium levels, *Plant Physiol.*, 111, 1271–1279, 1996.
47. Saijo, Y., Hata, S., Kyozuka, J., Shimamoto, K., and Izui, K., Overexpression of a single Ca^{2+}-dependent protein kinase confers both cold and salt/drought tolerance on rice plants, *Plant J.*, 23, 319–327, 2000.
48. Townley, H.E. and Knight, M.R., Calmodulin as a potential negative regulator of *Arabidopsis* COR gene expression, *Plant Physiol.*, 128, 1169–1172, 2002.
49. Kudla, J., Xu, Q., Harter, K., Gruissem, W., and Luan, S., Genes for calcineurin B-like proteins in *Arabidopsis* are differentially regulated by stress signals, *Proc. Natl. Acad. Sci. USA*, 96, 4718–4723, 1999.
50. Cheong, Y.H., Kim, K.N., Pandey, G.K., Gupta, R., Grant, J.J., and Luan, S., CBL1, a calcium sensor that differentially regulates salt, drought, and cold responses in *Arabidopsis*, *Plant Cell*, 15, 1833–1845, 2003.
51. Kim, K.N., Cheong, Y.H., Grant, J.J., Pandey, G.K., and Luan, S., CIPK3, a calcium sensor-associated protein kinase that regulates abscisic acid and cold signal transduction in *Arabidopsis*, *Plant Cell*, 15, 411–423, 2003.
52. Nakagami, H., Pitzschke, A., and Hirt, H., Emerging MAP kinase pathways in plant stress signaling, *Trends Plant Sci.*, 10, 339–346, 2005.
53. Shinozaki, K. and Yamaguchi-Shinozaki, K., Molecular responses to dehydration and low temperature: differences and cross-talk between two stress signaling pathways, *Curr. Opin. Plant Biol.*, 3, 217–223, 2000.
54. Heino, P. and Palva, E.T., Signal transduction in plant cold acclimation, in *Plant Responses to Abiotic Stress*, Vol. 4, Hirt, H. and Shinozaki, K., Eds., Springer-Verlag, Berlin, 2003, 151–186.
55. Seki, M., Narusaka, M., Abe, H., Kasuga, M., Yamaguchi-Shinozaki, K., Carninci, P., Hayashizaki, Y., and Shinozaki, K., Monitoring the expression pattern of 1300 *Arabidopsis* genes under drought and cold stresses by using a full-length cDNA microarray, *Plant Cell*, 13, 61–72, 2001.
56. Fowler, S. and Thomashow, M.F., *Arabidopsis* transcriptome profiling indicates that multiple regulatory pathways are activated during cold acclimation in addition to the CBF cold response pathway, *Plant Cell*, 14, 1675–1690, 2002.
57. Kreps, J.A., Wu, Y., Chang, H.S., Zhu, T., Wang, X. and Harper, J.F., Transcriptome changes for *Arabidopsis* in response to salt, osmotic, and cold stress, *Plant Physiol.*, 130, 2129–2141, 2002.
58. Thomashow, M.F., So what's new in the field of plant cold acclimation? Lots, *Plant Physiol.*, 125, 89–93, 2001.
59. Zhu, J.K., Cell signaling under salt, water and cold stresses, *Curr. Opin. Plant Biol.*, 4, 401–406, 2001.
60. Viswanathan, C. and Zhu, J.K., Molecular genetic analysis of cold-regulated gene transcription, *Philos. Trans. R. Soc. Lond. B. Biol. Sci.*, 357, 877–886, 2002.

61. Shinozaki, K., Yamaguchi-Shinozaki, K., and Seki, M., Regulatory network of gene expression in the drought and cold stress responses, *Curr. Opin. Plant Biol.*, 6, 410–417, 2003.
62. Ishitani, M., Xiong, L., Stevenson, B., and Zhu, J.K., Genetic analysis of osmotic and cold stress signal transduction in *Arabidopsis*: interactions and convergence of abscisic acid-dependent and abscisic acid-independent pathways, *Plant Cell*, 9, 1935–1949, 1997.
63. Chinnusamy, V., Schumaker, K., and Zhu, J.K., Molecular genetic perspectives on cross-talk and specificity in abiotic stress signalling in plants, *J. Exp. Bot.*, 55, 225–236, 2004.
64. Nordin, K., Heino, P., and Palva, E.T., Separate signal pathways regulate the expression of a low-temperature-induced gene in *Arabidopsis thaliana* (L.) Heynh, *Plant Mol. Biol.*, 16, 1061–1071, 1991.
65. Nordin, K., Vahala, T., and Palva, E.T., Differential expression of two related, low-temperature-induced genes in *Arabidopsis thaliana* (L.) Heynh, *Plant Mol. Biol.*, 21, 641–653, 1993.
66. Knight, H., Zarka, D.G., Okamoto, H., Thomashow, M.F., and Knight, M.R., Abscisic acid induces CBF gene transcription and subsequent induction of cold-regulated genes via the CRT promoter element, *Plant Physiol.*, 135, 1710–1717, 2004.
67. Gilmour, S.J., Zarka, D.G., Stockinger, E.J., Salazar, M.P., Houghton, J.M., and Thomashow, M.F., Low temperature regulation of the *Arabidopsis* CBF family of AP2 transcriptional activators as an early step in cold-induced COR gene expression, *Plant J.*, 16, 433–442, 1998.
68. Liu, Q., Kasuga, M., Sakuma, Y., Abe, H., Miura, S., Yamaguchi-Shinozaki, K., and Shinozaki, K., Two transcription factors, DREB1 and DREB2, with an EREBP/AP2 DNA binding domain separate two cellular signal transduction pathways in drought- and low-temperature-responsive gene expression, respectively, in *Arabidopsis*, *Plant Cell*, 10, 1391–1406, 1998.
69. Yamaguchi-Shinozaki, K. and Shinozaki, K., A novel *cis*-acting element in an *Arabidopsis* gene is involved in responsiveness to drought, low-temperature, or high-salt stress, *Plant Cell*, 6, 251–264, 1994.
70. Baker, S.S., Wilhelm, K.S., and Thomashow, M.F., The 5′-region of *Arabidopsis thaliana* cor15a has *cis*-acting elements that confer cold-, drought- and ABA-regulated gene expression, *Plant Mol. Biol.*, 24, 701–713, 1994.
71. Stockinger, E.J., Gilmour, S.J., and Thomashow, M.F., *Arabidopsis thaliana* CBF1 encodes an AP2 domain-containing transcriptional activator that binds to the C-repeat/DRE, a *cis*-acting DNA regulatory element that stimulates transcription in response to low temperature and water deficit, *Proc. Natl. Acad. Sci. USA*, 94, 1035–1040, 1997.
72. Kasuga, M., Liu, Q., Miura, S., Yamaguchi-Shinozaki, K., and Shinozak, K., Improving plant drought, salt, and freezing tolerance by gene transfer of a single stress-inducible transcription factor, *Nat. Biotechnol.*, 17, 287–291, 1999.
73. Haake, V., Cook, D., Riechmann, J.L., Pineda, O., Thomashow, M.F., and Zhang, J.Z., Transcription factor CBF4 is a regulator of drought adaptation in *Arabidopsis*, *Plant Physiol.*, 130, 639–648, 2002.
74. Medina, J., Bargues, M., Terol, J., Perez-Alonso, M., and Salinas, J., The *Arabidopsis* CBF gene family is composed of three genes encoding AP2 domain-containing proteins whose expression is regulated by low temperature but not by abscisic acid or dehydration, *Plant Physiol.*, 119, 463–469, 1999.

75. Chinnusamy, V., Ohta, M., Kanrar, S., Lee, B.H., Hong, X., Agarwal, M., and Zhu, J.K., ICE1: a regulator of cold-induced transcriptome and freezing tolerance in *Arabidopsis*, *Genes Dev.*, 17, 1043–1054, 2003.
76. Jaglo-Ottosen, K.R., Gilmour, S.J., Zarka, D.G., Schabenberger, O., and Thomashow, M.F., *Arabidopsis* CBF1 overexpression induces COR genes and enhances freezing tolerance, *Science*, 280, 104–106, 1998.
77. Gilmour, S.J., Sebolt, A.M., Salazar, M.P., Everard, J.D., and Thomashow, M.F., Overexpression of the *Arabidopsis* CBF3 transcriptional activator mimics multiple biochemical changes associated with cold acclimation, *Plant Physiol.*, 124, 1854–1865, 2000.
78. Seki, M., Narusaka, M., Ishida, J., Nanjo, T., Fujita, M., Oono, Y., Kamiya, A., Nakajima, M., Enju, A., Sakurai, T., Satou, M., Akiyama, K., Taji, T., Yamaguchi-Shinozaki, K., Carninci, P., Kawai, J., Hayashizaki, Y., and Shinozaki, K., Monitoring the expression profiles of 7000 *Arabidopsis* genes under drought, cold and high-salinity stresses using a full-length cDNA microarray, *Plant J.*, 31, 279–292, 2002.
79. Maruyama, K., Sakuma, Y., Kasuga, M., Ito, Y., Seki, M., Goda, H., Shimada, Y., Yoshida, S., Shinozaki, K., and Yamaguchi-Shinozaki, K., Identification of cold-inducible downstream genes of the *Arabidopsis* DREB1A/CBF3 transcriptional factor using two microarray systems, *Plant J.*, 38, 982–993, 2004.
80. Zhang, J.Z., Creelman, R.A., and Zhu, J.K., From laboratory to field. Using information from *Arabidopsis* to engineer salt, cold, and drought tolerance in crops, *Plant Physiol.*, 135, 615–621, 2004.
81. Vagujfalvi, A., Aprile, A., Miller, A., Dubcovsky, J., Delugu, G., Galiba, G., and Cattivelli, L., The expression of several Cbf genes at the Fr-A2 locus is linked to frost resistance in wheat, *Mol. Genet. Genomics*, 1–9, 2005.
82. Jaglo, K.R., Kleff, S., Amundsen, K.L., Zhang, X., Haake, V., Zhang, J.Z., Deits, T., and Thomashow, M.F., Components of the *Arabidopsis* C-repeat/dehydration-responsive element binding factor cold-response pathway are conserved in *Brassica napus* and other plant species, *Plant Physiol.*, 127, 910–917, 2001.
83. Choi, D.W., Rodriguez, E.M., and Close, T.J., Barley Cbf3 gene identification, expression pattern, and map location, *Plant Physiol.*, 129, 1781–1787, 2002.
84. Vagujfalvi, A., Galiba, G., Cattivelli, L., and Dubcovsky, J., The cold-regulated transcriptional activator Cbf3 is linked to the frost-tolerance locus Fr-A2 on wheat chromosome 5A, *Mol. Genet. Genomics*, 269, 60–67, 2003.
85. Xue, G.P., The DNA-binding activity of an AP2 transcriptional activator HvCBF2 involved in regulation of low-temperature responsive genes in barley is modulated by temperature, *Plant J.*, 33, 373–383, 2003.
86. Zhang, X., Fowler, S.G., Cheng, H., Lou, Y., Rhee, S.Y., Stockinger, E.J., and Thomashow, M.F., Freezing-sensitive tomato has a functional CBF cold response pathway, but a CBF regulon that differs from that of freezing-tolerant *Arabidopsis*, *Plant J.*, 39, 905–919, 2004.
87. Owens, C.L., Thomashow, M.F., Hancock, J.F., and Iezzoni, A.F., CBF1 orthologs in sour cherry and strawberry and the heterologous expression of CBF1 in strawberry, *J. Am. Soc. Hort. Sci.*, 127, 489–494, 2002.
88. Kitashiba, H., Ishizaka, T., Isuzugawa, K., Nishimura, K., and Suzuki, T., Expression of a sweet cherry DREB1/CBF ortholog in *Arabidopsis* confers salt and freezing tolerance, *J. Plant Physiol.*, 161, 1171–1176, 2004.
89. Puhakainen, T., Li, C., Boije-Malm, M., Kangasjarvi, J., Heino, P. and Palva, E.T., Short-day potentiation of low temperature-induced gene expression of a C-repeat-binding factor-controlled gene during cold acclimation in silver birch, *Plant Physiol.*, 136, 4299–4307, 2004.

90. Zhu, J.K., Salt and drought stress signal transduction in plants, *Annu. Rev. Plant Biol.*, 53, 247–273, 2002.
91. Verslues, P.E. and Zhu, J.K., Before and beyond ABA: upstream sensing and internal signals that determine ABA accumulation and response under abiotic stress, *Biochem. Soc. Trans.*, 33, 375–379, 2005.
92. Heino, P., Sandman, G., Lang, V., Nordin, K., and Palva, E.T., Abscisic-acid deficiency prevents development of freezing tolerance in *Arabidopsis thaliana* (L.) Heynh, *Theor. Appl. Genet.*, 79, 801–806, 1990.
93. Xiong, L., Ishitani, M., Lee, H., and Zhu, J.K., The *Arabidopsis* LOS5/ABA3 locus encodes a molybdenum cofactor sulfurase and modulates cold stress- and osmotic stress-responsive gene expression, *Plant Cell*, 13, 2063–2083, 2001.
94. Xiong, L., Schumaker, K.S., and Zhu, J.K., Cell signaling during cold, drought, and salt stress, *Plant Cell*, 14, S165–S183, 2002.
95. Rapacz, M., Waligorski, P., and Janowiak, F., ABA and gibberellin-like substances during prehardening, cold acclimation, de- and reacclimation of oilseed rape, *Acta Physiol. Plantarum*, 25, 151–161, 2003.
96. Scott, I.M., Clarke, S.M., Wood, J.E., and Mur, L.A., Salicylate accumulation inhibits growth at chilling temperature in *Arabidopsis*, *Plant Physiol.*, 135, 1040–1049, 2004.
97. Zhu, J., Verslues, P.E., Zheng, X., Lee, B.H., Zhan, X., Manabe, Y., Sokolchik, I., Zhu, Y., Dong, C.H., Zhu, J.K., Hasegawa, P.M., and Bressan, R.A., HOS10 encodes an R2R3-type MYB transcription factor essential for cold acclimation in plants, *Proc. Natl. Acad. Sci. USA*, 102, 9966–9971, 2005.
98. Choi, H., Hong, J., Ha, J., Kang, J., and Kim, S.Y., ABFs, a family of ABA-responsive element-binding factors, *J. Biol. Chem.*, 275, 1723–1730, 2000.
99. Uno, Y., Furihata, T., Abe, H., Yoshida, R., Shinozaki, K., and Yamaguchi-Shinozaki, K., *Arabidopsis* basic leucine zipper transcription factors involved in an abscisic acid-dependent signal transduction pathway under drought and high-salinity conditions, *Proc. Natl. Acad. Sci. USA*, 97, 11632–11637, 2000.
100. Yamaguchi-Shinozaki, K. and Shinozaki, K., Organization of *cis*-acting regulatory elements in osmotic- and cold-stress-responsive promoters, *Trends Plant Sci.*, 10, 88–94, 2005.
101. Kim, J.C., Lee, S.H., Cheong, Y.H., Yoo, C.M., Lee, S.I., Chun, H.J., Yun, D.J., Hong, J.C., Lee, S.Y., Lim, C.O., and Cho, M.J., A novel cold-inducible zinc finger protein from soybean, SCOF-1, enhances cold tolerance in transgenic plants, *Plant J.*, 25, 247–259, 2001.
102. Cakar, Z.P., Seker, U.O., Tamerler, C., Sonderegger, M., and Sauer, U., Evolutionary engineering of multiple-stress resistant *Saccharomyces cerevisiae*, *FEMS Yeast Res.*, 5, 569–578, 2005.
103. Guy, C.L., Niemi, K.J., and Brambl, R., Altered gene expression during cold acclimation of spinach, *Proc. Natl. Acad. Sci. USA*, 82, 3673–3677, 1985.
104. Thomashow, M.F., Role of cold-responsive genes in plant freezing tolerance, *Plant Physiol.*, 118, 1–8, 1998.
105. Stockinger, E.J., Mao, Y., Regier, M.K., Triezenberg, S.J., and Thomashow, M.F., Transcriptional adaptor and histone acetyltransferase proteins in *Arabidopsis* and their interactions with CBF1, a transcriptional activator involved in cold-regulated gene expression, *Nucleic Acids Res.*, 29, 1524–1533, 2001.
106. Kim, H.J., Hyun, Y., Park, J.Y., Park, M.J., Park, M.K., Kim, M.D., Kim, H.J., Lee, M.H., Moon, J., Lee, I., and Kim, J., A genetic link between cold responses and flowering time through FVE in *Arabidopsis thaliana*, *Nat. Genet.*, 36, 167–171, 2004.

107. Chiatante, D., Scippa, G.S., Maiuro, L., Onelli, E., and Patrignani, G., Modification of chromatin organization at low water potential in cultured cells of *Solarium tuberosum*: possible involvement of dehydrins, *Plant Biosyst.*, 136, 35–47, 2002.
108. Cremonini, R., Castiglione, M.R., Grif, V.G., Kotseruba, V.V., Punina, E.O., Rodionov, A.V., Muravenko, O.V., Popov, K.V., Samatadze, T.E., and Zelenin, A.V., Chromosome banding and DNA methylation patterns, chromatin organisation and nuclear DNA content in *Zingeria biebersteiniana*, *Biol. Plant.*, 46, 543–550, 2003.
109. Los, D.A., The effect of low-temperature-induced DNA supercoiling on the expression of the desaturase genes in synechocystis, *Cell. Mol. Biol.*, 50, 605–612, 2004.
110. Scippa, G.S., Di Michele, M., Onelli, E., Patrignani, G., Chiatante, D., and Bray, E.A., The histone-like protein H1-S and the response of tomato leaves to water deficit, *J. Exp. Bot.*, 55, 99–109, 2004.
111. Bray, E.A., Molecular responses to water deficit, *Plant Physiol.*, 103, 1035–1040, 1993.
112. Wise, M.J., LEAping to conclusions: a computational reanalysis of late embryogenesis abundant proteins and their possible roles, *BMC Bioinformatics*, 4, 52, 2003.
113. Wolkers, W.F., McCready, S., Brandt, W.F., Lindsey, G.G., and Hoekstra, F.A., Isolation and characterization of a D-7 LEA protein from pollen that stabilizes glasses *in vitro*, *Biochim. Biophys. Acta*, 1544, 196–206, 2001.
114. Ingram, J. and Bartels, D., The molecular basis for dehydration tolerance in plants, *Annu. Rev. Plant Physiol. Plant Mol. Biol.*, 47, 377–403, 1996.
115. Cuming, A.C., LEA proteins, in *Seed Proteins*, Shewry, P.R. and Casey, R., Eds., Kluwer Academic Publishers, The Netherlands, 1999, 753–780.
116. Palva, E.T. and Heino, P., Molecular mechanisms of plant cold acclimation and freezing tolerance, in *Plant Cold Hardiness: Molecular Biology, Biochemistry and Physiology,* Li, P.H. and Chen, T.H.H., Eds., Plenum Press, New York, 1998, 1–14.
117. Garay-Arroyo, A., Colmenero-Flores, J.M., Garciarrubio, A., and Covarrubias, A.A., Highly hydrophilic proteins in prokaryotes and eukaryotes are common during conditions of water deficit *J. Biol. Chem.*, 275, 5668–5674, 2000.
118. Swire-Clark, G.A. and Marcotte, W.R., Jr., The wheat LEA protein Em functions as an osmoprotective molecule in *Saccharomyces cerevisiae*, *Plant Mol. Biol.*, 39, 117–128, 1999.
119. Koag, M.C., Fenton, R.D., Wilkens, S., and Close, T.J., The binding of maize DHN1 to lipid vesicles. Gain of structure and lipid specificity, *Plant Physiol.*, 131, 309–316, 2003.
120. Soulages, J.L., Kim, K., Arrese, E.L., Walters, C., and Cushman, J.C., Conformation of a group 2 late embryogenesis abundant protein from soybean. Evidence of poly (L-proline)-type II structure, *Plant Physiol.*, 131, 963–975, 2003.
121. Allagulova, C., Gimalov, F.R., Shakirova, F.M., and Vakhitov, V.A., The plant dehydrins: structure and putative functions, *Biochemistry* (Moscow), 68, 945–951, 2003.
122. Goyal, K., Walton, L.J., and Tunnacliffe, A., LEA proteins prevent protein aggregation due to water stress, *Biochem. J.*, 388, 151–157, 2005.
123. Imai, R., Chang, L., Ohta, A., Bray, E.A., and Takagi, M., An LEA class gene of tomato confers salt and freezing tolerance when expressed in *Saccharomyces cerevisiae*, *Gene*, 170, 243–248, 1996.
124. Kim, H.S., Lee, J.H., Kim, J.J., Kim, C.H., Jun, S.S., and Hong, Y.N., Molecular and functional characterization of CaLEA6, the gene for a hydrophobic LEA protein from *Capsicum annuum*, *Gene*, 344, 115–123, 2005.
125. Wise, M.J. and Tunnacliffe, A., POPP the question: what do LEA proteins do? *Trends Plant Sci.*, 9, 13–17, 2004.

126. Hara, M., Terashima, S., Fukaya, T., and Kuboi, T., Enhancement of cold tolerance and inhibition of lipid peroxidation by citrus dehydrin in transgenic tobacco, *Planta*, 217, 290–298, 2003.
127. Xu, D., Duan, X., Wang, B., Hong, B., Ho, T., and Wu, R., Expression of a late embryogenesis abundant protein gene, HVA1, from barley confers tolerance to water deficit and salt stress in transgenic rice, *Plant Physiol.*, 110, 249–257, 1996.
128. Zhang, L., Ohta, A., Takagi, M., and Imai, R., Expression of plant group 2 and group 3 lea genes in *Saccharomyces cerevisiae* revealed functional divergence among LEA proteins, *J. Biochem.* (Tokyo), 127, 611–616, 2000.
129. Babu, R.C., Zhang, J., Blum, A., Ho, T.-.D., Wu, R., and Nguyen, H.T., HVA1, a LEA gene from barley confers dehydration tolerance in transgenic rice (*Oryza sativa* L.) via cell membrane protection, *Plant Sci.*, 166, 855–862, 2004.
130. Sutton, F., Ding, X., and Kenefick, D.G., Group 3 LEA gene HVA1 regulation by cold acclimation and deacclimation in two barley cultivars with varying freeze resistance, *Plant Physiol.*, 99, 338–340, 1992.
131. Sales, K., Brandt, W., Rumbak, E., and Lindsey, G., The LEA-like protein HSP 12 in *Saccharomyces cerevisiae* has a plasma membrane location and protects membranes against desiccation and ethanol-induced stress, *Biochim. Biophys. Acta*, 1463, 267–278, 2000.
132. Close, T.J., Dehydrins: emergence of a biochemical role of a family of plant dehydration proteins, *Physiol. Plantarum*, 97, 795–803, 1996.
133. Nylander, M., Svensson, J., Palva, E.T., and Welin, B.V., Stress-induced accumulation and tissue-specific localization of dehydrins in *Arabidopsis thaliana*, *Plant Molecular Biol.*, 45, 263–279, 2001.
134. Kruger, C., Berkowitz, O., Stephan, U.W., and Hell, R., A metal-binding member of the late embryogenesis abundant protein family transports iron in the phloem of *Ricinus communis* L., *J. Biol. Chem.*, 277, 25062–25069, 2002.
135. Heyen, B.J., Alsheikh, M.K., Smith, E.A., Torvik, C.F., Seals, D.F., and Randall, S.K., The calcium-binding activity of a vacuole-associated, dehydrin-like protein is regulated by phosphorylation, *Plant Physiol.*, 130, 675–687, 2002.
136. Puhakainen, T., Hess, M.W., Makela, P., Svensson, J., Heino, P., and Palva, E.T., Overexpression of multiple dehydrin genes enhances tolerance to freezing stress in *Arabidopsis*, *Plant Molecular Biol.*, 54, 743–753, 2004.
137. Iturriaga, G., Schneider, K., Salamini, F., and Bartels, D., Expression of desiccation-related proteins from the resurrection plant *Craterostigma plantagineum* in transgenic tobacco, *Plant Molecular Biol.*, 20, 555–558, 1992.
138. Park, S.H., Jun, S.S., An, G.H., Hong, Y.N., and Park, M.C., A comparative study on the protective role of trehalose and LEA proteins against abiotic stresses in transgenic Chinese cabbage (*Brassica campestris*) overexpressing CaLEA or otsA, *J. Plant Biol.*, 46, 277–286, 2003.
139. Artus, N.N., Uemura, M., Steponkus, P.L., Gilmour, S.J., Lin, C., and Thomashow, M.F., Constitutive expression of the cold-regulated *Arabidopsis thaliana* COR15a gene affects both chloroplast and protoplast freezing tolerance, *Proc. Natl. Acad. Sci. USA*, 93, 13404–13409, 1996.
140. McKersie, B.D., Murnaghan, J., and Bowley, S.R., Manipulating freezing tolerance in transgenic plants, *Acta Physiol. Plantarum*, 19, 485–495, 1997.
141. Lang, V. and Palva, E.T., The expression of a rab-related gene, rab18, is induced by abscisic acid during the cold acclimation process of *Arabidopsis thaliana* (L.) Heynh, *Plant Molecular Biol.*, 20, 951–962, 1992.

142. Welin, B.V., Olson, A., Nylander, M., and Palva, E.T., Characterization and differential expression of dhn/lea/rab-like genes during cold acclimation and drought stress in *Arabidopsis thaliana*, *Plant Molecular Biol.*, 26, 131–144, 1994.
143. Lang, V. The role of ABA and ABA-induced gene expression in cold acclimation of *Arabidopsis thaliana*, Dissertation, Swedish University of Agricultural Sciences, Uppsala Genetic Center, Uppsala, Sweden, 1993.
144. Steponkus, P.L., Role of the plasma membrane in freezing injury and cold acclimation, *Annu. Rev. Plant Physiol.*, 35, 543–584, 1984.
145. Xin, Z. and Browse, J., Cold comfort farm: the acclimation of plants to freezing temperatures, *Plant Cell Environ.*, 23, 893–902, 2000.
146. Kawamura, Y. and Uemura, M., Mass spectrometric approach for identifying putative plasma membrane proteins of *Arabidopsis* leaves associated with cold acclimation, *Plant J.*, 36, 141–154, 2003.
147. Steponkus, P.L., Uemura, M., Joseph, R.A., Gilmour, S.J., and Thomashow, M.F., Mode of action of the COR15a gene on the freezing tolerance of *Arabidopsis thaliana*, *Proc. Natl. Acad. Sci. USA*, 95, 14570–14575, 1998.
148. Uemura, M., Joseph, R.A., and Steponkus, P.L., Cold acclimation of *Arabidopsis thaliana*: effect on plasma membrane lipid composition and freeze-induced lesions, *Plant Physiol.*, 109, 15–30, 1995.
149. Newton, S.S. and Duman, J.G., An osmotin-like cryoprotective protein from the bittersweet nightshade *Solanum dulcamara*, *Plant Molecular Biol.*, 44, 581–589, 2000.
150. Hincha, D.K., Neukamm, B., Sror, H.A., Sieg, F., Weckwarth, W., Ruckels, M., Lullien-Pellerin, V., Schroder, W., and Schmitt, J.M., Cabbage cryoprotectin is a member of the nonspecific plant lipid transfer protein gene family, *Plant Physiol.*, 125, 835–846, 2001.
151. Sror, H.A.M., Tischendorf, G., Sieg, F., Schmitt, J.M., and Hincha, D.K., Cryoprotectin protects thylakoids during a freeze–thaw cycle by a mechanism involving stable membrane binding, *Cryobiology*, 47, 191–203, 2003.
152. Griffith, M. and Yaish, M.W., Antifreeze proteins in overwintering plants: a tale of two activities, *Trends Plant Sci.*, 9, 399–405, 2004.
153. Gilbert, J.A., Davies, P.L., and Laybourn-Parry, J., A hyperactive, Ca^{2+}-dependent antifreeze protein in an Antarctic bacterium, *FEMS Microbiol. Lett.*, 245, 67–72, 2005.
154. Bertrand, A. and Castonguay, Y., Plant adaptations to overwintering stresses and implications of climate change, *Can. J. Bot.—Rev. Can. Bot.*, 81, 1145–1152, 2003.
155. McKersie, B.D., Bowley, S.R., Harjanto, E., and Leprince, O., Water-deficit tolerance and field performance of transgenic alfalfa overexpressing superoxide dismutase, *Plant Physiol.*, 111, 1177–1181, 1996.
156. McKersie, B.D., Murnaghan, J., Jones, K.S., and Bowley, S.R., Iron-superoxide dismutase expression in transgenic alfalfa increases winter survival without a detectable increase in photosynthetic oxidative stress tolerance, *Plant Physiol.*, 122, 1427–1438, 2000.
157. van Camp, W., Capiau, K., Montagu, M.V., Inze, D., and Slooten, L., Enhancement of oxidative stress tolerance in transgenic tobacco plants overproducing Fe superoxide dismutase in chloroplasts, *Plant Physiol.*, 112, 1703–1714, 1996.
158. Samis, K., Bowley, S., and McKersie, B., Pyramiding Mn-superoxide dismutase transgenes to improve persistence and biomass production in alfalfa, *J. Exp. Bot.*, 53, 1343–1350, 2002.
159. Gupta, A.S., Webb, R.P., Holaday, A.S., and Allen, R.D., Overexpression of superoxide dismutase protects plants from oxidative stress: induction of ascorbate peroxidase in superoxide dismutase-overexpressing plants, *Plant Physiol.*, 103, 1067–1073, 1993.

160. Kingston-Smith, A.H. and Foyer, C.H., Overexpression of Mn-superoxide dismutase in maize leaves leads to increased monodehydroascorbate reductase, dehydroascorbate reductase and glutathione reductase activities, *J. Exp. Bot.*, 51, 1867–1877, 2000.
161. Mittler, R., Oxidative stress, antioxidants and stress tolerance, *Trends Plant Sci.*, 7, 405–410, 2002.
162. Shigeoka, S., Ishikawa, T., Tamoi, M., Miyagawa, Y., Takeda, T., Yabuta, Y., and Yoshimura, K., Regulation and function of ascorbate peroxidase isoenzymes, *J. Exp. Bot.*, 53, 1305–1319, 2002.
163. Yabuta, Y., Motoki, T., Yoshimura, K., Takeda, T., Ishikawa, T., and Shigeoka, S., Thylakoid membrane-bound ascorbate peroxidase is a limiting factor of antioxidative systems under photo-oxidative stress, *Plant J.*, 32, 915–925, 2002.
164. Murgia, I., Tarantino, D., Vannini, C., Bracale, M., Carravieri, S., and Soave, C., *Arabidopsis thaliana* plants overexpressing thylakoidal ascorbate peroxidase show increased resistance to paraquat-induced photo-oxidative stress and to nitric oxide-induced cell death, *Plant J.*, 38, 940–953, 2004.
165. Wang, Y.J., Wisniewski, M., Meilan, R., Cui, M.G., Webb, R., and Fuchigami, L., Overexpression of cytosolic ascorbate peroxidase in tomato confers tolerance to chilling and salt stress, *J. Am. Soc. Hort. Sci.*, 130, 167–173, 2005.
166. Funatsuki, H., Kurosaki, H., Murakami, T., Matsuba, S., Kawaguchi, K., Yumoto, S., and Sato, Y., Deficiency of a cytosolic ascorbate peroxidase associated with chilling tolerance in soybean, *Theor. Appl. Genet.*, 106, 494–502, 2003.
167. Ishikawa, T., Morimoto, Y., Madhusudhan, R., Sawa, Y., Shibata, H., Yabuta, Y., Nishizawa, A., and Shigeoka, S., Acclimation to diverse environmental stresses caused by a suppression of cytosolic ascorbate peroxidase in tobacco BY-2 cells, *Plant Cell Physiol.*, 46, 1264–1271, 2005.
168. Murata, N. and Los, D.A., Membrane fluidity and temperature perception, *Plant Physiol.*, 115, 875–879, 1997.
169. Mikami, K. and Murata, N., Membrane fluidity and the perception of environmental signals in cyanobacteria and plants, *Prog. Lipid Res.*, 42, 527–543, 2003.
170. Los, D.A. and Murata, N., Membrane fluidity and its roles in the perception of environmental signals, *Biochim. Biophys. Acta*, 1666, 142–157, 2004.
171. Inaba, M., Suzuki, I., Szalontai, B., Kanesaki, Y., Los, D.A., Hayashi, H., and Murata, N., Gene-engineered rigidification of membrane lipids enhances the cold inducibility of gene expression in synechocystis, *J. Biol. Chem.*, 278, 12191–12198, 2003.
172. Sung, D.Y., Kaplan, F., Lee, K.J., and Guy, C.L., Acquired tolerance to temperature extremes, *Trends Plant Sci.*, 8, 179–187, 2003.
173. Harwood, J.L., Fatty-acid metabolism, *Annu. Rev. Plant Physiol. Plant Molecular Biol.*, 39, 101–138, 1988.
174. Wallis, J.G. and Browse, J., Mutants of *Arabidopsis* reveal many roles for membrane lipids, *Prog. Lipid Res.*, 41, 254–278, 2002.
175. Iba, K., Acclimative response to temperature stress in higher plants: approaches of gene engineering for temperature tolerance, *Annu. Rev. Plant. Biol.*, 53, 225–245, 2002.
176. Kodama, H., Hamada, T., Horiguchi, G., Nishimura, M., and Iba, K., Genetic enhancement of cold tolerance by expression of a gene for chloroplast [omega]-3 fatty acid desaturase in transgenic tobacco, *Plant Physiol.*, 105, 601–605, 1994.
177. Orlova, I.V., Serebriiskaya, T.S., Popov, V., Merkulova, N., Nosov, A.M., Trunova, T.I., Tsydendambaev, V.D., and Los, D.A., Transformation of tobacco with a gene

for the thermophilic acyl-lipid desaturase enhances the chilling tolerance of plants, *Plant Cell Physiol.*, 44, 447–450, 2003.

178. Murakami, Y., Tsuyama, M., Kobayashi, Y., Kodama, H., and Iba, K., Trienoic fatty acids and plant tolerance of high temperature, *Science*, 287, 476–479, 2000.
179. Murata, N., Ishizakinishizawa, Q., Higashi, S., Hayashi, H., Tasaka, Y., and Nishida, I., Genetically engineered alteration in the chilling sensitivity of plants, *Nature*, 356, 710–713, 1992.
180. Ariizumi, T., Kishitani, S., Inatsugi, R., Nishida, I., Murata, N., and Toriyama, K., An increase in unsaturation of fatty acids in phosphatidylglycerol from leaves improves the rates of photosynthesis and growth at low temperatures in transgenic rice seedlings, *Plant Cell Physiol.*, 43, 751–758, 2002.
181. Sakamoto, A., Sulpice, R., Hou, C.X., Kinoshita, M., Higashi, S.I., Kanaseki, T., Nonaka, H., Moon, B.Y., and Murata, N., Genetic modification of the fatty acid unsaturation of phosphatidylglycerol in chloroplasts alters the sensitivity of tobacco plants to cold stress, *Plant Cell Environ.*, 27, 99–105, 2004.
182. Gomes, E., Jakobsen, M.K., Axelsen, K.B., Geisler, M., and Palmgren, M.G., Chilling tolerance in *Arabidopsis* involves ALA1, a member of a new family of putative aminophospholipid translocases, *Plant Cell*, 12, 2441–2454, 2000.
183. Ohta, S., Ishida, Y., and Usami, S., Expression of cold-tolerant pyruvate, orthophosphate dikinase cDNA, and heterotetramer formation in transgenic maize plants, *Transgenic Res.*, 13, 475–485, 2004.
184. Yancey, P.H., Clark, M.E., Hand, S.C., Bowlus, R.D., and Somero, G.N., Living with water stress: evolution of osmolyte systems, *Science*, 217, 1214–1222, 1982.
185. Yancey, P.H., Compatible and counteracting solutes: protecting cells from the Dead Sea to the deep sea, *Sci. Prog.*, 87, 1–24, 2004.
186. Hare, P.D., Cress, W.A., and Van Staden, J., Dissecting the roles of osmolyte accumulation during stress, *Plant Cell Environ.*, 21, 535–553, 1998.
187. Yancey, P.H., Compatible and counteracting solutes, in *Cellular and Molecular Physiology of Cell Volume Regulation*, Strange, K., Ed., CRC Press, Boca Raton, FL, 1994, 81–109.
188. Ahmad, I., Larher, F., and Stewart, G.R., Sorbitol, a compatible osmotic solute in *Plantago maritima*, *New Phytol.*, 82, 671–678, 1979.
189. Rhodes, D. and Hanson, A.D., Quaternary ammonium and tertiary sulfonium compounds in higher plants, *Annu. Rev. Plant Physiol. Plant Mol. Biol.*, 44, 357–384, 1993.
190. Bohnert, H.J., Nelson, D.E., and Jensen, R.G., Adaptations to environmental stresses, *Plant Cell*, 7, 1099–1111, 1995.
191. Cayley, S. and Record, M.T., Jr., Roles of cytoplasmic osmolytes, water, and crowding in the response of *Escherichia coli* to osmotic stress: biophysical basis of osmoprotection by glycine betaine, *Biochemistry*, 42, 12596–12609, 2003.
192. Larher, F.R., Aziz, A., Gibon, Y., Trotel-Aziz, P., Sulpice, R., and Bouchereau, A., An assessment of the physiological properties of the so-called compatible solutes using *in vitro* experiments with leaf discs, *Plant Physiol. Biochem.*, 41, 657–666, 2003.
193. Hare, P.D. and Cress, W.A., Metabolic implications of stress-induced proline accumulation in plants, *Plant Growth Regul.*, 21, 79–102, 1997.
194. Diamant, S., Rosenthal, D., Azem, A., Eliahu, N., Ben-Zvi, A.P., and Goloubinoff, P., Dicarboxylic amino acids and glycine-betaine regulate chaperone-mediated protein-disaggregation under stress, *Mol. Microbiol.*, 49, 401–410, 2003.

195. Wang, W., Vinocur, B., and Altman, A., Plant responses to drought, salinity and extreme temperatures: towards genetic engineering for stress tolerance, *Planta*, 218, 1–14, 2003.
196. Hanson, A.D., Rathinasabapathi, B., Rivoal, J., Burnet, M., Dillon, M.O., and Gage, D.A., Osmoprotective compounds in the Plumbaginaceae: a natural experiment in metabolic engineering of stress tolerance, *Proc. Natl. Acad. Sci. USA*, 91, 306–310, 1994.
197. McCue, K.F. and Hanson, A.D., Drought and salt tolerance: towards understanding and application, *Trends Biotechnol.*, 8, 358–362, 1990.
198. Rathinasabapathi, B., Metabolic engineering for stress tolerance: installing osmoprotectant synthesis pathways, *Ann. Bot.*, 86, 709–716, 2000.
199. McNeil, S.D., Nuccio, M.L., Ziemak, M.J., and Hanson, A.D., Enhanced synthesis of choline and glycine betaine in transgenic tobacco plants that overexpress phosphoethanolamine *N*-methyltransferase, *Proc. Natl. Acad. Sci. USA*, 98, 10001–10005, 2001.
200. Chen, T.H.H. and Murata, N., Enhancement of tolerance of abiotic stress by metabolic engineering of betaines and other compatible solutes, *Curr. Opin. Plant Biol.*, 5, 250–257, 2002.
201. Nanjo, T., Kobayashi, M., Yoshiba, Y., Kakubari, Y., Yamaguchi-Shinozaki, K., and Shinozaki, K., Antisense suppression of proline degradation improves tolerance to freezing and salinity in *Arabidopsis thaliana*, *FEBS Lett.*, 461, 205–210, 1999.
202. Flowers, T.J., Improving crop salt tolerance, *J. Exp. Bot.*, 55, 307–319, 2004.
203. Sakamoto, A. and Murata, N., The role of glycine betaine in the protection of plants from stress: clues from transgenic plants, *Plant Cell Environ.*, 25, 163–171, 2002.
204. Sakamoto, A. and Murata, N., The use of bacterial choline oxidase, a glycinebetaine-synthesizing enzyme, to create stress-resistant transgenic plants, *Plant Physiol.*, 125, 180–188, 2001.
205. Kumar, S., Dhingra, A., and Daniell, H., Plastid-expressed betaine aldehyde dehydrogenase gene in carrot cultured cells, roots, and leaves confers enhanced salt tolerance, *Plant Physiol.*, 136, 2843–2854, 2004.
206. Prasad, K.V.S.K. and Saradhi, P.P., Enhanced tolerance to photoinhibition in transgenic plants through targeting of glycinebetaine biosynthesis into the chloroplasts, *Plant Sci.*, 166, 1197–1212, 2004.
207. Konstantinova, T., Parvanova, D., Atanassov, A., and Djilianov, D., Freezing tolerant tobacco, transformed to accumulate osmoprotectants, *Plant Sci.*, 163, 157–164, 2002.
208. Park, E.J., Jeknic, Z., Sakamoto, A., DeNoma, J., Yuwansiri, R., Murata, N., and Chen, T.H., Genetic engineering of glycinebetaine synthesis in tomato protects seeds, plants, and flowers from chilling damage, *Plant J.*, 40, 474–487, 2004.
209. Holmstrom, K.O., Somersalo, S., Mandal, A., Palva, E.T., and Welin, B., Improved tolerance to salinity and low temperature in transgenic tobacco producing glycine betaine, *J. Exp. Bot.*, 51, 177–185, 2000.
210. Rontein, D., Rhodes, D., and Hanson, A.D., Evidence from engineering that decarboxylation of free serine is the major source of ethanolamine moieties in plants, *Plant Cell Physiol.*, 44, 1185–1191, 2003.
211. Tabuchi, T., Kawaguchi, Y., Azuma, T., Nanmori, T., and Yasuda, T., Similar regulation patterns of choline monooxygenase, phosphoethanolamine *N*-methyltransferase and S-adenosyl-L-methionine synthetase in leaves of the halophyte *Atriplex nummularia* L., *Plant Cell Physiol.*, 46, 505–513, 2005.
212. Nyyssola, A., Kerovuo, J., Kaukinen, P., von Weymarn, N., and Reinikainen, T., Extreme halophiles synthesize betaine from glycine by methylation, *J. Biol. Chem.*, 275, 22196–22201, 2000.

213. von Weymarn, N., Nyyssola, A., Reinikainen, T., Leisola, M., and Ojamo, H., Improved osmotolerance of recombinant *Escherichia coli* by *de novo* glycine betaine biosynthesis, *Appl. Microbiol. Biotechnol.*, 55, 214–218, 2001.
214. Waditee, R., Bhuiyan, M.N.H., Rai, V., Aoki, K., Tanaka, Y., Hibino, T., Suzuki, S., Takano, J., Jagendorf, A.T., Takabe, T., and Takabe, T., Genes for direct methylation of glycine provide high levels of glycinebetaine and abiotic-stress tolerance in *Synechococcus* and *Arabidopsis*, *Proc. Natl. Acad. Sci. USA*, 102, 1318–1323, 2005.
215. Huang, J., Hirji, R., Adam, L., Rozwadowski, K.L., Hammerlindl, J.K., Keller, W.A., and Selvaraj, G., Genetic engineering of glycinebetaine production toward enhancing stress tolerance in plants: metabolic limitations, *Plant Physiol.*, 122, 747–756, 2000.
216. Gibon, Y., Bessieres, M.A., and Larher, F., Is glycine betaine a noncompatible solute in higher plants that do not accumulate it? *Plant Cell Environ.*, 20, 329–340, 1997.
217. Sulpice, R., Gibon, Y., Bouchereau, A., and Larher, F., Exogenously supplied glycine betaine in spinach and rapeseed leaf discs: compatibility or noncompatibility? *Plant Cell Environ.*, 21, 1285–1292, 1998.
218. Sulpice, R., Gibon, Y., Cornic, G., and Larher, F.R., Interaction between exogenous glycine betaine and the photorespiratory pathway in canola leaf discs, *Physiol. Plantarum*, 116, 460–467, 2002.
219. Hanson, A.D. and Wyse, R., Biosynthesis, translocation, and accumulation of betaine in sugar-beet and its progenitors in relation to salinity, *Plant Physiol.*, 70, 1191–1198, 1982.
220. Raman, S.B. and Rathinasabapathi, B., Beta-alanine N-methyltransferase of *Limonium latifolium.* cDNA cloning and functional expression of a novel *N*-methyltransferase implicated in the synthesis of the osmoprotectant β-alanine betaine, *Plant Physiol.*, 132, 1642–1651, 2003.
221. Nolte, K.D., Hanson, A.D., and Gage, D.A., Proline accumulation and methylation to proline betaine in citrus: implications for genetic engineering of stress resistance, *J. Am. Soc. Hortic. Sci.*, 122, 8–13, 1997.
222. McNeil, S.D., Nuccio, M.L., and Hanson, A.D., Betaines and related osmoprotectants, targets for metabolic engineering of stress resistance, *Plant Physiol.*, 120, 945–949, 1999.
223. Klotke, J., Kopka, J., Gatzke, N., and Heyer, A.G., Impact of soluble sugar concentrations on the acquisition of freezing tolerance in accessions of *Arabidopsis thaliana* with contrasting cold adaptation—evidence for a role of raffinose in cold acclimation, *Plant Cell Environ.*, 27, 1395–1404, 2004.
224. Hoshida, H., Tanaka, Y., Hibino, T., Hayashi, Y., Tanaka, A., Takabe, T., and Takabe, T., Enhanced tolerance to salt stress in transgenic rice that overexpresses chloroplast glutamine synthetase, *Plant Mol. Biol.*, 43, 103–111, 2000.
225. Kavi Kishor, P.B., Sangam, S., Amrutha, R.N., Sri Laxmi, P., Naidu, K.R., Rao, K.R.S.S., Sreenath Rao, Reddy, K.J., Theriappan, P., and Sreenivasulu, N., Regulation of proline biosynthesis, degradation, uptake and transport in higher plants: its implications in plant growth and abiotic stress tolerance, *Curr. Sci.*, 88, 424–438, 2005.
226. Yoshiba, Y., Kiyosue, T., Katagiri, T., Ueda, H., Mizoguchi, T., Yamaguchi-Shinozaki, K., Wada, K., Harada, Y., and Shinozaki, K., Correlation between the induction of a gene for delta 1-pyrroline-5-carboxylate synthetase and the accumulation of proline in *Arabidopsis thaliana* under osmotic stress, *Plant J.*, 7, 751–760, 1995.
227. Wanner, L.A. and Junttila, O., Cold-induced freezing tolerance in *Arabidopsis*, *Plant Physiol.*, 120, 391–400, 1999.
228. Hur, J., Jung, K.H., Lee, C.H., and An, G.H., Stress-inducible OsP5CS2 gene is essential for salt and cold tolerance in rice, *Plant Sci.*, 167, 417–426, 2004.

229. Nayyar, H., Bains, T., and Kumar, S., Low temperature induced floral abortion in chickpea: relationship to abscisic acid and cryoprotectants in reproductive organs, *Environ. Exp. Bot.*, 53, 39–47, 2005.
230. Mani, S., Van De Cotte, B., Van Montagu, M., and Verbruggen, N., Altered levels of proline dehydrogenase cause hypersensitivity to proline and its analogs in *Arabidopsis*, *Plant Physiol.*, 128, 73–83, 2002.
231. Nanjo, T., Fujita, M., Seki, M., Kato, T., Tabata, S., and Shinozaki, K., Toxicity of free proline revealed in an *Arabidopsis* T-DNA-tagged mutant deficient in proline dehydrogenase, *Plant Cell Physiol.*, 44, 541–548, 2003.
232. Hendry, G.A.F., Evolutionary origin and natural functions of fructans—a climatological, biogeographic and mechanistic appraisal, *New Phytol.*, 123, 3–14, 1993.
233. McKown, R., Kuroki, G., and Warren, G., Cold responses of *Arabidopsis* mutants impaired in freezing tolerance, *J. Exp. Bot.*, 47, 1919–1925, 1996.
234. Castonguay, Y. and Nadeau, P., Enzymatic control of soluble carbohydrate accumulation in cold-acclimated crowns of alfalfa, *Crop Sci.*, 38, 1183–1189, 1998.
235. Taji, T., Ohsumi, C., Iuchi, S., Seki, M., Kasuga, M., Kobayashi, M., Yamaguchi-Shinozaki, K., and Shinozaki, K., Important roles of drought- and cold-inducible genes for galactinol synthase in stress tolerance in *Arabidopsis thaliana*, *Plant J.*, 29, 417–426, 2002.
236. Rohde, P., Hincha, D.K., and Heyer, A.G., Heterosis in the freezing tolerance of crosses between two *Arabidopsis thaliana* accessions (Columbia-0 and C24) that show differences in nonacclimated and acclimated freezing tolerance, *Plant J.*, 38, 790–799, 2004.
237. Zuther, E., Buchel, K., Hundertmark, M., Stitt, M., Hincha, D.K., and Heyer, A.G., The role of raffinose in the cold acclimation response of *Arabidopsis thaliana*, *FEBS Lett.*, 576, 169–173, 2004.
238. Abebe, T., Guenzi, A.C., Martin, B., and Cushman, J.C., Tolerance of mannitol-accumulating transgenic wheat to water stress and salinity, *Plant Physiol.*, 131, 1748–1755, 2003.
239. Sauter, J.J. and van Cleve, B., Biochemical and ultrastructural results during starch-sugar-conversion in ray parenchyma cells of populus during cold adaptation, *J. Plant Physiol.*, 139, 19–26, 1991.
240. Nagao, M., Minami, A., Arakawa, K., Fujikawa, S., and Takezawa, D., Rapid degradation of starch in chloroplasts and concomitant accumulation of soluble sugars associated with ABA-induced freezing tolerance in the moss *Physcomitrefla patens*, *J. Plant Physiol.*, 162, 169–180, 2005.
241. Kaplan, F. and Guy, C.L., β-Amylase induction and the protective role of maltose during temperature shock, *Plant Physiol.*, 135, 1674–1684, 2004.
242. Bohnert, H.J. and Sheveleva, E., Plant stress adaptations—making metabolism move, *Curr. Opin. Plant Biol.*, 1, 267–274, 1998.
243. Bae, H., Herman, E., and Sicher, R., Exogenous trehalose promotes nonstructural carbohydrate accumulation and induces chemical detoxification and stress response proteins in *Arabidopsis thaliana* grown in liquid culture, *Plant Sci.*, 168, 1293–1301, 2005.
244. Avonce, N., Leyman, B., Thevelein, J., and Iturriaga, G., Trehalose metabolism and glucose sensing in plants, *Biochem. Soc. Trans.*, 33, 276–279, 2005.
245. Sheveleva, E.V., Marquez, S., Chmara, W., Zegeer, A., Jensen, R.G., and Bohnert, H.J., Sorbitol-6-phosphate dehydrogenase expression in transgenic tobacco. High amounts of sorbitol lead to necrotic lesions, *Plant Physiol.*, 117, 831–839, 1998.
246. Cairns, A.J., Fructan biosynthesis in transgenic plants, *J. Exp. Bot.*, 54, 549–567, 2003.

247. Hisano, H., Kanazawa, A., Kawakami, A., Yoshida, M., Shimamoto, Y., and Yamada, T., Transgenic perennial ryegrass plants expressing wheat fructosyltransferase genes accumulate increased amounts of fructan and acquire increased tolerance on a cellular level to freezing, *Plant Sci.*, 167, 861–868, 2004.
248. Nuccio, M.L., Rhodes, D., McNeil, S.D., and Hanson, A.D., Metabolic engineering of plants for osmotic stress resistance, *Curr. Opin. Plant Biol.*, 2, 128–134, 1999.
249. Pennycooke, J.C., Jones, M.L., and Stushnoff, C., Down-regulating alpha-galactosidase enhances freezing tolerance in transgenic petunia, *Plant Physiol.*, 133, 901–909, 2003.
250. Rudolph, A.S. and Crowe, J.H., Membrane stabilization during freezing: the role of two natural cryoprotectants, trehalose and proline, *Cryobiology*, 22, 367–377, 1985.
251. Holmstrom, K.O., Mantyla, E., Welin, B., Mandal, A., and Palva, E.T., Drought tolerance in tobacco, *Nature*, 379, 683–684, 1996.
252. Garg, A.K., Kim, J.K., Owens, T.G., Ranwala, A.P., Choi, Y.D., Kochian, L.V., and Wu, R.J., Trehalose accumulation in rice plants confers high tolerance levels to different abiotic stresses, *Proc. Natl. Acad. Sci. USA*, 99, 15898–15903, 2002.
253. Tamminen, I., Puhakainen, T., Makela, P., Holmstrom, K., Mueller, J., Heino, P., and Palva, E.T., Engineering trehalose biosynthesis improves stress tolerance in *Arabidopsis*, in *Plant Cold Hardiness. Gene Regulation and Genetic Engineering*, Li, P.H. and Palva, E.T., Eds., Kluwer Academic/Plenum Publishers, New York, 2002, 249–257.
254. Eastmond, P.J. and Graham, I.A., Trehalose metabolism: a regulatory role for trehalose-6-phosphate? *Curr. Opin. Plant Biol.*, 6, 231–235, 2003.
255. Brodmann, D., Schuller, A., Ludwig-Muller, J., Aeschbacher, R.A., Wiemken, A., Boller, T., and Wingler, A., Induction of trehalase in *Arabidopsis* plants infected with the trehalose-producing pathogen *Plasmodiophora brassicae*, *Mol. Plant–Microbe Interact.*, 15, 693–700, 2002.
256. Eastmond, P.J., Glycerol-insensitive *Arabidopsis* mutants: gli1 seedlings lack glycerol kinase, accumulate glycerol and are more resistant to abiotic stress, *Plant J.*, 37, 617–625, 2004.
257. Yano, R., Nakamura, M., Yoneyama, T., and Nishida, I., Starch-related α-glucan/water dikinase is involved in the cold-induced development of freezing tolerance in *Arabidopsis*, *Plant Physiol.*, 138, 837–846, 2005.
258. Takagi, T., Nakamura, M., Hayashi, H., Inatsugi, R., Yano, R., and Nishida, I., The leaf-order-dependent enhancement of freezing tolerance in cold-acclimated *Arabidopsis* rosettes is not correlated with the transcript levels of the cold-inducible transcription factors of CBF/DREB1, *Plant Cell Physiol.*, 44, 922–931, 2003.
259. Browse, J. and Lange, B.M., Counting the cost of a cold-blooded life: metabolomics of cold acclimation, *Proc. Natl. Acad. Sci. USA*, 101, 14996–14997, 2004.
260. Lange, B.M. and Ghassemian, M., Comprehensive post-genomic data analysis approaches integrating biochemical pathway maps, *Phytochemistry*, 66, 413–451, 2005.
261. Steponkus, P.L., Cold acclimation of *Hedera helix*. Evidence for a two-phase process, *Plant Phys.*, 47, 175–180, 1971.
262. Kovtun, Y., Chiu, W.L., Tena, G., and Sheen, J., Functional analysis of oxidative stress-activated mitogen-activated protein kinase cascade in plants, *Proc. Natl. Acad. Sci. USA*, 97, 2940–2945, 2000.
263. Shou, H., Bordallo, P., Fan, J.B., Yeakley, J.M., Bibikova, M., Sheen, J., and Wang, K., Expression of an active tobacco mitogen-activated protein kinase kinase kinase enhances freezing tolerance in transgenic maize, *Proc. Natl. Acad. Sci. USA*, 101, 3298–3303, 2004.
264. Kovtun, Y., Chiu, W.L., Zeng, W.K., and Sheen, J., Suppression of auxin signal transduction by a MAPK cascade in higher plants, *Nature*, 395, 716–720, 1998.

265. Wang, C.Y., Bowen, J.H., Weir, I.E., Allan, A.C., and Ferguson, I.B., Heat-induced protection against death of suspension-cultured apple fruit cells exposed to low temperature, *Plant Cell Environ.*, 24, 1199–1207, 2001.
266. Teige, M., Scheikl, E., Eulgem, T., Doczi, R., Ichimura, K., Shinozaki, K., Dangl, J.L., and Hirt, H., The MKK2 pathway mediates cold and salt stress signaling in *Arabidopsis*, *Mol. Cell*, 15, 141–152, 2004.
267. Abbasi, F., Onodera, H., Toki, S., Tanaka, H., and Komatsu, S., OsCDPKI3, a calcium-dependent protein kinase gene from rice, is induced by cold and gibberellin in rice leaf sheath, *Plant Mol. Biol.*, 55, 541–552, 2004.
268. Xie, Z., Zhang, Z.L., Zou, X., Huang, J., Ruas, P., Thompson, D., and Shen, Q.J., Annotations and functional analyses of the rice WRKY gene superfamily reveal positive and negative regulators of abscisic acid signaling in aleurone cells, *Plant Physiol.*, 137, 176–189, 2005.
269. Kiyosue, T., Yamaguchi-Shinozaki, K., and Shinozaki, K., ERD15, a cDNA for a dehydration-induced gene from *Arabidopsis thaliana*, *Plant Physiol.*, 106, 1707, 1994.
270. Catala, R., Santos, E., Alonso, J.M., Ecker, J.R., Martinez-Zapater, J.M., and Salinas, J., Mutations in the $Ca^{2+}/H+$ transporter CAX1 increase CBF/DREB1 expression and the cold-acclimation response in *Arabidopsis*, *Plant Cell*, 15, 2940–2951, 2003.
271. Gong, Z., Lee, H., Xiong, L., Jagendorf, A., Stevenson, B., and Zhu, J.K., RNA helicase-like protein as an early regulator of transcription factors for plant chilling and freezing tolerance, *Proc. Natl. Acad. Sci. USA*, 99, 11507–11512, 2002.
272. Gong, Z., Dong, C.H., Lee, H., Zhu, J., Xiong, L., Gong, D., Stevenson, B., and Zhu, J.K., A DEAD box RNA helicase is essential for mRNA export and important for development and stress responses in *Arabidopsis*, *Plant Cell*, 17, 256–267, 2005.
273. Xin, Z. and Browse, J., Eskimo1 mutants of *Arabidopsis* are constitutively freezing tolerant, *Proc. Natl. Acad. Sci. USA*, 95, 7799–7804, 1998.
274. Tahtiharju, S. and Palva, T., Antisense inhibition of protein phosphatase 2C accelerates cold acclimation in *Arabidopsis thaliana*, *Plant J.*, 26, 461–470, 2001.
275. Naidu, S.L. and Long, S.P., Potential mechanisms of low-temperature tolerance of C4 photosynthesis in *Miscanthus* × *giganteus*: an *in vivo* analysis, *Planta*, 220, 145–155, 2004.
276. Bartels, D., Targeting detoxification pathways: an efficient approach to obtain plants with multiple stress tolerance? *Trends Plant Sci.*, 6, 284–286, 2001.
277. Kasuga, M., Miura, S., Shinozaki, K., and Yamaguchi-Shinozaki, K., A combination of the *Arabidopsis* DREB1A gene and stress-inducible rd29A promoter improved drought- and low-temperature stress tolerance in tobacco by gene transfer, *Plant Cell Physiol.*, 45, 346–350, 2004.
278. Pellegrineschi, A., Reynolds, M., Pacheco, M., Brito, R.M., Almeraya, R., Yamaguchi-Shinozaki, K., and Hoisington, D., Stress-induced expression in wheat of the *Arabidopsis thaliana* DREB1A gene delays water stress symptoms under greenhouse conditions, *Genome*, 47, 493–500, 2004.
279. Seki, M., Satou, M., Sakurai, T., Akiyama, K., Iida, K., Ishida, J., Nakajima, M., Enju, A., Narusaka, M., Fujita, M., Oono, Y., Kamei, A., Yamaguchi-Shinozaki, K., and Shinozaki, K., RIKEN *Arabidopsis* full-length (RAFL) cDNA and its applications for expression profiling under abiotic stress conditions, *J. Exp. Bot.*, 55, 213–223, 2004.
280. Sakamoto, H., Maruyama, K., Sakuma, Y., Meshi, T., Iwabuchi, M., Shinozaki, K., and Yamaguchi-Shinozaki, K., *Arabidopsis* Cys2/His2-type zinc-finger proteins function as transcription repressors under drought, cold, and high-salinity stress conditions, *Plant Physiol.*, 136, 2734–2746, 2004.

281. Daniell, H., Kumar, S., and Dufourmantel, N., Breakthrough in chloroplast genetic engineering of agronomically important crops, *Trends Biotechnol.*, 23, 238–245, 2005.
282. Yamamoto, A., Bhuiyan, M.N., Waditee, R., Tanaka, Y., Esaka, M., Oba, K., Jagendorf, A.T., and Takabe, T., Suppressed expression of the apoplastic ascorbate oxidase gene increases salt tolerance in tobacco and *Arabidopsis* plants, *J. Exp. Bot.*, 2005.
283. Till, B.J., Reynolds, S.H., Greene, E.A., Codomo, C.A., Enns, L.C., Johnson, J.E., Burtner, C., Odden, A.R., Young, K., Taylor, N.E., Henikoff, J.G., Comai, L., and Henikoff, S., Large-scale discovery of induced point mutations with high-throughput TILLING, *Genome Res.*, 13, 524–530, 2003.
284. Takken, F.L.W., van Wijk, R., Michielse, C.B., Houterman, P.M., Ram, A.F.J., and Cornelissen, B.J.C., A one-step method to convert vectors into binary vectors suited for Agrobacterium-mediated transformation, *Curr. Genet.*, 45, 242–248, 2004.
285. Altpeter, F., Baisakh, N., Beachy, R., Bock, R., Capell, T., Christou, P., Daniell, H., Datta, K., Datta, S., Dix, P.J., Fauquet, C., Huang, N., Kohli, A., Mooibroek, H., Nicholson, L., Nguyen, T.T., Nugent, G., Raemakers, K., Romano, A., Somers, D.A., Stoger, E., Taylor, N., and Visser, R., Particle bombardment and the genetic enhancement of crops: myths and realities, *Mol. Breed.*, 15, 305–327, 2005.
286. Murakeozy, E.P., Nagy, Z., Duhaze, C., Bouchereau, A., and Tuba, Z., Seasonal changes in the levels of compatible osmolytes in three halophytic species of inland saline vegetation in Hungary, *J. Plant Physiol.*, 160, 395–401, 2003.
287. Han, Q.M., Katahata, S., Kakubari, Y., and Mukai, Y., Seasonal changes in the xanthophyll cycle and antioxidants in sun-exposed and shaded parts of the crown of *Cryptomeria japonica* in relation to rhodoxanthin accumulation during cold acclimation, *Tree Physiol.*, 24, 609–616, 2004.
288. Marian, C.O., Eris, A., Krebs, S.L., and Arora, R., Environmental regulation of a 25-kDa dehydrin in relation to *Rhododendron* cold acclimation, *J. Am. Soc. Hort. Sci.*, 129, 354–359, 2004.
289. Welling, A., Rinne, P., Vihera-Aarnio, A., Kontunen-Soppela, S., Heino, P., and Palva, E.T., Photoperiod and temperature differentially regulate the expression of two dehydrin genes during overwintering of birch (*Betula pubescens* Ehrh.), *J. Exp. Bot.*, 55, 507–516, 2004.
290. Verhoeven, A.S., Swanberg, A., Thao, M., and Whiteman, J., Seasonal changes in leaf antioxidant systems and xanthophyll cycle characteristics in *Taxus* × *media* growing in sun and shade environments, *Physiol. Plantarum*, 123, 428–434, 2005.
291. Amiard, V., Morvan-Bertrand, A., Billard, J.P., Huault, C., Keller, F., and Prud'homme, M.P., Fructans, but not the sucrosyl-galactosides, raffinose and loliose, are affected by drought stress in perennial ryegrass, *Plant Physiol.*, 132, 2218–2229, 2003.
292. Tapernoux-Luthi, E.M., Bohm, A., and Keller, F., Cloning, functional expression, and characterization of the raffinose oligosaccharide chain elongation enzyme, galactan:galactan galactosyltransferase, from common bugle leaves, *Plant Physiol.*, 134, 1377–1387, 2004.
293. Ashraf, M. and Harris, P.J.C., Potential biochemical indicators of salinity tolerance in plants, *Plant Sci.*, 166, 3–16, 2004.
294. Sulpice, R., Tsukaya, H., Nonaka, H., Mustardy, L., Chen, T.H.H., and Murata, N., Enhanced formation of flowers in salt-stressed *Arabidopsis* after genetic engineering of the synthesis of glycine betaine, *Plant J.*, 36, 165–176, 2003.
295. Jing, Z.P., Gallardo, F., Pascual, M.B., Sampalo, R., Romero, J., de Navarra, A.T., and Canovas, F.M., Improved growth in a field trial of transgenic hybrid poplar overexpressing glutamine synthetase, *New Phytol.*, 164, 137–145, 2004.

296. Kaye, C., Neven, L., Hofig, A., Li, Q.B., Haskell, D., and Guy, C., Characterization of a gene for spinach CAP160 and expression of two spinach cold-acclimation proteins in tobacco, *Plant Physiol.*, 116, 1367–1377, 1998.
297. Houde, M., Dallaire, S., N'Dong, D., and Sarhan, F., Overexpression of the acidic dehydrin WCOR410 improves freezing tolerance in transgenic strawberry leaves, *Plant Biotechnol. J.*, 2, 381–387, 2004.
298. Honjoh, K., Shimizu, H., Nagaishi, N., Matsumoto, H., Suga, K., Miyamoto, T., Iio, M., and Hatano, S., Improvement of freezing tolerance in transgenic tobacco leaves by expressing the hiC6 gene, *Biosci. Biotechnol. Biochem.*, 65, 1796–1804, 2001.
299. NDong, C., Danyluk, J., Wilson, K.E., Pocock, T., Huner, N.P., and Sarhan, F., Cold-regulated cereal chloroplast late embryogenesis abundant-like proteins. Molecular characterization and functional analyses, *Plant Physiol.*, 129, 1368–1381, 2002.
300. Honjoh, K.I., Oda, Y., Takata, R., Miyamoto, T., and Hatano, S., Introduction of the hiC6 gene, which encodes a homologue of a late embryogenesis abundant (LEA) protein, enhances freezing tolerance of yeast, *J. Plant Physiol.*, 155, 509–512, 1999.
301. Grover, A., Aggarwal, P.K., Kapoor, A., Katiyar-Agarwal, S., Agarwal, M., and Chandramouli, A., Addressing abiotic stresses in agriculture through transgenic technology, *Curr. Sci.*, 84, 355–367, 2003.
302. Bartels, D. and Sunkar, R., Drought and salt tolerance in plants, *Crit. Rev. Plant Sci.*, 24, 23–58, 2005.
303. Chinnusamy, V., Jagendorf, A., and Zhu, J.K., Understanding and improving salt tolerance in plants, *Crop Sci.*, 45, 437–448, 2005.
304. Vinocur, B. and Altman, A., Recent advances in engineering plant tolerance to abiotic stress: achievements and limitations, *Curr. Opin. Biotechnol.*, 16, 123–132, 2005.
305. Ishizaki-Nishizawa, O., Fujii, T., Azuma, M., Sekiguchi, K., Murata, N., Ohtani, T., and Toguri, T., Low-temperature resistance of higher plants is significantly enhanced by a nonspecific cyanobacterial desaturase, *Nat. Biotechnol.*, 14, 1003–1006, 1996.
306. Shimizu, H., Furuya, N., Suga, K., Honjoh, K., Miyamoto, T., Hatano, S., and Masayoshi, I., Freezing tolerance of transgenic tobacco with increased content of unsaturated fatty acid by expressing the CvFad2 or CvFad3 gene, *J. Fac. Agric. Kyushu Univ.*, 47, 307–317, 2003.
307. Yokoi, S., Higashi, S., Kishitani, S., Murata, N., and Toriyama, K., Introduction of the cDNA for *Arabidopsis* glycerol-3-phosphate acyltransferase (GPAT) confers unsaturation of fatty acids and chilling tolerance of photosynthesis on rice, *Mol. Breed.*, 4, 269–275, 1998.
308. Yoshimura, K., Miyao, K., Gaber, A., Takeda, T., Kanaboshi, H., Miyasaka, H., and Shigeoka, S., Enhancement of stress tolerance in transgenic tobacco plants overexpressing *Chlamydomonas* glutathione peroxidase in chloroplasts or cytosol, *Plant J.*, 37, 21–33, 2004.
309. Gupta, A.S., Heinen, J.L., Holaday, A.S., Burke, J.J., and Allen, R.D., Increased resistance to oxidation stress in transgenic plants that overexpress chloroplastic Cu/Zn superoxide dismutase, *Proc. Natl. Acad. Sci. USA*, 90, 1629–1633, 1993.
310. Roxas, V.P., Smith, R.K., Jr., Allen, E.R., and Allen, R.D., Overexpression of glutathione S-transferase/glutathione peroxidase enhances the growth of transgenic tobacco seedlings during stress, *Nat. Biotechnol.*, 15, 988–991, 1997.
311. Kwon, S.Y., Choi, S.M., Ahn, Y.O., Lee, H.S., Lee, H.B., Park, Y.M., and Kwak, S.S., Enhanced stress tolerance of transgenic tobacco plants expressing a human dehydroascorbate reductase gene, *J. Plant Physiol.*, 160, 347–353, 2003.

312. Hu, Y., Jia, W., Wang, J., Zhang, Y., Yang, L., and Lin, Z., Transgenic tall fescue containing the *Agrobacterium tumefaciens* ipt gene shows enhanced cold tolerance, *Plant Cell Rep.*, 2004.
313. Deryabin, A.N., Dubinina, I.M., Burakhanova, E.A., Astakhova, N.V., Sabel'nikova, E.P., and Trunova, T.I., Influence of yeast-derived invertase gene expression in potato plants on membrane lipid peroxidation at low temperature, *J. Therm. Biol.*, 30, 73–77, 2005.
314. Jang, I.C., Oh, S.J., Seo, J.S., Choi, W.B., Song, S.I., Kim, C.H., Kim, Y.S., Seo, H.S., Choi, Y.D., Nahm, B.H., and Kim, J.K., Expression of a bifunctional fusion of the *Escherichia coli* genes for trehalose-6-phosphate synthase and trehalose-6-phosphate phosphatase in transgenic rice plants increases trehalose accumulation and abiotic stress tolerance without stunting growth, *Plant Physiol.*, 131, 516–524, 2003.
315. Prabhavathi, V., Yadav, J.S., Kumar, P.A., and Rajam, M.V., Abiotic stress tolerance in transgenic eggplant (*Solanum melongena* L.) by introduction of bacterial mannitol phosphodehydrogenase gene, *Mol. Breed.*, 9, 137–147, 2002.
316. Mukhopadhyay, A., Vij, S., and Tyagi, A.K., Overexpression of a zinc-finger protein gene from rice confers tolerance to cold, dehydration, and salt stress in transgenic tobacco, *Proc. Natl. Acad. Sci. USA*, 101, 6309–6314, 2004.
317. Qiao, J., Mitsuhara, I., Yazaki, Y., Sakano, K., Gotoh, Y., Miura, M., and Ohashi, Y., Enhanced resistance to salt, cold and wound stresses by overproduction of animal cell death suppressors Bcl-xL and Ced-9 in tobacco cells—their possible contribution through improved function of organella, *Plant Cell Physiol.*, 43, 992–1005, 2002.
318. McKersie, B.D., Bowley, S.R., and Jones, K.S., Winter survival of transgenic alfalfa overexpressing superoxide dismutase, *Plant Physiol.*, 119, 839–848, 1999.
319. Qi, Q.G., Huang, Y.F., Cutler, A.J., Abrams, S.R., and Taylor, D.C., Molecular and biochemical characterization of an aminoalcoholphosphotransferase (AAPT1) from *Brassica napus*: effects of low temperature and abscisic acid treatments on AAPT expression in *Arabidopsis* plants and effects of over-expression of BnAAPT1 in transgenic *Arabidopsis*, *Planta*, 217, 547–558, 2003.
320. Li, W.Q., Li, M.Y., Zhang, W.H., Welti, R., and Wang, X.M., The plasma membrane-bound phospholipase D delta enhances freezing tolerance in *Arabidopsis thaliana*, *Nat. Biotechnol.*, 22, 427–433, 2004.
321. Welti, R., Li, W.Q., Li, M.Y., Sang, Y.M., Biesiada, H., Zhou, H.E., Rajashekar, C.B., Williams, T.D., and Wang, X.M., Profiling membrane lipids in plant stress responses—role of phospholipase D alpha in freezing-induced lipid changes in *Arabidopsis*, *J. Biol. Chem.*, 277, 31994–32002, 2002.
322. Kasukabe, Y., He, L., Nada, K., Misawa, S., Ihara, I., and Tachibana, S., Overexpression of spermidine synthase enhances tolerance to multiple environmental stresses and upregulates the expression of various stress-regulated genes in transgenic *Arabidopsis thaliana*, *Plant Cell Physiol.*, 45, 712–722, 2004.
323. Alia, Hayashi, H., Chen, T.H.H., and Murata, N., Transformation with a gene for choline oxidase enhances the cold tolerance of *Arabidopsis* during germination and early growth, *Plant, Cell Environ.*, 21, 232–239, 1998.
324. Sakamoto, A., Alia, and Murata, N., Metabolic engineering of rice leading to biosynthesis of glycinebetaine and tolerance to salt and cold. [Erratum: May 1999, vol. 40(1), p. 195], *Plant Mol. Biol.*, 38, 1011–1019, 1998.
325. Fan, Y., Liu, B., Wang, H., Wang, S., and Wang, J., Cloning of an antifreeze protein gene from carrot and its influence on cold tolerance in transgenic tobacco plants, *Plant Cell Rep.*, 21, 296–301, 2002.

326. Vannini, C., Locatelli, F., Bracale, M., Magnani, E., Marsoni, M., Osnato, M., Mattana, M., Baldoni, E., and Coraggio, I., Overexpression of the rice Osmyb4 gene increases chilling and freezing tolerance of *Arabidopsis thaliana* plants, *Plant J.*, 37, 115–127, 2004.
327. Xiong, L. and Yang, Y., Disease resistance and abiotic stress tolerance in rice are inversely modulated by an abscisic acid-inducible mitogen-activated protein kinase, *Plant Cell*, 15, 745–759, 2003.
328. Xiong, L., Lee, H., Huang, R., and Zhu, J.K., A single amino acid substitution in the *Arabidopsis* FIERY1/HOS2 protein confers cold signaling specificity and lithium tolerance, *Plant J.*, 40, 536–545, 2004.
329. Kim, J.B., Kang, J.Y., and Kim, S.Y., Overexpression of a transcription factor regulating ABA-responsive gene expression confers multiple stress tolerance, *Plant Biotechnol. J.*, 2, 459–466, 2004.
330. Oh, S.J., Song, S.I., Kim, Y.S., Jang, H.J., Kim, S.Y., Kim, M., Kim, Y.K., Nahm, B.H., and Kim, J.K., *Arabidopsis* CBF3/DREB1A and ABF3 in transgenic rice increased tolerance to abiotic stress without stunting growth, *Plant Physiol.*, 138, 341–351, 2005.
331. Tamminen, I., Makela, P., Heino, P., and Palva, E.T., Ectopic expression of ABI3 gene enhances freezing tolerance in response to abscisic acid and low temperature in *Arabidopsis thaliana*, *Plant J.*, 25, 1–8, 2001.
332. Yi, S.Y., Kim, J.H., Joung, Y.H., Lee, S., Kim, W.T., Yu, S.H., and Choi, D., The pepper transcription factor CaPF1 confers pathogen and freezing tolerance in *Arabidopsis*, *Plant Physiol.*, 136, 2862–2874, 2004.
333. Qin, F., Sakuma, Y., Li, J., Liu, Q., Li, Y.Q., Shinozaki, K., and Yamaguchi–Shinozaki, K., Cloning and functional analysis of a novel DREB1/CBF transcription factor involved in cold-responsive gene expression in *Zea mays* L., *Plant Cell Physiol.*, 45, 1042–1052, 2004.
334. Hsieh, T.H., Lee, J.T., Yang, P.T., Chiu, L.H., Charng, Y.Y., Wang, Y.C., and Chan, M.T., Heterology expression of the *Arabidopsis* C-repeat/dehydration response element binding factor 1 gene confers elevated tolerance to chilling and oxidative stresses in transgenic tomato, *Plant. Physiol.*, 129, 1086–1094, 2002.
335. Lee, J.T., Prasad, V., Yang, P.T., Wu, J.F., Ho, T.H.D., Charng, Y.Y., and Chan, M.T., Expression of *Arabidopsis* CBF1 regulated by an ABA/stress inducible promoter in transgenic tomato confers stress tolerance without affecting yield, *Plant, Cell Environ.*, 26, 1181–1190, 2003.

Index

A

C

D

G

H

I

J

K

L

M

N

O

P

Q

R

S

U

V

W

X

Y

Z